Foundations of Geomagnetism

Foundations of Geomagnetism

GEORGE BACKUS

University of California, San Diego

ROBERT PARKER

University of California, San Diego

CATHERINE CONSTABLE

University of California, San Diego

CAMBRIDGE
UNIVERSITY PRESS

CAMBRIDGE UNIVERSITY PRESS
Cambridge, New York, Melbourne, Madrid, Cape Town, Singapore, São Paulo

Cambridge University Press
The Edinburgh Building, Cambridge CB2 2RU, UK

Published in the United States of America by Cambridge University Press, New York

www.cambridge.org
Information on this title: www.cambridge.org/9780521410069

First published 1996
This digitally printed first paperback version 2005

A catalogue record for this publication is available from the British Library

ISBN-13 978-0-521-41006-9 hardback
ISBN-10 0-521-41006-1 hardback

ISBN-13 978-0-521-01733-6 paperback
ISBN-10 0-521-01733-5 paperback

CONTENTS

PREFACE

The idea for this book was Philip Stark's: he suggested joining two of
the authors, Bob Parker and Cathy Constable, in a project to "tidy
up" George Backus's lecture notes and to make George a present of
them in book form in time for his sixtieth birthday. Before this project
got properly started, Philip left San Diego for Berkeley, and when the
two of us still here began to look at the job seriously, we soon realized
the timetable was not practical. Nonetheless, Cambridge University
Press thought the idea was worth supporting and encouraged us to
continue. The spirit of the enterprise was still that of offering George
Backus a gift, rather than giving him more to do.

Over the years 1962 to 1994, George Backus has taught graduate level classes covering geomagnetism, mathematical techniques for
geophysics, tensors in geophysics, and various other aspects of mathematical geophysics. Each time he taught, he would start from scratch,
composing for the students hundreds of pages of closely spaced, hand
written notes. He would devote himself entirely to the course each
time, putting aside research and administration while it was being
given. The level of the material was advanced, but the logical development, completeness, and consistency of notation were so compelling
that generations of students came away enriched. Philip Stark's idea
was to present something of this legacy to the world, but to spare
George the tedium of editing and collating.

George's notes, on close inspection, comprise the solid mathematical skeleton of the material, but he himself provided the flesh of
explanatory discourse and scientific context during his lectures. A
major part of the editors' job has been to re-create the interpolatory
text. Another part of our work was to decide what material to keep.
Since we are both primarily geomagnetists, we felt more comfortable
building around a geomagnetic theme rather than one of the other

topics on which George has lectured at length, such as continuum mechanics or tensors. Even then, we could not keep everything; for example, the material on the Haar measure for averaging over a sphere from great circle paths could not be fitted neatly into the book. Only in a few places, most notably the second half of Chapter 3, have we felt the need to write new text to fill a gap.

The subject matter of *Foundations of Geomagnetism* is the mathematical and physical basis of the science of geomagnetism; graduate students in the earth sciences are its intended audience. George Backus has always been passionately concerned with the logical foundation of scientific argument and mathematical rigor in quantitative developments. Thus when he taught about the decomposition of a magnetic field into its poloidal and toroidal parts, he would never begin, "It can be shown that"; a proper account must start with the demonstration that a unique decomposition of this kind is always possible. Chapter 5 opens on this point. To build the foundations of geomagnetism, George calls upon some unusual mathematical tools, some of his own invention. The earth is nearly spherical, and geophysicists continually need to treat vectors and solve differential equations in spherical geometry. George shows how it is possible to maintain the elegance of a coordinate-free notation and at the same time to preserve the simplicity and familiarity of a Gibbs-like vector calculus for operators on a spherical surface. In this way he avoids the heavy baggage of the currently fashionable coordinate-free differential geometry à la Cartan. George's notation is intuitively right for the problem.

The final form of the book consists of seven chapters. The first is a brief overview of the phenomena that are of interest in geomagnetism. It includes a sketch of the history of the subject, followed by a description of the geomagnetic field and its variability in time and space. Chapter 2 concerns the classical theory of electromagnetism based on Maxwell's equations. We cover the physical and mathematical ideas of sources for the electromagnetic field and how the vacuum form of Maxwell's equations is adapted for polarizable media. We discuss the mathematical basis for the practice of neglecting small terms in an equation, and in particular we justify the neglect in geomagnetic work of the displacement current in Maxwell's equations. The discussion of sources introduces the concept of the separation of the magnetic field into parts of internal and external origin.

Chapter 3 is devoted to spherical harmonics. The aim is to develop from first principles all the standard results. The perspective and methods used are not the traditional ones, however. In particular, the spherical harmonics of degree ℓ are exhibited as homogeneous harmonic polynomials in x, y, and z, which are treated as members of a finite-dimensional vector space. We introduce an inner product on the space and then consider an orthogonal basis for it. Various linear operators mapping this space onto itself are used to explore the symmetries of the system, and thereby to discover its properties, such as the Addition Theorem, and the existence of a self-reproducing kernel. Up to this point, the results have been independent of the particular coordinate system. Once a special axis system is chosen, we can develop explicit expressions for the traditional set of orthonormal functions. We investigate the asymptotic properties of the functions, derive recurrence relations among them, and describe a scheme for their practical computation.

Chapter 4 gives the application of spherical harmonics in the description of the main geomagnetic field. We study the question of the uniqueness of the coefficients in a spherical harmonic expansion containing internal and external parts. Other topics include the geomagnetic power spectrum, downward continuation to the core, and how little information about the sources is contained in the Gauss coefficients.

The subject of Chapter 5 is the Mie representation, which is the expression of a solenoidal field as a sum of poloidal and toroidal parts. Again the theory is developed from first principles, beginning with the Helmholtz representation for tangent vector fields on a spherical surface, a useful representation in its own right. The Mie representation is then applied to a variety of geomagnetic problems, including the generalization of Gauss' separation to regions containing sources, outward diffusion of the toroidal magnetic field of the core, geomagnetic sounding, and the free decay of magnetic fields in a stationary core.

The material of the preceding chapters is brought to bear in Chapter 6, where we consider the physical processes taking place in the core of the earth. After a quick look at a simplified dynamo model with only two degrees of freedom, we derive the full system of partial differential equations governing the interaction of a moving conducting fluid with an embedded magnetic field. To obtain the version of Ohm's law needed in a moving conductor, we must appeal to relativistic physics. The two descriptions of a continuum, Lagrangian and Eulerian, are discussed and their relationship exposed.

Two limiting problems are solved exactly: zero velocity and infinite conductivity. The latter is shown to be a useful approximation in the earth for time scales less than 100 years, and therefore of considerable interest in the interpretation of the secular variation. We derive the "frozen-flux" condition of Roberts and Scott, which, if valid, permits us to deduce from the magnetic field what the fluid velocity is on the lines where the radial field vanishes. The chapter closes with a sample of dynamo theory, a mixture of brief qualitative summaries of some important results together with a few topics laid out in mathematical detail, including Cowling's antidynamo theorem and a glimpse at mean field dynamos, currently so popular.

The Mie representation and, to a lesser extent, spherical harmonics are dependent on vector calculus on the surface of a sphere. Chapter 7 is a compendium of mathematical theorems and results needed elsewhere in the book. The chapter provides a general treatment of linear operators that act on scalar and vector fields. The general theory is specialized to the case of a spherical surface, and the properties of those most useful operators, surface gradient and surface curl, are then developed in some detail. Corresponding results for more general surfaces are touched upon. Other topics in Chapter 7 include surface forms of the integral theorems of Stokes and Gauss, and inclusion of jump discontinuities in those theorems and how this affects the Helmholtz Theorem.

The two junior authors have both learned an enormous amount by going over this material in detail. We can only hope some of the craftsmanship and the intellectual discipline demonstrated by this work has rubbed off on us. We are grateful to Philip Stark for suggesting the project. We would also like to express our thanks to Elaine Blackmore, who translated the hand written notes into TEX with great speed and accuracy, all the more amazing given that this was her first experience with TEX. We wish to express our gratitude to the Director of Scripps Institution of Oceanography for the financial help he provided as we were getting started. Cambridge University Press and its editors deserve our gratitude for their patience and understanding, as well as their enthusiasm for the idea. Once again, we want to thank the senior author, George Backus, for his example as a great scientist and a warm human being.

R. Parker
C. Constable

1

THE MAIN FIELD

1.1 A Whirlwind Tour

In this chapter we go on a whirlwind tour of the subject matter of geomagnetism. Far from being a comprehensive survey, this is an outline of the observations and phenomena that geomagnetism aims to understand. It is expected that every serious student of our science will be familiar with everything in this chapter, but those from outside the earth sciences may find the following summary helpful. The following references will help fill in the gaps.

For an accessible yet scholarly summary of the history of geomagnetism, one can find nothing that improves upon the final chapter in Volume II of *Geomagnetism* by Chapman and Bartels (1962). The short essay by Malin in the more recent *Geomagnetism*, edited by Jacobs (1987), is another useful source. Appendix B of the paper by Malin and Bullard (1981) gives a series of thumbnail sketches of the important players in geomagnetism in Great Britain from 1570 to 1900. A fascinating history of the discovery of the reversing nature of the main field can be found in *The Road to Jaramillo* (Glen, 1982).

A review of the spatial variation of the main field and the crustal magnetic anomalies can be found in Jacobs's *Geomagnetism* (1987). The practical techniques of magnetic data interpretation, particularly for fields originating in the crust, are thoroughly covered by Blakely (1995). A detailed account of the phenomena of the magnetosphere and the ionosphere is presented in *Physics of Geomagnetic Phenomena*, edited by Matsushita and Campbell (1967); a more modern but less detailed treatment is given by Parkinson (1983). These works are also good places to read about the externally caused time variations, but for detail of a purely descriptive kind, Volume I of Chapman and

Bartels (1962) remains the classic source. An excellent treatment of the longer-period time variations is to be found in the monograph by Merrill and McElhinny (1983), which also covers our other topics, though with somewhat less authority.

1.2 History

The existence of magnetic forces, through the tendency of lodestones to attract iron, has been known for perhaps 4000 years, having first been noted in China. Lodestones, which are naturally magnetic pieces of magnetite, are mentioned by Homer (ca 800 BC) and by later Greek and Roman writers such as Pliny the Elder (24–79 AD). The use of the magnetic field for navigational purposes cannot be unequivocally identified until 1088 in China and nearly 100 years later in Europe. Believing in the perfection of celestial phenomena, the early navigators assumed a compass would point exactly to geographic north. But of course this is mistaken, and the discrepancy, called variation in the original accounts but now referred to as *declination*, was generally recognized by the middle of the fifteenth century; its discovery in Europe is sometimes erroneously credited to Christopher Columbus. In 1581, Robert Norman (dates uncertain), a London instrument maker, reported the fact that the true direction of the field was not horizontal and that a compass needle, carefully balanced before being magnetized, would point downward, or dip. Thus magnetic *inclination*, as it is now called, was discovered.

It is time to introduce the first mathematical terms, the so-called geomagnetic elements that describe the magnetic field vector at a given point on the earth's surface. The following are the traditional names still in common use in the study of geomagnetism. We consider a local Cartesian coordinate system with x pointing to geographic north, y to the east, and z vertically downward. The magnetic elements X, Y, Z are the components of the magnetic field vector **B** in this frame. (We will define exactly what a magnetic field vector is in the next chapter.) Then the declination D is obtained from

$$\tan D = Y/X. \qquad (1.2.1)$$

The total intensity, variously referred to as T or F, is obviously

$$F = \sqrt{X^2 + Y^2 + Z^2}. \qquad (1.2.2)$$

The horizontal intensity H is just

$$H = \sqrt{X^2 + Y^2}. \tag{1.2.3}$$

Inclination, I, satisfies the equation

$$\tan I = Z/H. \tag{1.2.4}$$

Worldwide exploration by European navigators who measured declination and sometimes even inclination of the magnetic field enabled William Gilbert (1540–1603), chief physician to Queen Elizabeth I, to assemble a global picture of the field directions and to deduce that "The earth is a great magnet." He showed that a spherical lodestone, which he called a *terrella*, was surrounded by a magnetic field whose directional properties closely resembled those of the earth's field. His monumental book, *De Magnete* published in 1600, is widely regarded as the first scientific text, being entirely free of the appeals to heavenly causes and magic that were the common currency of explanation of the day.

Although it was probably suspected when Gilbert wrote his treatise, the fact that the earth's field changes in time was explicitly denied in *De Magnete*. In 1624, Edmund Gunter (1581–1626), professor of astronomy at Gresham College, had collected observations that pointed strongly to temporal variations in declination. His successor, Henry Gellibrand (1597–1636), completed the study after Gunter's death and published the discovery of secular variation in 1635. By 1680, Edmund Halley (1656–1742) produced an amazingly prescient model for the variation in terms of dipoles moving generally westward, deep within the earth, making a circuit every 700 years. The westward drift of small-scale features of the geomagnetic field is an important clue about fluid motions of the core; today the period is estimated to be about 2000 years.

An almost completely modern description of the main geomagnetic field was obtained by Karl Friedrich Gauss (1777–1855), the great German mathematician and geomagnetist. He completed the notion of the magnetic field as a vector by defining and determining its strength. Gauss invented spherical harmonics for the description of the field (a subject that will occupy us later on) and deduced that the origin of the main field is within the solid earth, not outside it, thus confirming Gilbert's early speculation. We will repeat an updated

form of Gauss' argument in Chapter 4. Gauss was responsible for setting up a worldwide system of magnetic observatories, some of which
have been running continuously up to the present day. Of course, the
cgs unit of magnetic induction is named for Gauss, although we adhere
to the Système Internationale (SI) convention in which 1 tesla = 10^4
gauss. Unlike the gauss, the tesla, named for the Polish-American
electrical engineer Nicola Tesla (1856–1943), is an inconvenient size
for geomagnetic use; generally the field intensities are referred to in
nanotesla (1 nT = 10^{-9} tesla) or microtesla (1 μT = 10^{-6} tesla). It
was Gauss who first observed that strength of the main field at the
surface varies from a maximum of about 60 μT (or 0.6 gauss) at the
poles to about 30 μT at the equator. His theoretical results predicted
that the intensity drops off approximately as the inverse cube of the
distance from the earth's center.

The major phenomenological discovery of the twentieth century
about the main geomagnetic field is the fact that the prominent
dipole has reversed polarity many hundreds of times during the earth's
geological history. The study of the magnetization of rocks in France
and Italy led Bernard Brunhes (1867–1910) to conclude in 1906 that
the ancient magnetic field had the opposite direction from today's.
In 1926 Paul Mercanton (1876–1963) came to the same conclusion
regarding rocks from Spitzbergen and Australia; Motonori Matuyama
(1884–1958) also deduced this from specimens taken in Japan and
Siberia. Yet for over half a century this evidence was not generally
regarded as conclusive for a variety of reasons. Most important, it was
not at all clear that rocks could retain magnetism gained millions of
years ago. Today much more about the physics of rock magnetism is
known and there is no real doubt that most rocks containing magnetic
minerals, both igneous and sedimentary, can record and preserve
almost indefinitely information about the field at the time when they
were formed. Stacey and Banerjee (1974) and O'Reilly (1984) provide
comprehensive treatments of the physics of rock magnetism. By 1963,
Cox, Doell, and Dalrymple had compelling evidence of reversals based
on radiometric ages; their measurements demonstrated conclusively
that the field was reversed all over the globe during a number of well-
defined epochs stretching back 4 million years; see Cox, et al. (1964).
In 1963 Vine and Matthews recognized the regular magnetic anomaly
stripes they had observed in the Indian Ocean as the record of the
reversing field in the seafloor rocks. Aside from the revolutionary
implications for geology, this great insight made it possible to establish

an accurate reversal chronology for the main field for over 100 million years; see Figure 1.4.4.

1.3 Spatial Variations

Now we imagine a journey from the outer reaches of the galaxy to the surface of the earth. At each stopping point we describe the environment and local magnetic field, averaging over a suitable volume. That averages in space are an essential part of a proper description is a theme in Chapter 2.

In interstellar space it has been observed from the polarization of light by magnetically oriented dust particles that there is a magnetic field with a magnitude of about 1 nT; in the arm of the galaxy the direction seems aligned with the arm.

Within the solar system the sun's magnetic field dominates interplanetary space. Streaming from the sun continuously at velocities from 350 to 500 km/s is a plasma of neutral hydrogen atoms (500 cm^{-3}), protons, and electrons (10 cm^{-3}). These particles compose the *solar wind;* they carry along a magnetic field whose strength is about 5 nT. All these figures refer to a point at the same distance from the sun as the earth is. Roughly speaking, the field is directed away from the sun, or toward it, in huge sectors. The equations of plasma physics govern the phenomena in the solar wind.

Immediately surrounding the earth, out to about 10 or 20 earth radii, is a region called the *magnetosphere.* Within this region the solar wind does not blow, and the magnetic field is that of the geomagnetic dipole, varying in intensity from 10 nT to 60 μT. The magnetosphere itself can be divided into a number of subregions: the boundary between the solar wind and the magnetosphere itself is a complex one in which a shock wave is continually being generated (since the speed of the solar wind exceeds the velocity of sound in the interplanetary plasma). Closer to the earth, highly energetic particles make up the *van Allen radiation belts,* in which protons and electrons move spirally along the magnetic field lines and then are reflected back and forth from the high-intensity fields near the poles. The outer belts are fed by the solar wind, and the equatorial circulation of the system constitutes an electrical current, the *ring current,* which generates a part of the magnetic field observed at the earth's surface. Modulation in the strength of the ring current is responsible for the longer-period recovery phase of magnetic storms (of which more later).

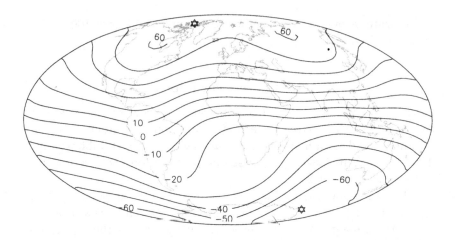

Figure 1.3.1: Magnetic element Z of the geomagnetic field in 1980, derived from the field model tabulated in Chapter 4; the map projection is Hammer equal-area elliptical; contour interval 10 μT.

From 100 km above the surface to about 10 earth radii, the magnetic field is well approximated by that of a dipole at the center of the earth, with its axis inclined at about 11° to the rotation axis. Of course the field is more complex than this, as can be seen in Figure 1.3.1, where we have plotted the Z component of a 1980 field model at the surface. The magnetic field model is one given by Langel et al. (1980) based on measurements made by the satellite MAGSAT; see section 4.4 for more details and a table of the corresponding coefficients. The so-called magnetic poles are the points on the earth's surface where the field is exactly vertical. According to the field model, in 1980 the magnetic north pole was in northern Canada at 77.4°N, 102.1°W and the south magnetic pole was in the Pacific Ocean off Antarctica at 65.4°S, 139.3°E; see Figure 1.3.1, where these points are marked by stars.

As we descend into the outer region of the earth's atmosphere we enter a layer called the *ionosphere*, so named because of the presence of high densities of charged particles (up to 10^6 cm^{-3}). The ionosphere lies between about 50 km and 600 km above the earth's surface and has been subdivided into layers: D from 50–90 km where the charged particles are NO^+, e^-, and O_2^-; E from 90–120 km (e^-, NO^+, O_2^+); F from 120–600 km (e^-, O_2^+). Because of the ultraviolet light from the sun that ionizes the atoms of the upper atmosphere, the

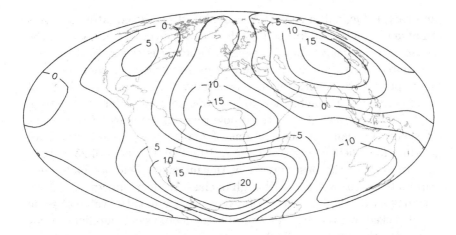

Figure 1.3.2: Field element Z of the nondipole field in 1980 from the same model field as in Figure 1.3.1; contour interval 5 μT.

sunlit hemisphere of the ionosphere is much more conducting than the nighttime hemisphere. Strong electric currents circulate in the sunlit hemisphere, driven by a dynamo process that derives its energy source from solar heating. The currents generate their own magnetic fields with intensities of up to 80 nT. An observer on the earth experiences a variation of the magnetic field once a day as the current system passes overhead following the sun. These daily variations are called quiet-solar variations, or S_q. The sun is not always quiet, and in periods of unrest during magnetic storms (see next 1.4) the daily variation is enhanced and frequently obscured by much more energetic magnetic activity.

At the earth's surface, as we mentioned in our brief historical notes, the departure of the geomagnetic field from that of a dipole is quite marked and has been known for over 300 years. A map of the vertical component of the nondipole field is provided in Figure 1.3.2. Those magnetic features with scales of several thousands of kilometers are associated with the main field. A notable feature, which has caused much speculation, is the very low nondipole field over the Pacific Ocean, something that may be associated with conditions at the base of the mantle. On shorter scales (say, less than 500 km) we find *magnetic anomalies* that are caused by the magnetization of crustal rocks. A huge anomaly of this kind, called the Bangui

anomaly, is located in central Africa, almost on the equator; it covers
an area of 250 by 700 km and reaches an amplitude of 500 nT. In the
oceans a prominent type of magnetic anomaly is the striped pattern
associated with seafloor spreading; this pattern is characterized by
an approximate constancy in one direction for many hundreds of
kilometers, and characteristic undulations perpendicular to the strike
that can be correlated with the dipole reversal sequence over millions
of years. On land, concentrations of magnetic ores can sometimes
produce local field changes as large as 10 μT in distances of 25 meters.

In each of the descriptions of the magnetic field we have been
careful to specify a scale. Although in classical electromagnetic theory
the magnetic field is a vector defined at every point, the value at some
point picked at random usually has little geophysical significance; for
example, the magnetic fields within an atom due to the intrinsic
moments and effective current circulation average to about 1 tesla
over the size of the atom (say, 10^{-10} m). Therefore, on this scale the
intensity fluctuates wildly between values that are millions of times
larger than those we typically think of as geomagnetic field strengths.

1.4 Time Variations

More so, perhaps, than variations in spatial scale, differences in characteristic time scales are tied to fundamental differences in the physical processes responsible. There is a remarkably large span of time-
scales on which the magnetic field fluctuates in the earth's environment, and a correspondingly rich collection of physical mechanisms.

Beginning at the smallest time scales, we note that at the earth's
surface in the daytime there is a strong, rapidly varying magnetic
field associated with the magnetic vector of sunlight; the time scale of
about 10^{-15} s is much too short for ordinary geomagnetic interests.
We calculate the magnetic field intensity of sunlight as follows: the
solar constant, 1360 W m^{-2}, gives the electromagnetic energy flux
arriving at the earth from the sun. This equals the average energy
density times the velocity of transport

$$\Phi = 2 \times \frac{<B^2>}{2\mu_0} \times c \qquad (1.4.1)$$

where the factor of 2 accounts for the fact that in an electromagnetic
wave the electric and magnetic energy densities are equal, c is the

velocity of light; the brackets $< \cdot >$ mean time average. Numerically, $\mu_0 = 4\pi \times 10^{-7}$ H/m, $c = 3 \times 10^8$ m/s; thus solving for the root-mean-square (rms) field

$$< B^2 >^{1/2} = 2.4 \times 10^{-6} \, \text{T} \quad = 2400 \, \text{nT}, \tag{1.4.2}$$

which is about 8 percent of the steady field at the equator.

Skipping through the electromagnetic spectrum, we come to the first geophysically interesting frequency at around 1 kHz or, equivalently, variations with periods of a millisecond or so. Sporadically at any place on the earth, the magnetic field oscillates with an amplitude of a few nanoteslas and a characteristic falling tone; since this is in the audible frequency range, the induced electrical signals can be fed into a loudspeaker. These transient signals are called *whistlers*. Each one is the result of a lightning stroke located on the other side of the earth approximately at the end of the magnetic field line that passes through the observer; the lightning sets off a wave in the ionospheric plasma that travels along the field line and disperses as it moves, so that high frequencies of the initial impulse arrive first.

At periods of 1 to 300 s there are quasi-periodic global variations called *micropulsations* that can last for several hours. They range in amplitude up to a few nanotesla, and are excitations by the solar wind of the resonances of the magnetosphere. During periods of high sunspot activity, the sun emits intense fluxes of particles in relatively narrow beams that travel to the earth with the solar wind, and collide with the magnetosphere. In a complex series of interactions these particles are responsible for *magnetic storms*, first identified as fluctuations with periods of minutes to days in the earth's field. If we look at the local X element of the field, storms have the following character: there is an initial phase in which X component is suddenly elevated. This lasts for a few minutes, after which the field drops almost as suddenly to a level below the mean value for the site; then a long period ensues of several days in which the field climbs back to its initial value, the *recovery phase* of the storm. Amplitudes of storms are usually a few tens of nanoteslas, but at high latitudes intense storms of 2000 nT or more have been recorded. In March 1989, a substantial fraction of the Canadian electrical power grid was shut down because of a surge caused by a particularly large magnetic storm, the X magnetogram of which is shown in Figure 1.4.1. The storm is unusual in a number of ways; for example, the drop after the

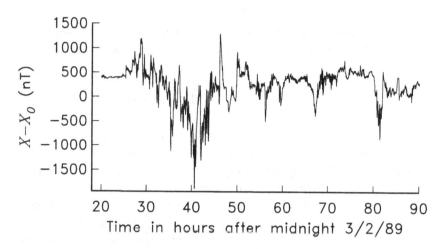

Figure 1.4.1: Magnetogram of the X element relative to a mean level measured at College, Alaska, showing the great magnetic storm of 1989.

initial rise is rather leisurely. Storms tend to recur with a period of about 27 days, this being the period of solar rotation.

, As mentioned in the previous section, the field also varies with a daily period because of the electric currents flowing in the magnetosphere.

All the variations in time that we have discussed so far owe their existence to some external cause; as such they will be of minor concern to us subsequently, except insofar as they can be exploited to provide information about the electrical properties of the earth's crust and mantle (see Chapter 5). Our major focus will be on the main geomagnetic field and the physics and mathematics of the processes that are sources within the earth. On a scale of about one year we encounter for the first time variation with internal origin. Because of *secular variation,* to be described in a later paragraph, the yearly averages of the magnetic field recorded at geomagnetic observatories increase or decrease at a steady rate of a few nanoteslas per year. This steady drift is often very accurately modeled by a quadratic polynomial in time. In 1969 there was observed in the records of European observatories a quite sudden change in the slope of the secular drift; "sudden" in this context means only that it took place in an interval of less than a year. This has come to be known as the *magnetic jerk* of 1969.

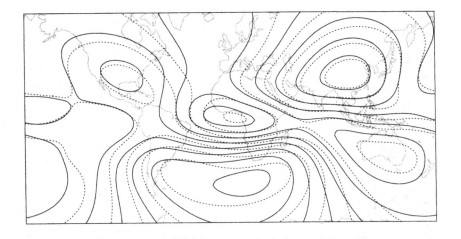

Figure 1.4.2: Westward drift of the nondipole field. The element Z is shown contoured at 5 μT intervals in 1945 (dashed) and in 1980 (solid line). The map projection is Mercator conformal.

The approximate 11-year periodicity in the sunspot population causes a long period modulation of the main field at that period through the frequency of occurrence of magnetic storms.

On time scales of 100 years or so the local declination and inclination of the field can undergo changes of 30° or more. This is a manifestation of the secular variation, caused by fluctuations in the source of the field itself. These changes have amplitudes at the earth's surface of the same size as the nondipole field, tens of microteslas. There is an overall tendency for the nondipole field to drift to the west, as we noted in section 1.2; this *westward drift* is illustrated in Figure 1.4.2, which shows the Z component of the nondipole field in 1945 and then 35 years later in 1980. The average rate of drift points to a complete circuit in 2000 years, although it is doubtful if any individual feature in the nondipole field survives for this length of time. Also there is a question of whether large features can cross the Pacific Ocean; the evidence from paleomagnetic and historical data suggests they do not.

The dipole part of the geomagnetic field is not static, of course. On the time scale of the secular variation, the dipole moment of the earth has been steadily decreasing during this century and is currently falling at a rate of nearly 10 percent per century (see Figure 1.4.3). Measurements of remanent magnetization of pottery and other

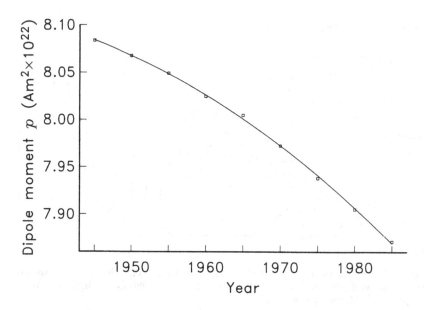

Figure 1.4.3: The earth's magnetic dipole moment as determined from some recent magnetic field models (IGRF 1945–1985).

archeomagnetic materials suggest that over periods of a thousand years and more the dipole strength may fluctuate by a factor of 2 or more. The most dramatic variations of the dipole are complete reversals, the interval between which appears to be random, but has been averaging about 10^5 years for the most recent reversals. Paleomagnetic measurements of magnetized sedimentary rocks and lavas point to a time scale of between 2000 and 10,000 years for the transition of the dipole field from one polarity to the other.

In the most recent epoch, as we mentioned, the dipole has been reversing about every 10^5 years, but a glance at the reversal history (Figure 1.4.4) shows that this has not always been the case even in fairly recent geological times. The reversal rate can undergo large changes: clearly 50 million years ago the interval between reversals was much longer. This suggests a possible physical phenomenon, as yet unidentified, with a characteristic time of nearly 10^8 years.

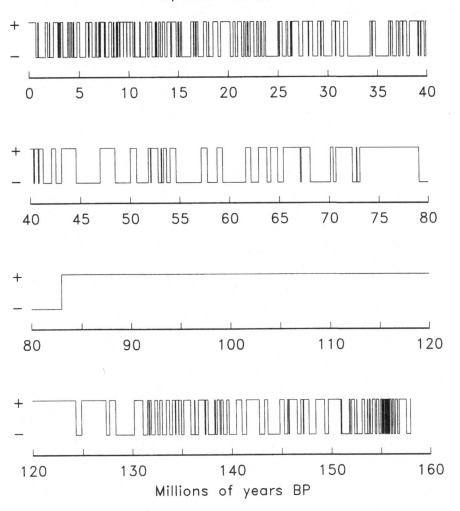

Figure 1.4.4: A geomagnetic polarity scale. Intervals of normal polarity are shown at the plus sign, reversed at the minus sign. The scale is that derived by Harland et al. (1989).

2

CLASSICAL ELECTRODYNAMICS

2.1 Helmholtz's Theorem and Maxwell's Equations

The basic tools for geomagnetism are the equations of classical electrodynamics in polarizable, moving media. Although a complete understanding of electromagnetism presumably involves the use of quantum theory, every phenomenon of interest to us is very accurately handled in the classical approximation. We develop a self-contained and self-consistent version of the classical theory. First we need an experimental definition of the electromagnetic field, one based on the idea that we can measure the force on an arbitrarily small charged test body.

The following is simultaneously a definition and means of measurement of the electric field $\mathbf{E}$ and magnetic field $\mathbf{B}$ relative to an observer in an inertial reference frame. An observer in an inertial frame can assign to all positions $\mathbf{r}$ and times t in vacuo two vectors, $\mathbf{E}(\mathbf{r},t)$ and $\mathbf{B}(\mathbf{r},t)$; and he can assign to each sufficiently small body in vacuo a number q, the charge on the body; if these assignments are done correctly, the force on a body with charge q located at position $\mathbf{r}$ at time t and moving with velocity $\mathbf{v}$ will be

$$\mathbf{f} = q\left[\mathbf{E}(\mathbf{r},t) + \mathbf{v} \times \mathbf{B}(\mathbf{r},t)\right]. \tag{2.1.1}$$

Thus in principle, an observer can, by measuring the forces on a small charged test particle, determine the local magnetic and electric field.

The basis of our development of electrodynamics is that Helmholtz's Theorem (McQuistan, 1965; Blakely, 1995) allows us uniquely to deduce a vector field from the known values of its divergence and curl, while Maxwell's equations define the values of the divergence

15

and curl of the fields in terms of sources, namely charges and electric currents. Thus, starting from the (presumably more primitive) sources, we can calculate the electric and magnetic fields throughout space. Complications naturally arise if the fields themselves affect the sources (for example, currents flow in response to electric fields in a conductor), and we will come to these in due course.

Helmholtz's Theorem states: a vector field $\mathbf{F}$ defined in R^3 (R^3 is all of Euclidean space) and continuously differentiable except for jump discontinuities across certain surfaces is uniquely determined by its divergence, its curl, and jump discontinuities if it approaches $\mathbf{0}$ at ∞; uniqueness is proved in subsection 7.3.3. Indeed, we may always write the field $\mathbf{F}$ as the sum of two parts:

$$\mathbf{F} = -\nabla V + \nabla \times \mathbf{A}, \tag{2.1.2}$$

and the two potentials can be explicitly computed from the integrals

$$V(\mathbf{r}) = \frac{1}{4\pi} \int d^3 \mathbf{s} \, \frac{\nabla \cdot \mathbf{F}(\mathbf{s})}{|\mathbf{r} - \mathbf{s}|} \tag{2.1.3}$$

$$\mathbf{A}(\mathbf{r}) = \frac{1}{4\pi} \int d^3 \mathbf{s} \, \frac{\nabla \times \mathbf{F}(\mathbf{s})}{|\mathbf{r} - \mathbf{s}|}. \tag{2.1.4}$$

We combine the theorem and its ability to provide explicit solutions, with Maxwell's equations in a vacuum; again classically we view charge density and current density as further vector fields, continuously differentiable if necessary.

Here are Maxwell's equations:

$$\nabla \times \mathbf{E} = -\partial_t \mathbf{B} \tag{2.1.5}$$

$$\nabla \cdot \mathbf{E} = \rho/\epsilon_0 \tag{2.1.6}$$

$$\nabla \times \mathbf{B} = \mu_0 \left(\mathbf{J} + \epsilon_0 \partial_t \mathbf{E} \right) \tag{2.1.7}$$

$$\nabla \cdot \mathbf{B} = 0 \tag{2.1.8}$$

where ρ = charge density, $\mathbf{J}$ = current density, μ_0 = permeability of vacuum, and ϵ_0 = capacitivity of vacuum. In SI units (see section 2.3) $\mathbf{B}$ is in teslas, $\mathbf{E}$ is in volts/m, ρ is in coulombs/m^3, $\mathbf{J}$ is in amperes/m^2, $\mu_0 = 4\pi \times 10^{-7}$ henries/m exactly, $\epsilon_0 = 10^7/4\pi c^2$ farads/m, where c is the velocity of light, and c = 299,792,458 m/s exactly.

Suppose S is a stationary oriented surface and $\hat{\mathbf{n}}$ is the unit normal on its positive side. If S carries a surface charge ρ_s coulombs/m, and a surface current $\mathbf{J}_s$ A/m, then there will be jump discontinuities in the electric field $[\mathbf{E}]_-^+ = \mathbf{E}^+ - \mathbf{E}^-$ and in the magnetic field $[\mathbf{B}]_-^+$ across S, whose amounts are

$$[\mathbf{E}]_-^+ = \hat{\mathbf{n}}(\rho_s/\epsilon_0) \qquad (2.1.9)$$

$$[\mathbf{B}]_-^+ = -\mu_0 \hat{\mathbf{n}} \times \mathbf{J}_s. \qquad (2.1.10)$$

If (2.1.2) is interpreted in terms of distribution theory, it implies these jump conditions automatically.

2.2 A Simple Solution: The Static Case

We imagine a system of stationary charges and steady current flows. As we saw in the first chapter, many phenomena of geomagnetism take place on a long time scale, and we might imagine the static approximation would be a good one; the validity of the approximation will be examined in section 2.4, along with a more complete solution to the equations. The electromagnetic fields can be computed for such a simple universe quite easily as follows. If $\partial_t \mathbf{B} = \mathbf{0}$, then from the first of Maxwell's equations (2.1.5) the curl of the electric field vanishes. Hence, by (2.1.2), the electric field may be written simply as the gradient of some scalar, ϕ, which of course is the electric potential: $\mathbf{E} = -\nabla\phi$. From the Maxwell equation (2.1.6) it follows that $\nabla^2\phi = \rho/\epsilon_0$, so, putting all this together with (2.1.3), we find

$$\phi(\mathbf{r}) = \frac{1}{4\pi\epsilon_0} \int d^3s \, \frac{\rho(\mathbf{s})}{|\mathbf{r}-\mathbf{s}|} \qquad (\rho \text{ includes } \rho_s). \qquad (2.2.1)$$

For the magnetic field, it follows from the Maxwell equation (2.1.8) together with (2.1.2–2.1.4) that we may always write $\mathbf{B} = \nabla \times \mathbf{A}$. The vector field $\mathbf{A}$ is called the magnetic vector potential. Then using the fact that in the steady state $\partial_t \mathbf{E} = \mathbf{0}$, we find from (2.1.4) and the Maxwell equation (2.1.7) that

$$\mathbf{A}(\mathbf{r}) = \frac{\mu_0}{4\pi} \int d^3s \, \frac{\mathbf{J}(\mathbf{s})}{|\mathbf{r}-\mathbf{s}|} \qquad (\mathbf{J} \text{ includes } \mathbf{J}_s). \qquad (2.2.2)$$

Since $\mathbf{J} = \boldsymbol{\nabla} \times \mathbf{B}/\mu_0$ in the static case, taking the divergence yields

$$\boldsymbol{\nabla} \cdot \mathbf{J} = 0, \qquad (2.2.3)$$

and this implies that the $\mathbf{A}$ of (2.2.2) satisfies

$$\boldsymbol{\nabla} \cdot \mathbf{A} = 0. \qquad (2.2.4)$$

We continue to study the case in which the currents and charges do not change in time. Suppose now we are interested in the $\mathbf{E}$ and $\mathbf{B}$ produced far outside a ball of radius a centered on the origin by a steady-state charge and current distribution contained entirely inside that ball. This is obviously a natural model for the fields produced within the earth seen by an observer outside. Let $B(a)$ denote the ball, $\partial B(a)$ its boundary. Then the ρ and $\mathbf{J}$ we study must vanish on $\partial B(a)$ and for all $\mathbf{r}$ with $|\mathbf{r}| \geq a$. We will look at $\mathbf{E}(\mathbf{r})$ and $\mathbf{B}(\mathbf{r})$ for $|\mathbf{r}| \gg a$. In (2.2.1) and (2.2.2), we can fix $\mathbf{r}$ and expand $1/|\mathbf{r} - \mathbf{s}|$ in a Taylor series in $-\mathbf{s}$ as

$$\frac{1}{|\mathbf{r} - \mathbf{s}|} = \frac{1}{r} - s_i \partial_i \left(\frac{1}{r}\right) + \frac{1}{2!} s_i s_j \partial_i \partial_j \left(\frac{1}{r}\right) - \frac{1}{3!} s_i s_j s_k \partial_i \partial_j \partial_k \left(\frac{1}{r}\right) + \cdots$$

$$(2.2.5)$$

where we remind the reader that the Einstein summation convention is in use; also recall that $\partial_i = \partial/\partial x_i$. Inserting this expansion in (2.2.1) gives

$$\phi(\mathbf{r}) = \frac{1}{4\pi\epsilon_0} \left[\frac{q}{r} - \mathbf{p} \cdot \boldsymbol{\nabla} \frac{1}{r} + O(qa^2/r^3)\right] \qquad (2.2.6)$$

where

$$q = \int_{B(a)} d^3\mathbf{s} \, \rho(\mathbf{s}) \qquad \text{(includes } \rho_s\text{)} \qquad (2.2.7)$$

$$\mathbf{p} = \int_{B(a)} d^3\mathbf{s} \, \mathbf{s}\rho(\mathbf{s}) \qquad \text{(includes } \rho_s\text{)} \qquad (2.2.8)$$

and $O(qa^2/r^3)$ represents an error term roughly equal to qa^2/r^3 or something less. The integral (2.2.7) is the total charge of the charge distribution in $B(a)$, and $\mathbf{p}$ is the electric dipole moment of that distribution relative to the origin. If $q \neq 0$, $\mathbf{p}$ depends on the choice of origin, but if $q = 0$, $\mathbf{p}$ is independent of where we put the origin; the reader is invited to verify this.

Suppose there is an externally produced field $\mathbf{E}^{(E)}$ that varies only slightly in $B(a)$. We compute the force and torque experienced by the body. By (2.1.1) the external field exerts a force on the local charge distribution whose amount is

$$\mathbf{f} = \int_{B(a)} d^3\mathbf{r}\ \rho(\mathbf{r})\mathbf{E}^{(E)}(\mathbf{r}). \qquad (2.2.9)$$

We expand $\mathbf{E}^{(E)}$ as

$$\mathbf{E}^{(E)}(\mathbf{r}) = \mathbf{E}^{(E)}(\mathbf{0}) + (\mathbf{r} \cdot \boldsymbol{\nabla})\mathbf{E}^{(E)}(\mathbf{0}) + \cdots, \qquad (2.2.10)$$

which is just another example of a Taylor series, this time expanded about the origin, and written in vector notation, rather than in components like (2.2.5). Then (2.2.9) becomes

$$\mathbf{f} = q\mathbf{E}^{(E)}(\mathbf{0}) + (\mathbf{p} \cdot \boldsymbol{\nabla})\mathbf{E}^{(E)}(\mathbf{0}) + \cdots. \qquad (2.2.11)$$

The torque about the origin $\mathbf{0}$ exerted on the local charge distribution by the external field is

$$\mathbf{T} = \int_{B(a)} d^3\mathbf{r}\ \mathbf{r} \times (\rho\mathbf{E}^{(E)}) = \mathbf{p} \times \mathbf{E}^{(E)}(\mathbf{0}) + \cdots. \qquad (2.2.12)$$

In (2.2.11) and (2.2.12) the dotted omissions are higher derivatives of $\mathbf{E}^{(E)}$, which are presumed negligible because $\mathbf{E}^{(E)}$ varies so little in $B(a)$.

Let us now turn to the equivalent set of results for the currents within the body. To deal with the magnetic field, insert (2.2.5) in (2.2.2), obtaining

$$\mathbf{A}(\mathbf{r}) = \frac{\mu_0}{4\pi} \left[\frac{1}{r} \int_{B(a)} d^3\mathbf{s}\ \mathbf{J}(\mathbf{s}) - \partial_i \left(\frac{1}{r}\right) \int_{B(a)} d^3\mathbf{s}\ s_i\mathbf{J}(\mathbf{s}) + O\left(\frac{1}{r^3}\right) \right].$$
$$(2.2.13)$$

Now

$$\partial_j(r_i J_j) = \delta_{ij} J_j + r_i \partial_j J_j = J_i + r_i(\boldsymbol{\nabla} \cdot \mathbf{J}) = J_i, \qquad (2.2.14)$$

so applying Gauss' Divergence Theorem

$$\int_{B(a)} d^3\mathbf{s}\, J_i(\mathbf{s}) = \int_{B(a)} d^3\mathbf{s}\, \partial_j(r_i J_j) = \int_{\partial B(a)} d^2\mathbf{s}\, n_j(r_i J_j)$$
$$= 0, \qquad (2.2.15)$$

because $n_i J_i = \hat{\mathbf{n}} \cdot \mathbf{J} = 0$, which states that no current flows in or out of $B(a)$.

Thus the term in $1/r$ in (2.2.13) vanishes. Moreover,

$$\partial_k(r_i r_j J_k) = \delta_{ik} r_j J_k + \delta_{jk} r_i J_k + r_i r_j \partial_k J_k$$
$$= r_j J_i + r_i J_j, \qquad (2.2.16)$$

so using Gauss' Theorem again

$$\int_{B(a)} d^3\mathbf{s}\, (s_i J_j + s_j J_i) = \int_{\partial B(a)} d^2\mathbf{s}\, n_k s_i s_j J_k = 0. \qquad (2.2.17)$$

Thus

$$\int_{B(a)} d^3\mathbf{s}\, s_i J_j = \tfrac{1}{2} \int_{B(a)} d^3\mathbf{s}\, (s_i J_j - s_j J_i) \qquad (2.2.18)$$

$$= (\delta_{ik}\delta_{j\ell} - \delta_{i\ell}\delta_{jk})\tfrac{1}{2} \int_{B(a)} d^3\mathbf{s}\, s_k J_\ell \qquad (2.2.19)$$

$$= \epsilon_{ijm}\epsilon_{k\ell m}\tfrac{1}{2} \int_{B(a)} d^3\mathbf{s}\, s_k J_\ell \qquad (2.2.20)$$

$$= -\epsilon_{jim} \left[\tfrac{1}{2} \int_{B(a)} d^3\mathbf{s}\, \mathbf{s} \times \mathbf{J} \right]_m. \qquad (2.2.21)$$

Then

$$\partial_i \left(\frac{1}{r} \right) \int_{B(a)} d^3\mathbf{s}\, s_i J_j = -\left(\boldsymbol{\nabla}\frac{1}{r} \times \mathbf{m} \right)_j \qquad (2.2.22)$$

where

$$\mathbf{m} = \tfrac{1}{2} \int_{B(a)} d^3\mathbf{s}\, (\mathbf{s} \times \mathbf{J}) \qquad (2.2.23)$$

and (2.2.13) becomes

$$\mathbf{A}(\mathbf{r}) = -\frac{\mu_0}{4\pi} \mathbf{m} \times \boldsymbol{\nabla}\frac{1}{r} + O\left(m\frac{a}{r^3} \right). \qquad (2.2.24)$$

The vector **m** is the magnetic dipole moment of the current distribution. Because $\int \mathbf{J}\, d^3s = 0$, **m** is independent of the choice of origin.

We note that

$$m_i = \tfrac{1}{2} \int_{B(a)} d^3 r \, \epsilon_{ijk} r_j J_k \tag{2.2.25}$$

$$m_i \epsilon_{i\ell m} = \tfrac{1}{2} \left(\delta_{\ell j} \delta_{mk} - \delta_{\ell k} \delta_{mj} \right) \int_{B(a)} d^3 r \, r_j J_k \tag{2.2.26}$$

$$= \tfrac{1}{2} \int_{B(a)} d^3 r \, \left(r_\ell J_m - r_m J_\ell \right) \tag{2.2.27}$$

$$= \int_{B(a)} d^3 r \, \left(r_\ell J_m \right) \quad \text{from (2.2.17).} \tag{2.2.28}$$

Thus

$$\int_{B(a)} d^3 s \, r_i J_j = \epsilon_{ijk} m_k. \tag{2.2.29}$$

We can use (2.2.29) to calculate the force and torque on the local current distribution due to an externally produced magnetic field $\mathbf{B}^{(E)}$, which changes only slightly in $B(a)$. Although the results can be expressed in a way exactly similar to those for electric charges and fields, a lot more work is required to obtain the answers because the sources are vector fields and the vector potential is more awkward than the scalar one. The force is

$$\mathbf{f} = \int_{B(a)} d^3 r \, \mathbf{J} \times \mathbf{B}^{(E)} \tag{2.2.30}$$

or in components

$$f_i = \int_{B(a)} d^3 r \, \epsilon_{ijk} J_j \left[B_k^{(E)}(\mathbf{0}) + r_\ell \partial_\ell B_k^{(E)}(\mathbf{0}) + \cdots \right]. \tag{2.2.31}$$

Since $\int_{B(a)} d^3 r \, J_j = 0$, the first term drops out, and

$$f_i = \epsilon_{ijk} \partial_\ell B_k^{(E)}(\mathbf{0}) \left(\int_{B(a)} d^3 r \, r_\ell J_j \right) + \cdots \tag{2.2.32}$$

$$= \epsilon_{ijk} \partial_\ell B_k^{(E)}(\mathbf{0}) (\epsilon_{\ell jm} m_m) + \cdots \tag{2.2.33}$$

$$= (\delta_{i\ell} \delta_{km} - \delta_{im} \delta_{k\ell}) m_m \partial_\ell B_k^{(E)}(\mathbf{0}) + \cdots \tag{2.2.34}$$

$$= m_k \partial_i B_k^{(E)}(\mathbf{0}) - m_i \partial_k B_k^{(E)}(\mathbf{0}) + \cdots . \tag{2.2.35}$$

Now $\partial_k B_k^{(E)} = \boldsymbol{\nabla} \cdot \mathbf{B}^{(E)} = 0$, and since $\mathbf{B}^{(E)}$ is produced by currents outside $B(a)$, $\boldsymbol{\nabla} \times \mathbf{B}^{(E)} = \mathbf{0}$. Thus $\partial_i B_k^{(E)}(0) = \partial_k B_i^{(E)}(0)$, and

$$f_i = m_k \partial_k B_i^{(E)}(0) + \cdots . \tag{2.2.36}$$

Therefore

$$\mathbf{f} = (\mathbf{m} \cdot \boldsymbol{\nabla})\mathbf{B}^{(E)}(\mathbf{0}) + \cdots . \tag{2.2.37}$$

This result differs from the equivalent electric-field expression (2.2.11) only because there are no isolated magnetic charges.

The torque exerted by $\mathbf{B}^{(E)}$ on the current inside $B(a)$ is

$$\mathbf{T} = \int_{B(a)} d^3\mathbf{r} \; \mathbf{r} \times (\mathbf{J} \times \mathbf{B}^{(E)}) \tag{2.2.38}$$

$$= \int_{B(a)} d^3\mathbf{r} \left[(\mathbf{r} \cdot \mathbf{B}^{(E)})\mathbf{J} - (\mathbf{r} \cdot \mathbf{J})\mathbf{B}^{(E)} \right] \tag{2.2.39}$$

$$T_i = \int_{B(a)} d^3\mathbf{r} \left(r_j B_j^{(E)} J_i - r_j J_j B_i^{(E)} \right). \tag{2.2.40}$$

In lowest order, $B_j^{(E)}$ is constant, so

$$T_i = B_j^{(E)}(0) \int_{B(a)} d^3\mathbf{r} \; r_j J_i - B_i^{(E)}(0) \int_{B(a)} d^3\mathbf{r} \; r_j J_j + \cdots .$$

From (5.1), $\int d^3\mathbf{r} \; r_j J_j = 0$, and so by (2.2.29)

$$T_i = B_j^{(E)}(0)\epsilon_{jik} m_k = \epsilon_{ikj} m_k B_j^{(E)}(0) + \cdots . \tag{2.2.41}$$

Thus

$$\mathbf{T} = \mathbf{m} \times \mathbf{B}^{(E)}(\mathbf{0}) + \cdots . \tag{2.2.42}$$

Again the dots in (2.2.41) and (2.2.42) represent terms involving higher spatial derivatives of $\mathbf{B}^{(E)}$ and are negligible if $\mathbf{B}^{(E)}$ is nearly constant in $B(a)$. The magnetic torque (the phenomenon that is responsible for the alignment of the compass needle, and so the primary reason why the subject of geomagnetism came into being!) is at first sight exactly like the electrical torque (2.2.12), but this is not really so. The electric dipole moment $\mathbf{p}$ depends on the location of the origin if the body is charged, while **m**, the magnetic dipole moment, is independent of the choice of origin and is therefore an intrinsic property of the current system.

2.3 Maxwell's Equations in a Polarized Medium

Maxwell's equations as we have written them in (2.1.5–2.1.8) are in the form that applies in a vacuum. In many applications we need equations for calculating the behavior of electromagnetic fields inside a material. What this really means is that we need equations for fields that have been smoothed on scales that are much larger than the spaces between the atoms, and which take into account the polarization of the atoms and molecules of which matter is built. Our purpose here is to obtain the necessary equations, taking a severely classical point of view.

Let $B(\mathbf{r}_0, a)$ be the ball of radius a centered at $\mathbf{r}_0$, i.e., the set of all $\mathbf{r}$ with $|\mathbf{r} - \mathbf{r}_0| < a$. Let $|B(\mathbf{r}_0, a)|$ denote the volume of the ball, $4\pi a^3/3$. For any function $f(\mathbf{r})$, let $\langle f \rangle_{B(\mathbf{r}_0,a)}$ denote the average of f over $B(\mathbf{r}_0, a)$:

$$\langle f \rangle_{B(\mathbf{r}_0,a)} = \frac{3}{4\pi a^3} \int_{B(\mathbf{r}_0,a)} d^3\mathbf{s} \; f(\mathbf{s}). \qquad (2.3.1)$$

Let ∂_i denote $\partial/\partial r_i$ or $\partial/\partial p_i$. We claim

$$\partial_i \langle f \rangle_{B(\mathbf{r}_0,a)} = \langle \partial_i f \rangle_{B(\mathbf{r}_0,a)} \qquad (2.3.2)$$

(i.e., differentiation commutes with averaging). To see this, note that

$$\langle f \rangle_{B(\mathbf{r}_0,a)} = \frac{3}{4\pi a^3} \int_{B(l0,a)} d^3\mathbf{s} \; f(\mathbf{s} + \mathbf{r}_0) \qquad (2.3.3)$$

$$\frac{\partial}{\partial r_i} \langle f \rangle_{B(\mathbf{r}_0,a)} = \frac{3}{4\pi a^3} \int_{B(l0,a)} d^3\mathbf{s} \; \partial_i f(\mathbf{s} + \mathbf{r}_0) \qquad (2.3.4)$$

$$= \frac{3}{4\pi a^3} \int_{B(\mathbf{r}_0,a)} d^3\mathbf{s} \; \partial_i f(\mathbf{s}). \qquad (2.3.5)$$

It is obvious that (2.3.3–2.3.5) also works if we replace ∂_i by ∂_t. Therefore, Maxwell's equations can be averaged over $B(\mathbf{r}, a)$, and the derivatives can be done after averaging instead of before, to give

$$\nabla \times \langle \mathbf{E} \rangle = -\partial_t \langle \mathbf{B} \rangle \tag{2.3.6}$$

$$\nabla \cdot \langle \mathbf{B} \rangle = \langle \rho \rangle / \epsilon_0 \tag{2.3.7}$$

$$\nabla \times \langle \mathbf{B} \rangle = \mu_0 (\langle \mathbf{J} \rangle + \epsilon_0 \partial_t \langle \mathbf{E} \rangle) \tag{2.3.8}$$

$$\nabla \cdot \langle \mathbf{B} \rangle = 0. \tag{2.3.9}$$

A point charge at $\mathbf{s}$ moving with velocity $\mathbf{v}$ and having charge q has charge density $\rho(\mathbf{r}) = q\delta(\mathbf{r} - \mathbf{s})$ and current density $\mathbf{J}(\mathbf{r}) = q\mathbf{v}\delta(\mathbf{r} - \mathbf{s})$ where δ is the Dirac delta function. Thus for a cloud of moving charges,

$$\langle \rho \rangle = \frac{1}{|B(\mathbf{r}, a)|} \sum_{B(\mathbf{r}, a)} q \tag{2.3.10}$$

$$\langle \mathbf{J} \rangle = \frac{1}{|B(\mathbf{r}, a)|} \sum_{B(\mathbf{r}, a)} q\mathbf{v}, \tag{2.3.11}$$

the sums being over all charges in $B(\mathbf{r}, a)$. It is in the form (2.3.6–2.3.9), (2.3.10, 2.3.11) that one usually works with Maxwell's equations in regions of macroscopic ($\gg 10^{-10}$ meters) size.

Suppose now that a certain region is filled with atoms rather than point charges. These atoms will have charge and current distributions of finite extent in space (length scale $\sim 10^{-10}$ meters) but at macroscopic distances, each looks like a point charge, an electric dipole, a magnetic dipole, plus corrections of higher order in (atomic size/distance), which we neglect. We can still define $\langle \rho \rangle$ and $\langle \mathbf{J} \rangle$, but calculating them is now more complicated than (2.3.10, 2.3.11) because $\partial B(a)$ will intersect some of the atoms. The calculation is done, for example, in Rosenfeld (1951). He assumes, as we will, that there is a range of values of a that are much greater than atomic size but so small that $\langle \rho \rangle$ and $\langle \mathbf{J} \rangle$ change very little if the center of $B(\mathbf{r}, a)$ is translated by a distance a. He defines

$$\mathbf{P}(\mathbf{r}) = \frac{1}{|B(\mathbf{r}, a)|} \sum_{B(\mathbf{r}, a)} \mathbf{p} \tag{2.3.12}$$

$$\mathbf{M}(\mathbf{r}) = \frac{1}{|B(\mathbf{r}, a)|} \sum_{B(\mathbf{r}, a)} \mathbf{m} \tag{2.3.13}$$

where the sums extend over all atoms whose centers of mass are in $B(\mathbf{r}, a)$. The fields $\mathbf{P}$ and $\mathbf{M}$ are called the electric polarization per unit volume and the magnetization, or magnetic polarization per unit volume; they are just the amount of the electric dipole $\mathbf{p}$ per unit volume and magnetic dipole moment $\mathbf{m}$ per unit volume, averaged within the ball. Both $\langle \rho \rangle$ and $\langle \mathbf{J} \rangle$ can be expanded in power series in (atomic size/a), and the lowest-order terms, found by a calculation given by Rosenfeld but too lengthy to give here, are

$$\langle \rho \rangle = \rho^{(F)} - \boldsymbol{\nabla} \cdot \mathbf{P} \qquad (2.3.14)$$

$$\langle \mathbf{J} \rangle = \mathbf{J}^{(F)} + \boldsymbol{\nabla} \times \mathbf{M} + \partial_t \mathbf{P} \qquad (2.3.15)$$

where the superscript (F) stands for *free* charges not bound into atoms: thus $\rho^{(F)}$ and $\mathbf{J}^{(F)}$ are calculated from any moving charges other than the orbital electrons in the atoms. They include net charge, if any, due to an excess of electrons over protons, and current produced by the drift of electrons or ions. They are calculated from equations (2.3.10, 2.3.11), approximating atoms, ions and electrons by point charges. The (perhaps rather unexpected) terms like $\boldsymbol{\nabla} \cdot \mathbf{P}$ arise from contributions in the boundary of the ball. If S is a stationary oriented surface with positive unit normal $\hat{\mathbf{n}}$ across which $\mathbf{M}$ and $\mathbf{P}$ have jump discontinuities, the jumps in $\langle \mathbf{E} \rangle$ and $\langle \mathbf{B} \rangle$ are

$$[\langle \mathbf{E} \rangle]_-^+ = \hat{\mathbf{n}} \left(\rho_S^{(F)} - \hat{\mathbf{n}} \cdot [\mathbf{P}]_-^+ \right) / \epsilon_0 \qquad (2.3.16)$$

$$[\langle \mathbf{B} \rangle]_-^+ = \mu_0 \left(\mathbf{J}_S^{(F)} + \hat{\mathbf{n}} \times [\mathbf{M}]_-^+ \right) \times \hat{\mathbf{n}} \qquad (2.3.17)$$

where $\rho_S^{(F)}$ and $\mathbf{J}_S^{(F)}$ stand for any net surface charge or surface current due to concentration of stationary or moving ions and electrons on S. These are treated as point charges in calculating the averages $\rho_S^{(F)}$ and $\mathbf{J}_S^{(F)}$. Now Maxwell's equations, averaged over $B(\mathbf{r}, a)$, become

$$\boldsymbol{\nabla} \times \langle \mathbf{E} \rangle = -\partial_t \langle \mathbf{B} \rangle \qquad (2.3.18)$$

$$\epsilon_0 \boldsymbol{\nabla} \cdot \langle \mathbf{E} \rangle = \rho^{(F)} - \boldsymbol{\nabla} \cdot \mathbf{P} \qquad (2.3.19)$$

$$\boldsymbol{\nabla} \times \langle \mathbf{B} \rangle / \mu_0 = \mathbf{J}^{(F)} + \partial_t \mathbf{P} + \boldsymbol{\nabla} \times \mathbf{M} + \epsilon_0 \partial_t \langle \mathbf{E} \rangle \qquad (2.3.20)$$

$$\boldsymbol{\nabla} \cdot \langle \mathbf{B} \rangle = 0. \qquad (2.3.21)$$

The jump conditions on $\langle \mathbf{E} \rangle$ and $\langle \mathbf{B} \rangle$ are (2.3.16, 2.3.17). To solve these equations for $\langle \mathbf{E} \rangle$ and $\langle \mathbf{B} \rangle$ we need to know $\rho^{(F)}$, $\mathbf{J}^{(F)}$, $\rho_S^{(F)}$, $\mathbf{J}_S^{(F)}$, $\mathbf{M}$, and $\mathbf{P}$. Both $\rho^{(F)}$ and $\rho_S^{(F)}$ are easy. If we add the time derivative of (2.3.19) to the divergence of (2.3.20) we get

$$\partial_t \rho^{(F)} + \boldsymbol{\nabla} \cdot \mathbf{J}^{(F)} = 0. \qquad (2.3.22)$$

That is, the charge not bound into atoms is conserved. If we know $\rho^{(F)}$ at $t = 0$, and $\mathbf{J}^{(F)}$ for all $\mathbf{r}$ and t, (2.3.22) gives $\rho^{(F)}$ for $t > 0$. From (2.3.16) and (2.3.20)

$$\partial_t \rho_S^{(F)} + \boldsymbol{\nabla}_S \cdot \mathbf{J}_S^{(F)} = -\hat{\mathbf{n}} \cdot \lfloor \mathbf{J}^{(F)} \rfloor_-^+ \qquad (2.3.23)$$

where $\boldsymbol{\nabla}_S$ is the surface gradient on S (see subsection 7.4.3). That is, charge is conserved on S if one includes flow of charge onto S from space. If we know $\rho_S^{(F)}$ at $t = 0$, we can find it at all later times.

Thus, to solve (2.3.18–2.3.21) and (2.3.16, 2.3.17) for $\langle \mathbf{E} \rangle$ and $\langle \mathbf{B} \rangle$ all we really need are $\mathbf{J}^{(F)}$, $\mathbf{J}_S^{(F)}$, $\mathbf{P}$, and $\mathbf{M}$. These can be given *a priori* or determined by $\langle \mathbf{E} \rangle$ and $\langle \mathbf{B} \rangle$ themselves. The simplest example of the latter situation is

$$\mathbf{J}^{(F)} = \sigma \langle \mathbf{E} \rangle \qquad (2.3.24)$$

$$\mathbf{J}_S^{(F)} = \sigma_S [\langle \mathbf{E} \rangle - \hat{\mathbf{n}} \hat{\mathbf{n}} \cdot \langle \mathbf{E} \rangle] = \sigma_S \mathbf{E}_S \qquad (2.3.25)$$

$$\mathbf{P} = \epsilon_0 \chi_E \langle \mathbf{E} \rangle \qquad (2.3.26)$$

$$\mathbf{M} = \beta \langle \mathbf{B} \rangle / \mu_0 \qquad (2.3.27)$$

where σ, σ_S, χ_E, and β are constants characteristic of the material. These relations do not have the same standing as Maxwell's equations: laws like (2.3.24–2.3.27) must be determined empirically or calculated from a detailed knowledge of the atomic arrangement in the material. They are *constitutive relations*. We recognize (2.3.24) as Ohm's law with σ as the electrical conductivity. In many materials met in paleomagnetism, for example, the constitutive relation connecting magnetization to average field $\mathbf{B}$ is much more complex than (2.3.27); not only is the magnetization a complicated nonlinear function of the field, it usually depends upon the magnetic field history of the material through time. Notice that the constitutive relation (2.3.27) for magnetization is usually written in terms of $\mathbf{H}$ defined in (2.3.30):

$$\mathbf{M} = \chi_M \mathbf{H} \qquad (2.3.28)$$

where the dimensionless number χ_M is called the magnetic susceptibility.

Maxwell's equations(2.3.18–2.3.21) and (2.3.16, 2.3.17) can be given a much simpler appearance if we *define* fields $\mathbf{H}$ and $\mathbf{D}$ by

$$\mathbf{D} = \epsilon_0 \langle \mathbf{E} \rangle + \mathbf{P} \qquad (2.3.29)$$

$$\mathbf{H} = \langle \mathbf{B} \rangle / \mu_0 - \mathbf{M}. \qquad (2.3.30)$$

The field $\mathbf{D}$ is called the *electric displacement vector*. Unfortunately, particularly in view of its secondary role in the theory, $\mathbf{H}$ has traditionally been called the magnetic field, while the fundamental vector $\mathbf{B}$ was called the flux intensity, or the magnetic induction. In this book we call $\mathbf{B}$ the magnetic field vector, and by analogy with $\mathbf{D}$ we invent a new name for $\mathbf{H}$ – the *magnetic displacement vector*. In the new field variables (2.3.19, 2.3.20) are considerably simplified:

$$\boldsymbol{\nabla} \cdot \mathbf{D} = \rho^{(F)} \qquad (2.3.31)$$

$$\boldsymbol{\nabla} \times \mathbf{H} = \mathbf{J}^{(F)} + \partial_t \mathbf{D}. \qquad (2.3.32)$$

And the jump conditions (2.3.16, 2.3.17) can be written in terms of components normal and parallel to S:

$$\hat{\mathbf{n}} \times [\mathbf{E}]_-^+ = \mathbf{0} \qquad (2.3.33)$$

$$\hat{\mathbf{n}} \cdot [\mathbf{D}]_-^+ = \rho_S^{(F)} \qquad (2.3.34)$$

$$\hat{\mathbf{n}} \times [\mathbf{H}]_-^+ = \mathbf{J}_S^{(F)} \qquad (2.3.35)$$

$$\hat{\mathbf{n}} \cdot [\mathbf{B}]_-^+ = 0. \qquad (2.3.36)$$

Notice that we have removed the brackets denoting averaging in the equations containing $\mathbf{E}$ and $\mathbf{B}$. We will be dealing only with regions much larger than atoms, so we will treat only $\langle \mathbf{E} \rangle$ and $\langle \mathbf{B} \rangle$, not the highly oscillatory $\mathbf{E}$ and $\mathbf{B}$. Therefore, henceforth we abbreviate $\langle \mathbf{E} \rangle$

and $\langle \mathbf{B} \rangle$ as $\mathbf{E}$ and $\mathbf{B}$. Thus finally the form in which we will use Maxwell's equations is

$$\boldsymbol{\nabla} \times \mathbf{E} = -\partial_t \mathbf{B} \tag{2.3.37}$$

$$\boldsymbol{\nabla} \cdot \mathbf{D} = \rho^{(F)} \tag{2.3.38}$$

$$\boldsymbol{\nabla} \times \mathbf{H} = \mathbf{J}^{(F)} + \partial_t \mathbf{D} \tag{2.3.39}$$

$$\boldsymbol{\nabla} \cdot \mathbf{B} = 0 \tag{2.3.40}$$

where

$$\mathbf{D} = \epsilon_0 \mathbf{E} + \mathbf{P} \tag{2.3.41}$$

$$\mathbf{H} = \mathbf{B}/\mu_0 - \mathbf{M}. \tag{2.3.42}$$

The table below shows the comparison of units in the International System (SI) and the Electrostatic (ESU) and Electromagnetic (EMU) systems of units. In the table, $\bar{c} = 2.99792458 =$ speed of light

Quantity	1 SI unit	ESU	EMU
Length	meter	10^2 cm	10^2 cm
Mass	kilogram	10^3 gram	10^3 g
Time	second	1 second	1 s
Force	newton	10^5 dynes	10^5 dynes
Energy	joule	10^7 ergs	10^7 ergs
Power	watt	10^7 ergs/s	10^7 ergs/s
Electric Charge	coulomb	$10^9 \times \bar{c}$ statC	10^{-1} abC
Electric Potential	volt	$10^{-2} \times 1/\bar{c}$ statV	10^8 abV
Electric Field $\mathbf{E}$	volt/m	$10^{-4} \times 1/\bar{c}$ statV/cm	10^6 abV/cm
Electric Displacement $\mathbf{D}$	coulomb/m^2	$10^5 \times 4\pi\bar{c}$ statC/cm^2	$4\pi \times 10^{-5}$ abC/cm^2
Electric Polarization $\mathbf{P}$	coulomb/m^2	$10^5 \times \bar{c}$ statC/cm^2	$4\pi \times 10^{-5}$ abC/cm^2
Current Density $\mathbf{J}$	ampere/m^2	$10^5 \times \bar{c}$ statA/cm^2	10^{-5} abA/cm^2
Conductivity σ	siemens/m	$10^9 \times \bar{c}^2$ statΩ^{-1} cm^{-1}	10^{-11} abΩ^{-1} cm^{-1}
Magnetic Field $\mathbf{B}$	tesla	$10^{-6} \times 1/\bar{c}$	10^4 gauss
Magnetic Displacement $\mathbf{H}$	ampere/m	$10^7 \times 4\pi\bar{c}$	$4\pi \times 10^{-3}$ oersted
Magnetization $\mathbf{M}$	ampere/m	$10^{-3} \times 1/\bar{c}$	10^{-3}
Magnetic Dipole Moment $\mathbf{m}$	ampere m^2	$10^3 \times 1/\bar{c}$	10^3
Magnetic Susceptibility χ_M	1	$10^{-20} \times 1/4\pi\bar{c}^2$	$1/4\pi$

in units of 10^8 m/s. The subunit 1 nanotesla = 1 nT = 10^{-9} tesla = 10^{-5} gauss = 1 gamma = 1 γ. In SI units, $\mu_0 = 4\pi \times 10^{-7}$ henries/m; $\epsilon_0 = 10^{-9}/(4\pi \bar{c}^2)$ farads/m; $(\mu_0 \epsilon_0)^{-1/2} = c$, the speed of light. Both μ_0 and c are defined as exact numbers in SI units, and so therefore ϵ_0 is a defined constant too. Note the naming rules in SI: all units named for a person are spelled out as lower-case words, for example, tesla, ampere, siemens, watt; each unit is abbreviated to the initial letter of the full name, and is written uppercase: hence T, A, S, W. Exceptions to the abbreviation rule are Pa for pascal, Ω for ohm, and °C for celsius degrees.

2.4 On Judicious Neglect of Terms in Equations

In this section we discuss approximation by dropping terms that are suspected of being small from the full system of equations; one application is the derivation of the pre-Maxwell equations in which we neglect the displacement current, $\partial_t \mathbf{D}$ in (2.3.39), or $\epsilon_0 \partial_t \mathbf{E}$ in (2.1.7). Suppose A and B are expressions involving some physical quantities we would like to calculate, and the physical law governing those quantities takes the form

$$A = B. \tag{2.4.1}$$

For example: $A = \boldsymbol{\nabla} \times \mathbf{E}$, $B = -\partial_t \mathbf{B}$. Suppose that without any law to guide us, we would expect *a priori* that the order of magnitude of A was about α and the order of magnitude of B was about β. If $\beta \ll \alpha$, then the essential content of (2.4.1) is that the physical law makes A much smaller than we expected, and in lowest approximation (2.4.1) can be replaced by

$$A = 0. \tag{2.4.2}$$

If (2.4.2) has solutions, we can see whether these solutions really do make B of order β, and if so we may start a perturbation expansion for A and B in powers of β/α by beginning with (2.4.2). If (2.4.2) has no solutions, we must reexamine our *a priori* expectations about α and β and see what went wrong.

If the physical law has the form $A_1 + \cdots + A_N = 0$, and we have *a priori* estimates about the orders of magnitude of some but not all the A_n's, say $A_n \approx \alpha_n$, then for any m, if there is an n with $\alpha_m \ll \alpha_n$, we can omit A_m in solving the equations.

As an example, consider the exact Maxwell equations (2.1.5–2.1.8), where field quantities are exact, not averages over polarized media. Rearranging (2.1.7, 2.1.8) somewhat, we have

$$\mathbf{\nabla} \times \mathbf{E} = -\partial_t \mathbf{B} \qquad (2.4.3)$$

$$\mathbf{\nabla} \times \mathbf{B} = \mu_0 \mathbf{J} + \frac{1}{c^2}\, \partial_t \mathbf{E}. \qquad (2.4.4)$$

Recall $c^2 = 1/\mu_0 \epsilon_0$. Suppose we are interested in fields whose typical length scale is L and whose typical time scale is T. Then *a priori* (i.e., without any physical laws to guide us) we might expect $\mathbf{\nabla} \cdot \mathbf{E} \approx E/L$, $|\mathbf{\nabla} \times \mathbf{E}| \approx E/L$, $|\partial_t \mathbf{E}| \approx E/T$, $|\partial_t \mathbf{B}| \approx B/T$, and $|\mathbf{\nabla} \times \mathbf{B}| \approx B/L$. One way to see this is to imagine the decomposition of the fields into their Fourier spectra, in either frequency or wavenumber, whichever is appropriate. Now (2.4.3) and (2.4.4) tell us that these various magnitudes are not independent; according to (2.4.3) $E/L \approx B/T$ and therefore $|\partial_t \mathbf{E}| \approx LB/T^2$. It further follows that

$$\frac{1}{c^2}\, |\partial_t \mathbf{E}| \ / \ |\mathbf{\nabla} \times \mathbf{B}| \approx \frac{(L/T)^2}{c^2}. \qquad (2.4.5)$$

Thus $|\partial_t \mathbf{E}|/c^2 \ll |\mathbf{\nabla} \times \mathbf{B}|$ as long as $(L/T) \ll c$. This condition implies that time scales are long compared with the time needed for light to cross a feature with a typical length in the system, surely a valid restriction in geomagnetism. It is apparently a good approximation, therefore, to drop the second term on the right of (2.4.4), which is called the displacement current. Maxwell's equations without the displacement current $\partial_t \mathbf{E}/c^2$ will be called the *pre-Maxwell equations* because this omission returns classical electromagnetism to its state in the times before Maxwell.

Some housekeeping remains. Do the pre-Maxwell equations have solutions? Obviously, yes: all electrostatic fields, magnetostatic fields, and inductively generated electric fields of elementary electromagnetism. Having found such a solution, we must verify *a posteriori* that it satisfies the order-of-magnitude estimates that justified neglecting the displacement current.

Logically, the situation is a little unsatisfactory. By neglecting $\partial_t \mathbf{E}/c^2$, we find solutions in which $\partial_t \mathbf{E}/c^2$ is negligible. But are these solutions of the pre-Maxwell equations really close to genuine solutions of the Maxwell equations? Most physical theories are too complicated to permit a rigorous answer to this question. Fortunately, the Maxwell equations are exactly soluble in terms of retarded potentials, so it can be shown for them that our approximation scheme always works. If $L/T \ll$ c, the exact solution is close to a pre-Maxwell solution.

The argument is as follows. First we find a closed-form solution to the full Maxwell equations, like the one we found for the static case (2.2.1, 2.2.2). We can then check the pre-Maxwell solution against the correct answer. Since $\mathbf{\nabla} \cdot \mathbf{B} = 0$, we can find many vector potentials $\mathbf{A}'$ with

$$\mathbf{B} = \mathbf{\nabla} \times \mathbf{A}'. \qquad (2.4.6)$$

Then (2.4.3) can be written $\mathbf{\nabla} \times (\mathbf{E} + \partial_t \mathbf{A}') = \mathbf{0}$, and so for any choice of an $\mathbf{A}'$ satisfying (2.4.6) there is a scalar potential ϕ', unique up to an additive function of time alone, such that

$$\mathbf{E} = -\mathbf{\nabla}\phi' - \partial_t \mathbf{A}'. \qquad (2.4.7)$$

Suppose we choose an arbitrary function ψ and define $\mathbf{A} = \mathbf{A}' + \mathbf{\nabla}\psi$. Then $\mathbf{B} = \mathbf{\nabla} \times \mathbf{A}$, and $\mathbf{E} = -\mathbf{\nabla}\phi' - \partial_t(\mathbf{A} - \mathbf{\nabla}\psi) = -\mathbf{\nabla}(\phi' - \partial_t\psi) - \partial_t \mathbf{A}$. Thus if $\mathbf{A}'$ and ϕ' give $\mathbf{B}$ and $\mathbf{E}$ correctly via (2.4.6) and (2.4.7), so do $\mathbf{A}$ and ϕ, where

$$\mathbf{A} = \mathbf{A}' + \mathbf{\nabla}\psi \qquad (2.4.8)$$

$$\phi = \phi' - \partial_t\psi. \qquad (2.4.9)$$

The replacement of $\mathbf{A}'$ and ϕ' by $\mathbf{A}$ and ϕ is a *gauge transformation*. We make use of this freedom of choice to construct potentials that are decoupled from each other and in which the source for one is charge density ρ, and for the other current density $\mathbf{J}$.

So far we have accounted for two of Maxwell's equations, $\mathbf{\nabla} \cdot \mathbf{B} = 0$ and (2.4.3). The two other Maxwell equations written in terms of these potentials are

$$-\nabla^2\phi' - \partial_t(\mathbf{\nabla} \cdot \mathbf{A}') = \rho/\epsilon_0 \qquad (2.4.10)$$

$$\mathbf{\nabla} \times (\mathbf{\nabla} \times \mathbf{A}') = \mu_0\mathbf{J} - \mathbf{\nabla}\left(\frac{1}{c^2}\partial_t\phi'\right) - \frac{1}{c^2}\partial_t^2\mathbf{A}'. \qquad (2.4.11)$$

But by the vector identity $\mathbf{\nabla} \times (\mathbf{\nabla} \times \mathbf{A}') = \mathbf{\nabla}(\mathbf{\nabla} \cdot \mathbf{A}') - \nabla^2\mathbf{A}'$, so these equations can be written

$$\frac{1}{c^2}\partial_t^2\phi' - \nabla^2\phi' - \partial_t\left(\mathbf{\nabla} \cdot \mathbf{A}' + \frac{1}{c^2}\partial_t\phi'\right) = \rho/\epsilon_0 \qquad (2.4.12)$$

$$\frac{1}{c^2} \partial_t^2 \mathbf{A}' - \nabla^2 \mathbf{A}' = \mu_0 \mathbf{J} - \nabla \left(\nabla \cdot \mathbf{A}' + \frac{1}{c^2} \partial_t \phi' \right). \qquad (2.4.13)$$

These equations are also satisfied by $\mathbf{A}$ and ϕ so that ψ is still arbitrary. Now we can choose ψ so that

$$\nabla \cdot \mathbf{A} + \frac{1}{c^2} \partial_t \phi = 0. \qquad (2.4.14)$$

From (2.4.8, 2.4.9) this requires

$$\nabla \cdot \mathbf{A}' + \nabla^2 \psi + \frac{1}{c^2} \partial_t \phi' - \frac{1}{c^2} \partial_t^2 \psi = 0 \qquad (2.4.15)$$

or

$$\frac{1}{c^2} \partial_t^2 \psi - \nabla^2 \psi = \frac{1}{c^2} \partial_t \phi' + \nabla \cdot \mathbf{A}'. \qquad (2.4.16)$$

Having chosen ψ in order to solve (2.4.16), we define $\mathbf{A}$ and ϕ by (2.4.8, 2.4.9) and then we have

$$\mathbf{B} = \nabla \times \mathbf{A} \qquad (2.4.17)$$

$$\mathbf{E} = -\nabla \phi - \partial_t \mathbf{A} \qquad (2.4.18)$$

where the potentials ϕ and $\mathbf{A}$ satisfy (2.4.14) and

$$\frac{1}{c^2} \partial_t^2 \phi - \nabla^2 \phi = \rho/\epsilon_0 \qquad (2.4.19)$$

$$\frac{1}{c^2} \partial_t^2 \mathbf{A} - \nabla^2 \mathbf{A} = \mu_0 \mathbf{J}. \qquad (2.4.20)$$

Suppose first that we are interested only in the $\mathbf{E}$ and $\mathbf{B}$ produced by charges and currents inside a certain bounded region V. In this case we can solve the above wave equations explicitly in terms of the sources ρ and $\mathbf{J}$: ϕ and $\mathbf{A}$ are the *retarded potentials*.

$$\phi(\mathbf{r}, t) = \frac{1}{4\pi\epsilon_0} \int_V d^3\mathbf{s} \, \frac{\rho(\mathbf{s}, t - |\mathbf{r} - \mathbf{s}|/c)}{|\mathbf{r} - \mathbf{s}|} \qquad (2.4.21)$$

$$\mathbf{A}(\mathbf{r}, t) = \frac{\mu_0}{4\pi} \int_V d^3\mathbf{s} \, \frac{\mathbf{J}(\mathbf{s}, t - |\mathbf{r} - \mathbf{s}|/c)}{|\mathbf{r} - \mathbf{s}|}. \qquad (2.4.22)$$

The reader may check that these are solutions by substitution into the differential equations. Of course we need a uniqueness theorem to be sure there are no other solutions, and indeed there are unless further assumptions are made: we must demand that the fields decay to zero at infinity and require that physical signals arrive at an observer only after the initial event within V. So-called advanced potentials, with the sign of the delay term reversed, also satisfy (2.4.19, 2.4.20), and in classical electromagnetic theory they must be rejected.

Returning to the question of approximation, suppose that T is the shortest period present with appreciable amplitude in the time variations of ρ and $\mathbf{J}$. Then (ignoring 2π's)

$$|\partial_t \rho| \le |\rho|/T, \quad |\partial_t \mathbf{J}| \le |\mathbf{J}|/T. \tag{2.4.23}$$

Now suppose that all the positions $\mathbf{r}$ at which we want to know $\mathbf{E}$ and $\mathbf{B}$ satisfy

$$|\mathbf{r} - \mathbf{s}| \ll cT \tag{2.4.24}$$

for every $\mathbf{s}$ in V. Then at every such $\mathbf{r}$, and for all $\mathbf{s}$ in V,

$$\frac{|\mathbf{r} - \mathbf{s}|}{c} |\partial_t \rho| \ll |\rho|, \quad \frac{|\mathbf{r} - \mathbf{s}|}{c} |\partial_t \mathbf{J}| \ll |\mathbf{J}|. \tag{2.4.25}$$

A Taylor series in t shows then that with good accuracy we can omit the time delays $|\mathbf{r}-\mathbf{s}|/c$ in (2.4.21, 2.4.22). But then ϕ and $\mathbf{A}$ give, via (2.4.17, 2.4.18), the standard solution of the pre-Maxwell equations. The exact solutions are closely approximated by solutions of the pre-Maxwell equations.

Next, suppose we are interested in $\mathbf{E}$ and $\mathbf{B}$ produced by very distant matter. If this $\mathbf{E}$ and $\mathbf{B}$ are static, of course we can neglect displacement current; it vanishes. If $\mathbf{E}$ and $\mathbf{B}$ are not static, then we observe them as electromagnetic waves. The relation between their length and time scales is $L/T = c$. If we are interested only in variations with $L/T \ll c$, we will filter out these waves, perhaps by low-pass filtering in time. Our decision to study only signals with $L/T \ll c$ filters out high frequencies (assuming L is given, e.g., as the radius of the earth or the thickness of an interesting layer of sediment). This time-averaging removes the displacement current and returns us to the pre-Maxwell equations.

As a second application of the idea of approximation of exact equations, we discuss the appropriate equations governing the main geomagnetic field (and the slowly varying parts of the externally generated field). The atmosphere is only very slightly polarizable magnetically, and we can set $\mathbf{M} = \mathbf{0}$. Hence (2.3.42) states $\mathbf{B} = \mu_0\mathbf{H}$. We know that $|\mathbf{B}| \approx 30$ to $60\mu T$ and so $|\mathbf{H}| \approx 25$ to 50 A/m. Another observational fact we call upon is that $\mathbf{E}$ at the surface is ≈ 100 V/m downward and it decreases with altitude. The electric polarization $\mathbf{P}$ is negligible, so by (2.3.41) $\mathbf{D} = \epsilon_0\mathbf{E}$ and $|\mathbf{D}| \approx 10^{-9}$ C/m^2. If we are interested in length scales of the order of a, the radius of the earth, and time scales of the order of T, then we expect *a priori* that

$$|\partial_t \mathbf{D}| \approx |\mathbf{D}|/T = 10^{-9}/T \qquad (2.4.26)$$

$$|\nabla \times \mathbf{H}| \approx |\mathbf{H}|/a \approx 25/a \qquad (2.4.27)$$

and dividing these two we find

$$\frac{|\partial_t \mathbf{D}|}{|\nabla \times \mathbf{H}|} \approx \frac{4a}{T} \times 10^{-11} \approx \frac{2.5 \times 10^{-4}}{T}. \qquad (2.4.28)$$

Thus if $T \gg 2.5 \times 10^{-4}$ s, then for global-scale magnetic fields we can neglect the displacement current in the atmosphere and write the averaged Maxwell equation (2.3.39) as

$$\nabla \times \mathbf{H} = \mathbf{J}^{(F)}. \qquad (2.4.29)$$

We can go further. The conductivity of the atmosphere near the ground can be measured (it is due to ions produced by cosmic rays and radon decay products). It is about $\sigma \approx 10^{-13}$ S/m, so by Ohm's law

$$J^{(F)} = \sigma E \approx 10^{-11} \text{ A/m}^2. \qquad (2.4.30)$$

But *a priori*

$$|\nabla \times \mathbf{H}| \approx 25/a = 4 \times 10^{-6} \text{ A/m}^2, \qquad (2.4.31)$$

which is obviously much more than the current density in (2.4.30). Thus the essential content of (2.4.29) in the atmosphere is $\nabla \times \mathbf{H} = \mathbf{0}$. Recall we already observed that $\mathbf{M} = \mathbf{0}$ and that this implies $\mathbf{H} = \mathbf{B}/\mu_0$; it follows that

$$\nabla \times \mathbf{B} = \mathbf{0} \qquad \text{in the atmosphere.} \qquad (2.4.32)$$

The other equation for **B** in the atmosphere is the Maxwell equation (2.3.40)

$$\nabla \cdot \mathbf{B} = 0. \tag{2.4.33}$$

It follows from (2.4.32) and Helmholtz's Theorem that in the atmosphere there is a scalar potential Ω such that

$$\mathbf{B} = -\nabla\Omega \tag{2.4.34}$$

and by (2.4.33) Ω is harmonic, which means it obeys Laplace's equation:

$$\nabla^2\Omega = 0. \tag{2.4.35}$$

Harmonic functions are very smooth: they are infinitely differentiable at all points away from the boundaries and are even analytic, which means they have locally convergent power series about each interior point.

2.5 Internal and External Fields

We turn briefly again to the idea of sources for the earth's magnetic field and their location. Recall that Helmholtz's Theorem (subsection 7.3.3) asserts that if **v** is a vector field in R^3 that vanishes at infinity and is continuously differentiable except for possible jump discontinuities across an oriented surface S, then **v** is determined everywhere if we know $\nabla \times \mathbf{v}$ and $\nabla \cdot \mathbf{v}$ outside S and $[\mathbf{v}]_-^+$ on S. The three quantities $\nabla \times \mathbf{v}$, $\nabla \cdot \mathbf{v}$, and $[\mathbf{v}]_-^+$ are the sources of **v**. The discussion of Maxwell's equations in the previous section leads us to conclude that, to an excellent approximation, **B** has no sources in the atmosphere.

In Gauss' theory of the earth's magnetic field (which we consider in much more detail in Chapter 4), sources are located above the atmosphere or within the earth. It is essential to the theory that a vector field depends *linearly* on its sources. Suppose we have two vector fields $\mathbf{v}_1$ and $\mathbf{v}_2$, and we write their sources as

$$\mathbf{j}_i = \nabla \times \mathbf{v}_i, \quad \rho_i = \nabla \cdot \mathbf{v}_i, \quad \mathbf{s}_i = [\mathbf{v}_i]_-^+$$

where i is either 1 or 2. Now let c_1 and c_2 be constants; obviously the vector field $\mathbf{v} = c_1\mathbf{v}_1 + c_2\mathbf{v}_2$ will have as its sources $\mathbf{j} = c_1\mathbf{j}_1 + c_2\mathbf{j}_2$, $\rho = c_1\rho_1 + c_2\rho_2$, and $\mathbf{s} = c_1\mathbf{s}_1 + c_2\mathbf{s}_2$. But since a field is uniquely

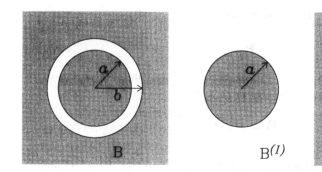

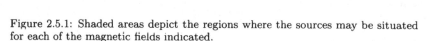

Figure 2.5.1: Shaded areas depict the regions where the sources may be situated for each of the magnetic fields indicated.

determined by its sources, $\mathbf{v}$ is the vector field whose sources are $\mathbf{j}$, ρ, and $\mathbf{s}$.

It follows that the sources of the magnetic field in the atmosphere can be divided into two parts: those outside the atmosphere (in the ionosphere and above it) and those inside the earth. The field produced by internal sources is often called the *internal field* (!) and written $\mathbf{B}^{(I)}$. Similarly, the field produced by sources above the atmosphere is called the *external field* and is written $\mathbf{B}^{(E)}$. The total field is

$$\mathbf{B} = \mathbf{B}^{(E)} + \mathbf{B}^{(I)}. \qquad (2.5.1)$$

Figure 2.5.1 is a sketch that locates the source regions for each of the fields $\mathbf{B}$, $\mathbf{B}^{(E)}$, and $\mathbf{B}^{(I)}$. The radius of the solid earth is a, and the inner radius of the ionosphere is $b > a$.

Maxwell's equations show us that

$$\boldsymbol{\nabla} \cdot \mathbf{B}^{(I)} = \boldsymbol{\nabla} \cdot \mathbf{B}^{(E)} = 0 \qquad \text{everywhere} \qquad (2.5.2)$$

$$\boldsymbol{\nabla} \times \mathbf{B}^{(I)} = 0 \qquad \text{if } r > a \qquad (2.5.3)$$

$$\boldsymbol{\nabla} \times \mathbf{B}^{(E)} = 0 \qquad \text{if } r < b \qquad (2.5.4)$$

In $r < a$ we have

$$\boldsymbol{\nabla} \times \mathbf{B}^{(I)} = \mu_0[\mathbf{J}^{(F)} + \partial_t \mathbf{D} + \boldsymbol{\nabla} \times \mathbf{M}]$$

and in $r > b$

$$\boldsymbol{\nabla} \times \mathbf{B}^{(E)} = \mu_0[\mathbf{J}^{(F)} + \partial_t \mathbf{D} + \boldsymbol{\nabla} \times \mathbf{M}].$$

It follows by combining (2.5.2) and (2.5.3) that everywhere underneath the ionosphere, namely in $r < b$, we may write

$$\mathbf{B}^{(I)} = -\boldsymbol{\nabla}\psi^{(I)} \quad \text{with} \quad \nabla^2\psi^{(I)} = 0.$$

And from (2.5.2) and (2.5.4), everywhere above the solid earth, that is, when $r > a$, we have

$$\mathbf{B}^{(E)} = -\boldsymbol{\nabla}\psi^{(E)} \quad \text{with} \quad \nabla^2\psi^{(E)} = 0.$$

The scalar potentials $\psi^{(I)}$ and $\psi^{(E)}$ have different kinds of spherical harmonic representations, as we will see in section 4.2.

2.6 Solving Maxwell's Equations as an Initial Value Problem

We briefly describe a program for solving Maxwell's equations, starting from initial conditions. Rewrite the averaged Maxwell's equations of section 2.3 in the following order:

$$\partial_t \mathbf{D} = \boldsymbol{\nabla} \times \mathbf{H} - \mathbf{J}^{(F)} \tag{2.6.1}$$

$$\partial_t \mathbf{B} = -\boldsymbol{\nabla} \times \mathbf{E} \tag{2.6.2}$$

$$\mathbf{H} = \mu_0^{-1}\mathbf{B} - \mathbf{M} \tag{2.6.3}$$

$$\mathbf{E} = \epsilon_0^{-1}(\mathbf{D} - \mathbf{P}) \tag{2.6.4}$$

$$\boldsymbol{\nabla} \cdot \mathbf{D} = \rho^{(F)} \tag{2.6.5}$$

$$\boldsymbol{\nabla} \cdot \mathbf{B} = 0 \tag{2.6.6}$$

$$\partial_t \rho^{(F)} + \boldsymbol{\nabla} \cdot \mathbf{J}^{(F)} = 0. \tag{2.6.7}$$

Equations (2.6.3, 2.6.4) simply define $\mathbf{H}$ and $\mathbf{E}$ as auxiliary quantities. When the expressions for $\mathbf{H}$ and $\mathbf{E}$ are substituted into (2.6.1, 2.6.2), we have equations for $\partial_t\mathbf{D}$ and $\partial_t\mathbf{B}$ in terms of $\mathbf{D}$, $\mathbf{B}$, $\mathbf{J}^{(F)}$, and $\mathbf{P}$ and $\mathbf{M}$. We treat $\mathbf{J}^{(F)}$, $\mathbf{P}$, and $\mathbf{M}$ as sources for the fields that must be specified; if they are known for all $\mathbf{r}$ and t, and if initial conditions are known for $\mathbf{D}$ and $\mathbf{B}$ (i.e., if $\mathbf{D}$ and $\mathbf{B}$ are known for all $\mathbf{r}$ at $t = 0$), then the resulting equations for $\partial_t\mathbf{D}$ and $\partial_t\mathbf{B}$ can be integrated forward (and backward) in time to give $\mathbf{D}$ and $\mathbf{B}$ for all $\mathbf{r}$ and t.

Example 1:

    M given (a permanent magnet)

    P given (an electret)

    $\mathbf{J}^{(F)}$ given (a superconductor or known current source)

Example 2: Alternatively, instead of expressing the sources explicitly, we can specify them in terms of other fields through constitutive relations. Practical considerations essentially limit solution to linear constitutive relations. Then

    $\mathbf{M} = \beta\mathbf{B}/\mu_0$ (a magnetically polarizable material)

    $\mathbf{P} = \epsilon_0\chi_E\mathbf{E}$ (an electrically polarizable material)

    $\mathbf{J}^{(F)} = \sigma\mathbf{E}$ (a motionless Ohmic conductor)

Having found **B** and **D** for all **r** and t, we can calculate **H**, **E**, $\rho^{(F)}$ from (2.6.3–2.6.5). But what role do (2.6.6) and (2.6.7) play? Equation (2.6.7) is not new information. It will be satisfied automatically, because it follows from (2.6.1) and (2.6.5). Equation (2.6.6) occupies an intermediate position. It does convey the new mathematical information that, although we can choose **D** arbitrarily at time $t = 0$, we cannot do this with **B**. We must choose the initial **B** so that $\boldsymbol{\nabla} \cdot \mathbf{B} = 0$. Once we have done this, however, the solution **D**, **B** of (2.6.1–2.6.4) will automatically satisfy (2.6.6) at all other times because $\partial_t(\boldsymbol{\nabla} \cdot \mathbf{B}) = \boldsymbol{\nabla} \cdot (\partial_t\mathbf{B}) = -\boldsymbol{\nabla} \cdot (\boldsymbol{\nabla} \times \mathbf{E}) = 0.$

Exercises

1. A volume V with smooth boundary ∂V and outward unit normal $\hat{\mathbf{n}}$ contains matter with a uniform magnetization density **M** amp/meter. No other charges or currents are present anywhere. The magnetic field produced by **M** is **B** teslas. Assume you know **B**.

   (a) Find the surface conduction current $\mathbf{J}_s^{(F)}$ on ∂V that would produce the same **B** if **M** were zero.

   (b) Find the magnetic displacement **H** in the two cases that **B** is produced by **M** or $\mathbf{J}_s^{(F)}$.

   (c) Find the free surface charge $\rho_s^{(F)}$ on ∂V that would produce everywhere the electric field $\sqrt{\mu_0/\epsilon_0}\mathbf{H}$, **H** being the magnetic displacement produced by **M**.

2. In exercise 1, suppose that V is the spherical ball of radius a centered at **0**. Choose Cartesian coordinates so that **M** points in the positive z

direction, and use the notation of (7.2.96) for spherical polar coordinates. Let C and D be unspecified constants and define a scalar field Ψ in R^3 by requiring $\Psi = Cz$ in V, $\Psi = Dz/r^3$ in R^3 outside V.

(a) Show that $|\nabla\Psi| \to 0$ ar $|\mathbf{r}| \to \infty$ and that C and D can be chosen so that $-\nabla\Psi$ has the same sources as the $\mathbf{B}$ in exercise 1. It follows from Helmholtz's source theorem that with such a C and D, $\mathbf{B} = -\nabla\Psi$.

(b) Sketch the lines of force of $\mathbf{B}$, $\mathbf{H}$, and $\mathbf{J}_s^{(F)}$ in exercise 1, with arrows to show their directions.

(c) If a is the radius of the earth, how large must $\mathbf{M}$ or $\mathbf{J}_s^{(F)}$ be to produce a dipole field as strong as the earth's?

3. Some paleomagnetic investigators suggest there is evidence that in the northern hemisphere at certain past epochs the magnetic declination may have been consistently about 5° east of north. Accepting this hypothesis, and assuming that the changes in $\mathbf{B}$ have not been large, *estimate* what was the average radial current density in the cap between the north pole and colatitude θ. Find the resulting average radial electric field in the air just above the surface of the cap, assuming that the electrical conductivity of air there had its present value, $\sigma \approx 10^{-13}$ S/meter.

3

SPHERICAL HARMONICS

In his wonderful book *Fourier Analysis* T. W. Körner (1988) has a
short chapter on the future of mathematics as it appeared to the
French mathematicians of 1800: there was general agreement that
they had exhausted the methods of calculus available to them for
solving physical problems, which were based in the main on power
series and polynomials. As Körner remarks, just ahead lay the
developments of Fourier, which toppled most of the barriers facing the
eighteenth-century analysts; in physical terms mathematical physics
was about to master wave motion. Spherical harmonics represent
an unusually rich intertwining of both kinds of analysis: the Fourier
description and the polynomial (see section 3.6, for example). Yet
the traditional training of a geophysicist presents the topic from its
Fourier, or wave-like, aspect as the spherical version of Fourier series
or the eigensolutions arising from separation of variables, these in
turn often motivated by the free oscillations of the earth (standing
waves) or for students from physics, the energy levels of hydrogen
atoms (more ineffable waves). Thus it is that most geophysicists
never realize that spherical harmonics are also harmonic polynomials
and that studying them as polynomials provides an avenue for the
discovery of all of their properties. This is a particularly appealing
notion in geomagnetism or classical gravity, where wave motion is not
central but potentials are.

Spherical harmonics are the functions most commonly used to
represent scalar fields on a spherical surface, and they are used exten-
sively in the geomagnetic applications involving Laplace's equation,
discussed in Chapters 4 and 5. Properties of spherical harmonics
and their representation will occupy us for the rest of this chapter.

41

An important property of any representation of a scalar field is
that it should be able to describe any member of a large class of
functions. Thus we begin by demonstrating that on a spherical surface
of unit radius the spherical harmonics provide an arbitrarily good
approximation for any differentiable function. Before beginning the
serious work of the chapter, we note that the Einstein summation
convention has been turned off.

3.1 Completeness on $S(1)$

The key to the completeness question is the following result, known
as the *Weierstrass Approximation Theorem*. Suppose V is a bounded
open set; we denote its boundary by ∂V, and its closure (V
together with ∂V) by $\bar{V}$. If f is any function N times continuously
differentiable on $\bar{V}$, then for any $\epsilon > 0$ there is a polynomial P_ϵ with
the property that $\mid f - P_\epsilon \mid < \epsilon$ in $\bar{V}$, and moreover any partial
derivative of f, up to order N, lies within ϵ of the same partial
derivative of P_ϵ, everywhere in V. A proof can be found in Courant
and Hilbert (1953, vol. I), or in Körner (1988, chaps. 4, 5). The
theorem suggests that a good foundation for approximating arbitrary
continuous functions would be the set of polynomials.

3.1.1 Homogeneous and Harmonic Polynomials

Some typical homogeneous polynomials of degree 3 in x, y, and z are

$$xyz, \quad x^3 + y^3 + z^3 + 3x^2y + 3xy^2, \quad x^2z + y^2z - z^3.$$

The term *homogeneous* refers to the fact that the sums of powers of
each term are the same, in this case 3, which is the *degree*. If the
variables are thought of as coordinates of a point, the polynomials
are scalar fields in R^3. We can ask which, if any, of these polynomials
satisfy Laplace's equation, that is, which of them is harmonic. In
the list above the first and third are indeed harmonic homogeneous
polynomials. But it is not immediately clear how to construct such
functions for a general degree, or how many essentially different
polynomials there are in each degree. These are some of the questions
that will concern us in this chapter.

We denote the set of all homogeneous polynomials in **r** of degree ℓ
by $\mathcal{P}_\ell$. Every polynomial in **r** is a sum of homogeneous polynomials of

various degrees, and if $p_\ell(\mathbf{r})$ is a homogeneous polynomial of degree l, then $\sum_{\ell=0}^{L} p_\ell(\mathbf{r}) = 0$ vanishes identically only if each p_l vanishes separately. Moreover, $\nabla^2 p_\ell$ is a homogeneous polynomial of degree $\ell - 2$ so $\nabla^2 \sum_{\ell=0}^{L} p_\ell(\mathbf{r}) = 0$ only if $\nabla^2 p_\ell = 0$ for all ℓ. This leads us to look at homogeneous harmonic polynomials of various degrees: we denote the set of homogeneous harmonic polynomials of degree ℓ by $\mathcal{H}_\ell$. Thus $\mathcal{H}_\ell$ is the set of all $h_\ell \in \mathcal{P}_\ell$ such that $\nabla^2 h_\ell = 0$. Members of $\mathcal{H}_\ell$ are called *solid spherical harmonics* of degree ℓ. *Surface spherical harmonics* are the restriction of h_ℓ to a spherical surface; a typical example will be written $p_\ell(\hat{\mathbf{r}})$, which means of course that the solid spherical harmonic is evaluated on the unit sphere $S(1)$ centered on the origin, $\mathbf{0}$. In the geophysical literature, the term *spherical harmonic* commonly denotes the surface spherical harmonic, but here we always mean the solid kind.

In this section, we discuss the formula

$$f(\hat{\mathbf{r}}) = \sum_{\ell=0}^{\infty} h_\ell(\hat{\mathbf{r}}), \qquad (3.1.1)$$

which says that every scalar-valued function on $S(1)$ is an infinite sum of spherical harmonics on $S(1)$. We will not prove the formula completely, but will give references, and discuss the limitation required on f and the meaning of the convergence. Our main goal in section 3.1 is to make the spherical harmonic representation (3.1.1) seem obvious. We start by outlining some properties of $\mathcal{P}_\ell$, and then specialize to harmonic polynomials on a sphere. Our first objective is to show that any polynomial on the sphere can be uniquely represented by a sum of harmonic polynomials; the interesting point is of course that in R^3 there are many more homogeneous polynomials than there are harmonic ones, yet if we evaluate such functions only on $S(1)$, the harmonic subset is large enough to represent them. Then we can invoke the Weierstrass Approximation Theorem to justify (3.1.1), which states that harmonic polynomials are able to represent general functions evaluated on a sphere.

The set $\mathcal{P}_\ell$ of all homogeneous polynomials of degree ℓ with complex coefficients is a complex linear vector space (Halmos, 1958). We denote its dimension by dim $\mathcal{P}_\ell$. We can choose $\hat{\mathbf{x}}$, $\hat{\mathbf{y}}$, $\hat{\mathbf{z}}$ as an normal basis for R^3 and write $\mathbf{r} = \hat{\mathbf{x}}x + \hat{\mathbf{y}}y + \hat{\mathbf{z}}z$. Then any p_ℓ in $\mathcal{P}_\ell$

can be written as

$$p_\ell(\mathbf{r}) = \sum_{a+b+c=\ell} C_{a,b,c} x^a y^b z^c, \tag{3.1.2}$$

where $C_{a,b,c}$ is a complex constant; a, b, and c are nonnegative integers; and the sum is over all triples (a, b, c) with $a + b + c = \ell$. The monomials $x^a y^b z^c$ form a basis for $\mathcal{P}_l$, so counting them gives us dim $\mathcal{P}_l$. To do this we note that c can take any value from 0 to ℓ. Having chosen c, b can take on any value from 0 to $\ell - c$, but a is then determined. Thus for any value of c there are $\ell - c + 1$ monomials; hence

$$\dim \mathcal{P}_\ell = \sum_{c=0}^{\ell} (\ell - c + 1) = \frac{(\ell + 1)(\ell + 2)}{2}. \tag{3.1.3}$$

Thus the dimension of the space $\mathcal{P}_l$ has been relatively easy to obtain. But the calculation of the dimension of $\mathcal{H}_\ell$, the space of harmonic homogeneous polynomials, is deferred until we study orthogonality in section 3.2.

3.1.2 The Laplacian of a Certain Homogeneous Polynomial

The object of this subsection is to obtain the Laplacian of $r^{n+2} h_\ell(\mathbf{r})$, which will be the crucial result for showing the existence of an expansion of arbitrary homogeneous polynomials in terms of harmonic polynomials. We begin by deducing some properties of derivatives of members of $\mathcal{P}_l$ and $\mathcal{H}_l$.

It also follows from equation (3.1.2) that if $p_\ell \in \mathcal{P}_\ell$ then

$$p_\ell(\mathbf{r}) = r^\ell p_\ell(\hat{\mathbf{r}}). \tag{3.1.4}$$

To see this, write $\mathbf{r} = r\hat{\mathbf{r}} = r(\hat{\mathbf{x}} x/r + \hat{\mathbf{y}} y/r + \hat{\mathbf{z}} z/r)$. Then

$$\begin{aligned}
p_\ell(\mathbf{r}) &= \sum_{a+b+c=\ell} C_{a,b,c} r^{a+b+c} (x/r)^a (y/r)^b (z/r)^c \\
&= r^l \sum_{a+b+c=\ell} C_{a,b,c} (x/r)^a (y/r)^b (z/r)^c \\
&= r^l p_\ell(\hat{\mathbf{r}}).
\end{aligned}$$

Equation (3.1.4) naturally implies

$$h_\ell(\mathbf{r}) = r^\ell h_\ell(\hat{\mathbf{r}}),$$

a relation that we will use repeatedly in the next few pages. Furthermore

$$(r\partial_r)p_\ell = \ell p_\ell. \tag{3.1.5}$$

Proof: We introduce the spherical coordinates r, θ, λ. Then $\hat{\mathbf{r}}$ depends only on θ and λ. Thus in (3.1.4), $\partial_r p_\ell(\mathbf{r}) = \ell r^{\ell-1} p_\ell(\hat{\mathbf{r}})$ and $r\partial_r p_\ell(\mathbf{r}) = \ell r^\ell p_\ell(\hat{\mathbf{r}}) = \ell p_\ell(\mathbf{r})$. This proves (3.1.5). $\triangleleft$

Next we call on a result from Chapter 7, the mathematical appendix (7.2.13):

$$r^2 \nabla^2 = r\partial_r(r\partial_r + 1) + \nabla_1^2, \tag{3.1.6}$$

where in any system of spherical polar coordinates, the surface Laplacian operator is

$$\nabla_1^2 = \frac{1}{\sin\theta}\partial_\theta \sin\theta\, \partial_\theta + \frac{1}{\sin^2\theta}\partial_\lambda^2. \tag{3.1.7}$$

Applying this to $h_\ell \in \mathcal{H}_\ell$ gives

$$r^2 \nabla^2 h_\ell = r\partial_r(r\partial_r + 1)h_\ell + \nabla_1^2 h_\ell = 0.$$

But $r\partial_r$ is linear, so $r\partial_r(r\partial_r + 1)h_\ell = \ell(\ell+1)h_\ell$ by (3.1.5). Therefore, for every $h_\ell \in \mathcal{H}_\ell$

$$\nabla_1^2 h_\ell = -\ell(\ell+1)h_\ell. \tag{3.1.8}$$

Thus we see that the eigenvalues of the ∇_1^2 operator are $-\ell(\ell+1)$ and the corresponding eigenfunctions are surface spherical harmonics.

It is obvious by applying the representation (3.1.7) that

$$\nabla_1^2 r^n h_\ell(\mathbf{r}) = r^n \nabla_1^2 h_\ell(\mathbf{r})$$
$$= -\ell(\ell+1)r^n h_\ell(\mathbf{r})$$
$$= -\ell(\ell+1)r^{n+\ell} h_\ell(\hat{\mathbf{r}}).$$

Also

$$r\partial_r(r\partial_r + 1)r^n h_\ell(\mathbf{r}) = r\partial_r(r\partial_r + 1)r^{\ell+n} h_\ell(\hat{\mathbf{r}})$$
$$= (\ell+n)(\ell+n+1)r^{\ell+n} h_\ell(\hat{\mathbf{r}}).$$

We now apply equation (3.1.6), whose terms we have been busy assembling:

$$r^2 \nabla^2 r^n h_\ell(\mathbf{r}) = [(\ell + n)(\ell + n + 1) - \ell(\ell + 1)] r^{\ell+n} h_\ell(\hat{\mathbf{r}}),$$

so

$$\nabla^2 r^n h_\ell(\mathbf{r}) = n(2\ell + n + 1) r^{n-2} h_\ell(\mathbf{r})$$

or, on substituting $n + 2$ for n,

$$r^n h_\ell(\mathbf{r}) = \nabla^2 \left[r^{n+2} \frac{h_\ell(\mathbf{r})}{(n+2)(2\ell + n + 3)} \right]. \qquad (3.1.9)$$

This is the key result for the next subsection.

3.1.3 An Expansion in Harmonic Polynomials

Before exploiting (3.1.9) we introduce the following notation. If ℓ is an integer,

$$\sigma(\ell) = 0, \quad \ell \text{ even}$$
$$= 1, \quad \ell \text{ odd}.$$

Also $r^n \mathcal{H}_\ell$ denotes the set of all functions of the form $r^n h_\ell(\mathbf{r})$ with $h_\ell \in \mathcal{H}_\ell$. Now we are ready to show that any p_ℓ can be uniquely represented in a series of harmonic polynomials.

Theorem: If $p_\ell \in \mathcal{P}_\ell$, there are unique harmonic polynomials $h_\ell \in \mathcal{H}_\ell$, $h_{\ell-2} \in \mathcal{H}_{\ell-2}$, $h_{\ell-4} \in \mathcal{H}_{\ell-4}, \ldots, h_{\sigma(\ell)} \in \mathcal{H}_{\sigma(\ell)}$ such that

$$p_\ell(\mathbf{r}) = h_\ell(\mathbf{r}) + r^2 h_{\ell-2}(\mathbf{r}) + r^4 h_{\ell-4}(\mathbf{r}) + \cdots + r^{\ell-\sigma(\ell)} h_{\sigma(\ell)}(\mathbf{r}). \quad (3.1.10)$$

Proof: We proceed by induction on ℓ. The theorem is trivial for $\ell = 0$ or $\ell = 1$ because then $\mathcal{P}_\ell = \mathcal{H}_\ell$. Therefore we assume $\ell \geq 2$. Then $\nabla^2 p_\ell \in \mathcal{P}_{\ell-2}$, so, by induction, we can assume the theorem is true for $\nabla^2 p_\ell$. Note that $\sigma(\ell - 2) = \sigma(\ell)$. Hence there are polynomials $\tilde{h}_{\ell-2} \in \mathcal{H}_{\ell-2}$, $\tilde{h}_{\ell-4} \in \mathcal{H}_{\ell-4}, \ldots, \tilde{h}_{\sigma(\ell)} \in \mathcal{H}_{\sigma(\ell)}$ such that for all $\mathbf{r} \in R^3$

$$\nabla^2 p_\ell = \tilde{h}_{\ell-2} + r^2 \tilde{H}_{\ell-4} + r^4 \tilde{H}_{\ell-6} + \cdots + r^{\ell-2-\sigma(\ell)} \tilde{h}_{\sigma(\ell)}. \quad (3.1.11)$$

Now (3.1.9) permits us to write

$$\nabla^2 p_\ell = \nabla^2 \left[r^2 \frac{\tilde{h}_{\ell-2}}{2(2\ell-1)} + r^4 \frac{\tilde{h}_{\ell-4}}{4(2\ell-3)} + \cdots \right.$$

$$\left. + r^{\ell-\sigma(\ell)} \frac{\tilde{h}_{\sigma(\ell)}}{[\ell-\sigma(\ell)][\ell+\sigma(\ell)+1]} \right]. \qquad (3.1.12)$$

Define

$$h_\ell = p_\ell - r^2 \frac{\tilde{h}_{\ell-2}}{2(2\ell-1)} - r^4 \frac{\tilde{h}_{\ell-4}}{4(2\ell-3)} - \cdots$$

$$- r^{\ell-\sigma(\ell)} \frac{\tilde{h}_{\sigma(\ell)}}{[\ell-\sigma(\ell)][\ell+\sigma(\ell)+1]}$$

$$h_{\ell-2} = \tilde{h}_{\ell-2} \,/\, 2(2\ell-1)$$

$$h_{\ell-4} = \tilde{h}_{\ell-4} \,/\, 4(2\ell-3)$$

$$\vdots$$

$$h_{\sigma(\ell)} = \tilde{h}_{\sigma(\ell)} \,/\, [\ell-\sigma(\ell)][\ell+\sigma(\ell)+1].$$

Then (3.1.12) shows $\nabla^2 h_\ell = 0$. Moreover, by its definition, $h_\ell \in \mathcal{P}_\ell$. Thus $h_\ell \in \mathcal{H}_\ell$. This proves that p_ℓ has an expression of the form (3.1.10). To prove uniqueness, we note that (3.1.5) implies, with (3.1.9), that

$$\nabla^2 p_\ell = 2(2\ell-1)h_{\ell-2} + 4(2\ell-3)r^2 h_{\ell-4} + \cdots$$

$$+ [\ell-\sigma(\ell)][\ell+\sigma(\ell)+1]r^{\ell-2-\sigma(\ell)}h_{\sigma(\ell)}.$$

But then by induction the polynomials $h_{\ell-2},\ h_{\ell-4}, \ldots, h_{\sigma(\ell)}$ are uniquely determined by p_ℓ. Then equation (3.1.10) determines h_ℓ uniquely.   ◁

Now we define

$$\tilde{\mathcal{P}}_\ell = \mathcal{P}_\ell + \mathcal{P}_{\ell-1} + \mathcal{P}_{\ell-2} + \cdots + \mathcal{P}_0. \qquad (3.1.13)$$

In other words, $\tilde{\mathcal{P}}_\ell$ is the set of *all* polynomials in **r** of degree ℓ, whether homogeneous or not.

We are finally in a position to use the Weierstrass Approximation Theorem. In any rectangular subset U of R^3, a continuous scalar-valued function f can be uniformly approximated arbitrarily well by a member of $\tilde{\mathcal{P}}_\ell$. Thus for any $\epsilon > 0$ there is an L so large that there are polynomials $p_0, p_1, \ldots, p_L$ with the properties $p_\ell \in \mathcal{P}_\ell$ and

$$\left| f - \sum_{\ell=0}^{L} p_\ell \right| < \epsilon \quad \text{if} \quad |r_i| \le 1 \quad \text{for} \quad i = 1, 2, 3. \qquad (3.1.14)$$

But then

$$\left| f(\hat{\mathbf{r}}) - \sum_{\ell=0}^{L} p_\ell(\hat{\mathbf{r}}) \right| < \epsilon$$

for all $\hat{\mathbf{r}} \in S(1)$, and on $S(1)$ (3.1.10) says

$$p_\ell(\hat{\mathbf{r}}) = h_\ell(\hat{\mathbf{r}}) + r^2 h_{\ell-2}(\hat{\mathbf{r}}) + r^4 h_{\ell-4}(\hat{\mathbf{r}}) + \cdots + r^{\ell - \sigma(\ell)} h_{\sigma(\ell)}(\hat{\mathbf{r}}). \quad (3.1.15)$$

Thus there are polynomials $\tilde{h}_\ell \in \mathcal{H}_\ell$ such that

$$\left| f(\hat{\mathbf{r}}) - \sum_{\ell=0}^{L} \tilde{h}_\ell(\hat{\mathbf{r}}) \right| < \epsilon \quad \text{for all} \quad \hat{\mathbf{r}} \in S(1). \qquad (3.1.16)$$

This completes our discussion of completeness.

3.2 Orthogonality in $\mathcal{P}_\ell$

An important property of the spherical harmonics is their orthogonality. So far in our development we have shown that a wide class of functions on $S(1)$ can be expressed by an expansion in terms of spherical harmonics, but we have not yet presented a way to construct such an expansion given a particular function. Among other things, orthogonality provides the necessary mechanism. Also, we are able to discover the dimension of $\mathcal{H}_\ell$.

Let us define the average value of a scalar field on the surface $S(1)$ by

$$\langle f \rangle = \frac{1}{4\pi} \int_{S(1)} d^2 \hat{\mathbf{s}} \, f(\hat{\mathbf{s}}).$$

An important application for the average is the *inner product* of two complex scalar fields f, g on $S(1)$ denoted by $\langle f|g \rangle$. It is defined as follows:

$$\langle f|g \rangle = \langle \bar{f}g \rangle = \frac{1}{4\pi} \int_{S(1)} d^2\hat{s} \ \overline{f(\hat{s})} \, g(\hat{s})$$

where the bar denotes complex conjugate. We note the following obvious facts about the inner product:

$$\langle f|g \rangle = \overline{\langle g|f \rangle}$$
$$\langle f|ag + bh \rangle = a\langle f|g \rangle + b\langle f|h \rangle$$
$$\langle f|f \rangle > 0 \text{ unless } f = 0 \text{ on } S(1)$$

where f, g, h are functions on $S(1)$ and a and b are complex constants. Two functions f and g are said to be *orthogonal* if $\langle f|g \rangle = 0$. Also the *2-norm* of f is defined by $\|f\|_2 = \sqrt{\langle f|f \rangle}$; the norm (we usually omit the subscript 2 when there is no ambiguity in doing so) of a function is a quantity analogous to the length of a vector. The orthogonality relationship for spherical harmonics is expressed in the following theorem.

Theorem 1: If $\ell \neq \ell'$ and $h_\ell \in \mathcal{H}_\ell$ and $h_{\ell'} \in \mathcal{H}_{\ell'}$ then $\langle h_\ell|h_{\ell'} \rangle = 0$.

Proof: We apply a result derived in the mathematical appendix for a general smooth closed oriented surface (Corollary 3.1 of section 7.4), namely that the Beltrami operator on a closed surface is self-adjoint. On the spherical surface of unit radius this gives us

$$\langle \overline{h_\ell}(\nabla_1^2 h_{\ell'}) \rangle = \langle (\overline{\nabla_1^2 h_\ell}) h_{\ell'} \rangle.$$

But from the eigenvalue property (3.1.8)

$$\nabla_1^2 h_{\ell'} = -\ell'(\ell'+1)h'_\ell$$

and

$$\nabla_1^2 \bar{h}_\ell = \overline{\nabla_1^2 h_\ell} = -\ell(\ell+1)\bar{h}_\ell$$

so

$$-\ell'(\ell'+1)\langle \bar{h}_\ell h_{\ell'} \rangle = -\ell(\ell+1)\langle \bar{h}_\ell h_{\ell'} \rangle$$

and

$$(\ell-\ell')(\ell+\ell'+1)\langle \bar{h}_\ell h_{\ell'} \rangle = 0.$$

Thus if $\ell \neq \ell'$, $\langle \bar{h}_\ell h_{\ell'} \rangle = 0$. But $\langle \bar{h}_\ell h_{\ell'} \rangle = \langle h_\ell | h_{\ell'} \rangle$.   ◁

Another way of stating this result is to say $\mathcal{H}_\ell \perp \mathcal{H}_{\ell'}$, that is, the spaces $\mathcal{H}_\ell$ and $\mathcal{H}_{\ell'}$ are orthogonal, if $\ell \neq \ell'$. A further result from this theorem is for $r^n \mathcal{H}_\ell$, the set of all functions of the form $r^n h_\ell$ with $h_\ell \in \mathcal{H}_\ell$. If n is even, $r^n \mathcal{H}_\ell$ consists of polynomials in $\mathcal{P}_{\ell+n}$. Then $r^n \mathcal{H}_\ell \perp r^{n'} \mathcal{H}_{\ell'}$ if $\ell \neq \ell'$. This follows from the fact that inner products involve only the values of the functions on $S(1)$.

An immediate consequence of this result and (3.1.10) is that $\mathcal{P}_\ell$ may be constructed by the union of the following orthogonal spaces

$$\mathcal{P}_\ell = \mathcal{H}_\ell \oplus r^2 \mathcal{H}_{\ell-2} \oplus r^4 \mathcal{H}_{\ell-4} \oplus \cdots \oplus r^{\ell-\sigma(\ell)} \mathcal{H}_{\sigma(\ell)}.$$

Then

$$\dim \mathcal{P}_\ell = \dim \mathcal{H}_\ell + \dim \mathcal{H}_{\ell-2} + \dim \mathcal{H}_{\ell-4} + \cdots + \dim \mathcal{H}_{\sigma(\ell)}.$$

From this result we can derive the dimension of the space $\mathcal{H}_\ell$. We have

$$\dim \mathcal{P}_{\ell-2} = \dim \mathcal{H}_{\ell-2} + \dim \mathcal{H}_{\ell-4} + \cdots + \dim \mathcal{H}_{\sigma(\ell)}$$

so

$$\dim \mathcal{P}_\ell - \dim \mathcal{P}_{\ell-2} = \dim \mathcal{H}_\ell.$$

But $\dim \mathcal{P}_\ell = (\ell+1)(\ell+2)/2$, a result derived in the last section, thus we have $\dim \mathcal{H}_\ell = 2\ell + 1$.

A useful result, which we will need later, is contained in the following theorem, which is another way of saying that the spherical harmonics are complete on $S(1)$.

Theorem 2: Suppose $\|f\|_2 < \infty$, that is, f is a square-integrable function on $S(1)$. If f is orthogonal to all surface harmonics then $\|f\|_2 = 0$. We do not prove this theorem.

Now we discuss the expansion of an arbitrary square-integrable function f on $S(1)$ in spherical harmonics. In each $\mathcal{H}_\ell$ we can choose an orthonormal basis, which means every member of the basis is of unit norm, and is orthogonal to every other element. Because the dimension of $\mathcal{H}_\ell$ is $2\ell + 1$ the basis set must contain $2\ell + 1$ members, so we can label them β_ℓ^m with $m = -\ell, -\ell+1, \ldots, \ell-1, \ell$. Note that here the index m is nothing but a label for the basis functions β_ℓ^m; it does not refer to any particular angular dependence of the function β_ℓ^m. Orthonormality of the basis in $\mathcal{H}_\ell$ means that $\langle \beta_\ell^m | \beta_\ell^n \rangle = \delta_{mn}$. Similarly orthogonality between spaces means that if $\ell \neq \ell'$ then $\langle \beta_\ell^m | \beta_{\ell'}^n \rangle = 0$.

Theorem 3: Suppose f is a scalar field on $S(1)$ such that $\|f\|_2 < \infty$, and $\beta_\ell^{-\ell}, \ldots, \beta_\ell^\ell$ are any orthonormal bases for $\mathcal{H}_\ell$. Then there are unique scalars f_ℓ^m such that on $S(1)$

$$f(\hat{\mathbf{r}}) = \sum_{\ell=0}^{\infty} \sum_{m=-\ell}^{\ell} f_\ell^m \beta_\ell^m(\hat{\mathbf{r}}) \tag{3.2.1}$$

and those scalars are given by

$$f_\ell^m = \langle \beta_\ell^m | f \rangle. \tag{3.2.2}$$

Proof: The orthonormality condition for the bases can be summarized by

$$\langle \beta_l^m | \beta_{l'}^{m'} \rangle = \delta_{\ell\ell'} \delta_{mm'} .$$

For notational convenience rename the β_ℓ^m as $(\gamma_1, \gamma_2, \gamma_3, \ldots)$, the sequence of γ's being simply all the β_ℓ^m in any order. Then

$$\langle \gamma_m | \gamma_n \rangle = \delta_{mn} . \tag{3.2.3}$$

With this nomenclature, equation (3.2.1) becomes:

$$f = \sum_{n=1}^{\infty} f_n \gamma_n . \tag{3.2.4}$$

Suppose that f is a square-integrable function on $S(1)$; that is, $\|f\|_2 < \infty$. Then the integral $\langle \gamma_n | f \rangle$ exists, because whenever $\|f\|_2 < \infty$ and $\|g\|_2 < \infty$, $\langle f | g \rangle$ exists, which follows from *Schwarz's inequality:* $|\langle f | g \rangle| \leq \|f\|_2 \|g\|_2$ (Debnath and Mikusiński, 1990). Whenever there is an infinite sum, as in (3.2.4), we must define what is meant, since obviously some kind of limiting process is involved. We elect to interpret (3.2.4) for the purposes of the theorem as follows: we wish to show that, if $f_1, f_2, \ldots$, are complex constants,

$$\lim_{N \to \infty} \left\| f - \sum_{n=1}^{N} f_n \gamma_n \right\|_2 = 0 \tag{3.2.5}$$

if and only if

$$f_n = \langle \gamma_n | f \rangle . \tag{3.2.6}$$

Equation (3.2.5) says that the size of the discrepancy between f and the approximation in a finite expansion in the basis functions, measured by the norm, can be made as small as we like. First, observe that

$$\left\| f - \sum_{n=1}^{N} f_n \gamma_n \right\|_2^2$$

$$= \left\langle f - \sum_{n=1}^{N} f_n \gamma_n \ \middle| \ f - \sum_{m=1}^{N} f_m \gamma_m \right\rangle$$

$$= \langle f | f \rangle - \left\langle f \middle| \sum_{m=1}^{N} f_m \gamma_m \right\rangle - \left\langle \sum_{n=1}^{N} f_n \gamma_n \middle| f \right\rangle$$

$$+ \left\langle \sum_{n=1}^{N} f_n \gamma_n \middle| \sum_{m=1}^{N} f_m \gamma_m \right\rangle$$

$$= \| f \|_2^2 - \sum_{m=1}^{N} f_m \langle f | \gamma_m \rangle - \sum_{n=1}^{N} \bar{f}_n \langle \gamma_n | f \rangle + \sum_{n,m=1}^{N} \bar{f}_n f_m \langle \gamma_n | \gamma_m \rangle.$$

Now from (3.2.6) $\langle \gamma_n | f \rangle = f_n$ and $\langle f | \gamma_m \rangle = \bar{f}_m$; from (3.2.3) $\langle \gamma_n | \gamma_m \rangle = \delta_{mn}$. Substituting these relations we find

$$\left\| f - \sum_{n=1}^{N} f_n \gamma_n \right\|_2^2 = \| f \|_2^2 - \sum_{n=1}^{N} |f_n|^2 . \tag{3.2.7}$$

The left side of (3.2.7) is ≥ 0, so $\sum_{n=1}^{N} |f_n|^2 \leq \| f \|_2^2$. Since this is true for every N, $\sum_{n=1}^{\infty} |f_n|^2$ converges and we obtain Bessel's inequality:

$$\sum_{n=1}^{\infty} |f_n|^2 \leq \| f \|_2^2. \tag{3.2.8}$$

By the Riesz–Fischer Theorem (Riesz and Nagy, 1955), (3.2.8) implies that there is a square-integrable function g such that

$$\lim_{N \to \infty} \left\| g - \sum_{n=1}^{N} f_n \gamma_n \right\|_2 = 0 \tag{3.2.9}$$

and also

$$\langle \gamma_n | g \rangle = f_n . \qquad (3.2.10)$$

But then if f_n is given by (3.2.6), we have $\langle \gamma_n | f - g \rangle = 0$ for all n. By Theorem 2, $\|f - g\|_2 = 0$. Now

$$\left\| f - \sum_{n=1}^{N} f_n \gamma_n \right\|_2 = \left\| f - g + g - \sum_{n=1}^{N} f_n \gamma_n \right\|_2$$

$$\leq \|f - g\|_2 + \left\| g - \sum_{n=1}^{N} f_n \gamma_n \right\|_2$$

$$\leq \left\| g - \sum_{n=1}^{N} f_n \gamma_n \right\|_2 .$$

Therefore (3.2.9) implies (3.2.5). Hence (3.2.6) implies (3.2.5). Now we address the issue of uniqueness. Suppose (3.2.5) is true for some $f_1, f_2, \ldots$. Let $f_n' = \langle \gamma_n | f \rangle$. We want to prove $f_n = f_n'$. We have

$$\sum_{n=1}^{N} |f_n - f_n'|^2 = \left\| \sum_{n=1}^{N} (f_n - f_n') \gamma_n \right\|_2^2$$

$$= \left\| \sum_{n=1}^{N} f_n \gamma_n - \sum_{n=1}^{N} f_n' \gamma_n \right\|_2^2$$

$$= \left\| \sum_{n=1}^{N} f_n \gamma_n - f + f - \sum_{n=1}^{N} f_n' \gamma_n \right\|_2^2$$

$$\leq \left(\left\| \sum_{n=1}^{N} f_n \gamma_n - f \right\|_2 + \left\| f - \sum_{n=1}^{N} f_n' \gamma_n \right\|_2 \right)^2 .$$

$$(3.2.11)$$

We have already proved $\lim_{N \to \infty} \|f - \sum_{n=1}^{N} f_n' \gamma_n\|_2 = 0$, and now we have the hypothesis that also $\lim_{N \to \infty} \|f - \sum_{n=1}^{N} f_n \gamma_n\|_2 = 0$. Then (3.2.11) implies $\sum_{n=1}^{\infty} |f_n - f_n'|^2 = 0$, so $f_n = f_n'$. Recalling the equivalence of f_n to f_l^m and γ_n to β_l^m completes the proof of the theorem as expressed in (3.2.1) and (3.2.2).   ◁

Some comments are required about (3.2.1). We chose to interpret this equation in the sense of (3.2.5), convergence under the 2-norm.

Such convergence follows from the rather weak hypothesis that the 2-norm of f is finite; if that is all we know about f, then there is little more to be said about the ability of the expansion to approximate the field on $S(1)$. If, on the other hand, we know that f is continuously differentiable, it can be shown that the series converges uniformly on $S(1)$, by which we mean that the maximum absolute difference between the approximation and f tends to zero as $\ell \to \infty$. This is the same sense of convergence as the Weierstrass Approximation Theorem provides. However, if f is square-integrable but discontinuous, the series converges only in a weaker sense: for example, it can be shown to have a subsequence point-wise convergent almost everywhere on $S(1)$. Stated explicitly, this means there is a subsequence of 1, 2, 3, ..., say, $L_1, L_2, L_3, \ldots$, such that

$$\lim_{n \to \infty} \sum_{\ell=0}^{L_n} \sum_{m=-\ell}^{\ell} f_\ell^m \beta_\ell^m(\hat{\mathbf{r}}) = f(\hat{\mathbf{r}})$$

everywhere on $S(1)$ except in a subset of $S(1)$ whose total area is zero.

3.3 The Self-Reproducing Kernel on $\mathcal{H}_\ell$

We continue to explore the properties of spherical harmonics, making use of the inner products defined earlier. In this section we show that any linear functional on a finite-dimensional vector space possessing an inner product can be written uniquely in terms of the inner product on the space. This result is applied to $\mathcal{H}_\ell$ to find the self-reproducing kernel $\tilde{q}_\ell$, a harmonic polynomial whose inner product with another evaluates it at a fixed point in space. The invariance of $\mathcal{H}_\ell$ under rigid rotations on $S(1)$ is exploited to derive the Spherical Harmonic Addition Theorem.

3.3.1 Inner Product Spaces

In the previous section we encountered the notion of an inner product, a complex-valued function obtained by averaging whose arguments were in that case two complex scalar fields on $S(1)$. We now make the obvious generalization to an abstract inner product space, whose structure is that of a linear vector space, with an inner product between pairs of elements; the range of the inner product is the set of scalars on the vector space. Suppose V is such a finite-dimensional space with real or complex scalars; we need both in later applications. The set of scalars is denoted F, for the algebraic "field." We make first a few elementary remarks. We provide the general inner product with the properties seen in the previous section: The inner product is *hermitian*, which means

$$\langle u|v \rangle = \overline{\langle v|u \rangle} \tag{3.3.1}$$

where the overbar is complex conjugate; this would have no effect when the scalars are real. It is *positive definite*:

$$\langle v|v \rangle > 0 \text{ unless } v = \mathbf{0}. \tag{3.3.2}$$

Further, it is linear in its second argument: this means if $b_1, \ldots, b_n \in F$ and $v_1, \ldots, v_n \in V$

$$\left\langle u \Big| \sum_{i=1}^{n} b_i v_i \right\rangle = \sum_{i=1}^{n} b_i \langle u|v_i \rangle. \tag{3.3.3}$$

From (3.3.1) and (3.3.3) we deduce

$$\left\langle \left(\sum_{i=1}^{n} b_i v_i \right) \Big| u \right\rangle = \overline{\left\langle u \Big| \left(\sum_{i=1}^{n} b_i v_i \right) \right\rangle} = \sum_{i=1}^{n} \overline{b_i \langle u|v_i \rangle} = \sum_{i=1}^{n} \overline{b_i} \langle v_i|u \rangle$$

so that

$$\left\langle \left(\sum_{i=1}^{n} b_i v_i \right) \Big| u \right\rangle = \sum_{i=1}^{n} \overline{b_i} \langle v_i|u \rangle. \tag{3.3.4}$$

Assume now that dim $V = n$. If $\beta_1, \ldots, \beta_n$ is an orthonormal basis for V then for every $v \in V$ we have

$$v = \sum_{i=1}^{n} v_i \beta_i \tag{3.3.5}$$

$$\langle \beta_j|\beta_i \rangle = \delta_{ij} \tag{3.3.6}$$

$$v_j = \langle \beta_j|v \rangle. \tag{3.3.7}$$

That (3.3.5) is possible follows from the fact that $\{\beta_1, \ldots, \beta_n\}$ spans V. Equation (3.3.6) is a property of any orthonormal basis. Then (3.3.7) follows from (3.3.4) and (3.3.6).

Now suppose $p, q \in V$, and expand them in terms of the orthonormal basis $\{\beta_1, \ldots, \beta_n\}$. Write

$$p = \sum_{i=1}^{n} p_i \beta_i, \qquad q = \sum_{j=1}^{n} q_j \beta_j. \qquad (3.3.8)$$

Then

$$\langle p|q \rangle = \sum_{i=1}^{n} \overline{p_i}\, q_i. \qquad (3.3.9)$$

To see this, expand p and q in the basis and then use (3.3.4) and (3.3.7).

3.3.2 Inner Products and Linear Functionals

We come next to linear functionals and their representation as inner products. If a function Q on domain V assigns to each $v \in V$ a value $Q(v)$ in some set W, we say Q maps V into W, and we write $Q : V \to W$. When $Q : V \to F$, Q is called a *functional* on V. If, for any $b_1, \ldots, b_n \in F$ and $v_i, \ldots, v_n \in V$, we have

$$\sum_{i=1}^{n} Q(b_i v_i) = \sum_{i=1}^{n} b_i Q(v_i) \qquad (3.3.10)$$

then Q is a *linear functional*.

Suppose V is an n-dimensional inner product space with scalar field F. For any $q \in V$ we can define a linear functional, Q, on V by requiring for each $v \in V$ that

$$Q(v) = \langle q|v \rangle. \qquad (3.3.11)$$

Obviously because inner products are scalar-valued, $Q : V \to F$, and its linearity follows from (3.3.3). The interesting thing is not this trivial remark but its converse.

Theorem: With V as in the preceding remark, suppose $Q : V \to F$ and Q is linear. Then there is a unique $q \in V$ such that (3.3.11) holds for all $v \in V$.

Proof: Suppose there is a $q \in V$ such that (3.3.11) works for all $v \in V$. We show that this is the only such q. Let $\beta_1, \ldots, \beta_n$ be an orthonormal basis for V. Then we can write

$$q = \sum_{j=1}^{n} q_j \beta_j \tag{3.3.12}$$

and from (3.3.11) with $v = \beta_j$ we have

$$Q(\beta_j) = \langle q | \beta_j \rangle = \overline{\langle \beta_j | q \rangle} = \overline{q_j},$$

so

$$q_j = \overline{Q(\beta_j)}. \tag{3.3.13}$$

Thus the only q that can possibly work in (3.3.11) for all $v \in V$ is the q given by (3.3.12), (3.3.13). Any other q would fail to work for β_j.

Now we address the issue of whether any such q exists. There is no point trying any q other than the one just constructed, so let us see if it does work. Choose any $v \in V$, and write it $v = \sum_{i=1}^{n} v_i \beta_i$. Then

$$Q(v) = \sum_{i=1}^{n} v_i Q(\beta_i) = \sum_{i=1}^{n} \overline{q_i} v_i = \langle q | v \rangle.$$

Therefore it works!   ◁

We note as a matter of mathematical culture that the ability to represent a linear functional uniquely by an inner product also holds true on infinite-dimensional inner product spaces provided certain conditions are satisfied: first, the linear functional must be bounded (which is automatic on finite-dimensional spaces); second, the space must be complete — it is then called a *Hilbert space*. Completeness of a space is the property that certain sequences of elements, called Cauchy sequences, converge to a member of the space. Every finite-dimensional space is complete. The general result that every bounded linear functional in a Hilbert space can be written uniquely as an inner product is the *Riesz Representation Theorem* (Debnath and Mikusiński, 1990).

3.3.3 A Special Linear Functional on $\mathcal{H}_\ell$

Now let us return to the space of homogeneous harmonic polynomials, $\mathcal{H}_\ell$. We invest this space with the inner product used throughout

section 3.2. Choose a particular $\hat{\mathbf{r}}$ in $S(1)$ and *fix it*. Consider the
scalar $h_\ell(\hat{\mathbf{r}})$, which is the evaluation of some polynomial at the fixed
point $\hat{\mathbf{r}}$. For every $h_\ell \in \mathcal{H}_\ell$ this has a value in F. And if a and b are
scalars and h_ℓ and $g_\ell \in \mathcal{H}_\ell$, then by the definition of $ah_\ell + bg_\ell$ we
have

$$(ah_\ell + bg_\ell)(\hat{\mathbf{r}}) = a[h_\ell(\hat{\mathbf{r}})] + b[g_\ell(\hat{\mathbf{r}})]. \tag{3.3.14}$$

Suppose we define the functional $Q_{\hat{\mathbf{r}}} : \mathcal{H}_\ell \to F$ by requiring

$$Q_{\hat{\mathbf{r}}}(h_\ell) = h_\ell(\hat{\mathbf{r}}) \tag{3.3.15}$$

for every $h_\ell \in \mathcal{H}_\ell$; in other words $Q_{\hat{\mathbf{r}}}$ takes any polynomial in $\mathcal{H}_\ell$ and
evaluates it at $\hat{\mathbf{r}}$. Then (3.3.14) shows that $Q_{\hat{\mathbf{r}}}$ is a linear functional
on $\mathcal{H}_\ell$. Therefore, by the theorem of the previous subsection, there is
a $q_{\hat{\mathbf{r}}} \in \mathcal{H}_\ell$ such that

$$Q_{\hat{\mathbf{r}}}(h_\ell) = \langle q_{\hat{\mathbf{r}}} | h_\ell \rangle \tag{3.3.16}$$

for every $h_\ell \in \mathcal{H}_\ell$. But then $q_{\hat{\mathbf{r}}}(\mathbf{s})$ is a harmonic polynomial in $\mathbf{s}$ of
degree ℓ, and $\langle q_{\hat{\mathbf{r}}} | h_\ell \rangle = \langle \overline{q_{\hat{\mathbf{r}}}}\, h_\ell \rangle$, so (3.3.16) says

$$h_\ell(\hat{\mathbf{r}}) = \frac{1}{4\pi} \int_{S(1)} d^2\hat{\mathbf{s}}\, \overline{q_{\hat{\mathbf{r}}}}\,(\hat{\mathbf{s}})\, h_\ell(\hat{\mathbf{s}}) \tag{3.3.17}$$

for every $h_\ell \in \mathcal{H}_\ell$. Furthermore, the theorem says there is only one
function $q_{\hat{\mathbf{r}}} \in \mathcal{H}_\ell$, which makes (3.3.17) true for all $h_\ell \in \mathcal{H}_\ell$. We
are curious about this function and we spend the rest of this section
deducing some of its properties.

First, we show the polynomial $q_{\hat{\mathbf{r}}}$ is real. We can use the above
argument when the field of scalars $F = R$, the real numbers. In that
case, $q_{\hat{\mathbf{r}}}$ is a real-valued polynomial, and the complex conjugation
can be omitted in (3.3.17). But this real-valued polynomial makes
(3.3.17) work for the real and imaginary parts of any complex-valued
h_ℓ. Therefore the real $q_{\hat{\mathbf{r}}}$ works in (3.3.17) even for complex h_ℓ.
Because $q_{\hat{\mathbf{r}}}$ is unique, it is the appropriate member of $\mathcal{H}_\ell$ to use in
(3.3.17) even when $F = C$. The same harmonic polynomial $q_{\hat{\mathbf{r}}}$ works
in (3.3.17) whether h_ℓ is real or complex, and that $q_{\hat{\mathbf{r}}}$ is itself real.

In order to show explicitly the dependence of $q_{\hat{\mathbf{r}}}(\mathbf{s})$ on ℓ as well
as $\hat{\mathbf{r}}$ and $\mathbf{s}$, we write it

$$q_{\hat{\mathbf{r}}}(\mathbf{s}) = \tilde{q}_\ell(\hat{\mathbf{r}}, \mathbf{s}).$$

For each fixed $\hat{\mathbf{r}} \in S(1)$, $\tilde{q}_\ell(\hat{\mathbf{r}}, \mathbf{s})$ is a real, harmonic homogeneous polynomial in $\mathbf{s}$ of degree ℓ (i.e., a member of $\mathcal{H}_\ell$) and for every $h_\ell \in \mathcal{H}_\ell$

$$h_\ell(\hat{\mathbf{r}}) = \frac{1}{4\pi} \int_{S(1)} d^2\hat{\mathbf{s}} \; \tilde{q}_\ell(\hat{\mathbf{r}}, \hat{\mathbf{s}}) h_\ell(\hat{\mathbf{s}}) \,. \qquad (3.3.18)$$

Moreover, any function of $\hat{\mathbf{r}}$ and $\hat{\mathbf{s}}$ that makes (3.3.18) work for all $h_\ell \in \mathcal{H}_\ell$ is this particular $\tilde{q}_\ell$. It is uniquely determined by (3.3.18). For obvious reasons, $\tilde{q}_\ell$ is called the *self-reproducing kernel* on $\mathcal{H}_\ell$. As an aside we note that not every inner product space possesses self-reproducing kernels: those that do are called *reproducing kernel spaces*, and they play a major role in the modern theory of interpolation, for example.

3.3.4 A Rotational Symmetry of $\tilde{q}_\ell$

An important property of $\tilde{q}_\ell$ can be deduced by considering the rigid rotations of $S(1)$. We can show that $\tilde{q}_\ell$ depends only on $\hat{\mathbf{r}} \cdot \hat{\mathbf{s}}$, and not on the specific locations of $\hat{\mathbf{r}}$ and $\hat{\mathbf{s}}$. Put another way, this means $\tilde{q}_\ell$ is invariant under rigid rotations about the axis $\hat{\mathbf{r}}$.

If $\Re$ is the rotation by angle ϕ about axis $\hat{\mathbf{w}}$, then $\Re\hat{\mathbf{s}} \in S(1)$ whenever $\hat{\mathbf{s}} \in S(1)$. If f is a scalar field on R^3, we *rotate* f into a new scalar field $\Re f$ by requiring

$$(\Re f)(\Re \mathbf{s}) = f(\mathbf{s}) \qquad (3.3.19)$$

for all $\mathbf{s} \in R^3$. Let $\Re^{-1}$ denote the rotation inverse to $\Re$, which carries $\Re \mathbf{s}$ back to $\mathbf{s}$. If $\Re \mathbf{s} = \mathbf{r}$ then $\mathbf{s} = \Re^{-1}\mathbf{r}$, so

$$(\Re f)(\mathbf{r}) = f(\Re^{-1}\mathbf{r}) \,. \qquad (3.3.20)$$

This equation defines the scalar field $\Re f$ on R^3, and hence defines $\Re$ as a scalar operator on R^3. As such $\Re$ is obviously linear. Moreover, $\Re$ is also an operator on $S(1)$. Finally, ∇^2 and $\Re$ commute; that is

$$\nabla^2 \Re = \Re \nabla^2 \,. \qquad (3.3.21)$$

Proof. Consider an arbitrary scalar field f on R^3. Choose a Cartesian basis $\hat{\mathbf{x}}_1, \hat{\mathbf{x}}_2, \hat{\mathbf{x}}_3$, and write $\mathbf{r} = \sum r_i \hat{\mathbf{x}}_i$. Let $\mathbf{r}' = \Re \mathbf{r} =$

$\sum r_i' \hat{\mathbf{x}}_i$, and $f' = \Re f$. Then $f'(\mathbf{r}') = f(\mathbf{r})$. Let $\partial_i = \partial/\partial r_i$ and $\partial_i' = \partial/\partial r_i'$. Then by (7.2.85)

$$\nabla^2 f'(\mathbf{r}') = \sum_{i=1}^{3} \partial_i' \partial_i' f'(\mathbf{r}') = \sum_{i=1}^{3} \partial_i \partial_i f'(\mathbf{r}') = \sum_{i=1}^{3} \partial_i \partial_i f(\mathbf{r}) = \nabla^2 f(\mathbf{r}).$$

Thus $[\nabla^2(\Re f)](\Re \mathbf{r}) = (\nabla^2 f)(\mathbf{r})$, so

$$[\nabla^2(\Re f)](\mathbf{s}) = (\nabla^2 f)(\Re^{-1}\mathbf{s}) = [\Re(\nabla^2 f)](\mathbf{s}).$$

Thus $\nabla^2(\Re f) = \Re(\nabla^2 f)$, so $(\nabla^2 \Re)f = (\Re\nabla^2)f$. Hence (3.3.21). ◁

From (3.3.21) it follows that

$$\Re\mathcal{H}_\ell \to \mathcal{H}_\ell \qquad\qquad (3.3.22)$$

because if $h_\ell \in \mathcal{H}_\ell$ then $\nabla^2 \Re h_\ell = \Re\nabla^2 h_\ell = 0$.

We can now apply the rotation operator to the fundamental property of the self-reproducing kernel, equation (3.3.18). This equation implies

$$(\Re h_\ell)(\mathbf{r}) = \frac{1}{4\pi} \int_{S(1)} d^2\hat{\mathbf{s}} \; \tilde{q}_\ell(\hat{\mathbf{r}}, \hat{\mathbf{s}})(\Re h_\ell)(\hat{\mathbf{s}})$$

or

$$h_\ell(\Re^{-1}\hat{\mathbf{r}}) = \frac{1}{4\pi} \int_{S(1)} d^2\hat{\mathbf{s}} \; \tilde{q}_\ell(\hat{\mathbf{r}}, \hat{\mathbf{s}}) h_\ell(\Re^{-1}\hat{\mathbf{s}}).$$

Suppose we replace $\hat{\mathbf{r}}$ by $\Re\hat{\mathbf{r}}$ in this equation. Then

$$h_\ell(\hat{\mathbf{r}}) = \frac{1}{4\pi} \int_{S(1)} d^2\hat{\mathbf{s}} \; \tilde{q}_\ell(\Re\hat{\mathbf{r}}, \hat{\mathbf{s}}) h_\ell(\Re^{-1}\hat{\mathbf{s}}).$$

Now suppose we change the variable of integration from $\hat{\mathbf{s}}$ to $\hat{\mathbf{s}}' = \Re^{-1}\hat{\mathbf{s}}$. Then, clearly, $d^2\hat{\mathbf{s}}' = d^2\hat{\mathbf{s}}$, since $\Re$ preserves areas. Moreover, $\hat{\mathbf{s}} = \Re\hat{\mathbf{s}}'$, so

$$h_\ell(\hat{\mathbf{r}}) = \frac{1}{4\pi} \int_{S(1)} d^2\hat{\mathbf{s}}' \; \tilde{q}_\ell(\Re\hat{\mathbf{r}}, \Re\hat{\mathbf{s}}') h_\ell(\hat{\mathbf{s}}').$$

Now relabel the variable of integration as $\hat{s}$, and we obtain for all $h_\ell \in \mathcal{H}$

$$h_\ell(\hat{r}) = \frac{1}{4\pi} \int_{S(1)} d^2\hat{s}\; \tilde{q}_\ell(\Re\hat{r}, \Re\hat{s}) h_\ell(\hat{s}). \qquad (3.3.23)$$

Comparing (3.3.18) with (3.3.23) and using the fact that (3.3.18) determines $\tilde{q}_\ell(\hat{r}, \hat{s})$ uniquely, we conclude that for all $\hat{r}, \hat{s} \in S(1)$ and all rotations $\Re$

$$\tilde{q}_\ell(\Re\hat{r},\ \Re\hat{s}) = \tilde{q}_\ell(\hat{r},\ \hat{s}). \qquad (3.3.24)$$

Now suppose that $\hat{r}, \hat{s}, \hat{r}', \hat{s}' \in S(1)$ and $\hat{r} \cdot \hat{s} = \hat{r}' \cdot \hat{s}'$. Then the angle θ between $\hat{r}$ and $\hat{s}$ is the same as the angle θ' between $\hat{r}'$ and $\hat{s}'$, so there is a rotation $\Re$ such that $\hat{r}' = \Re\hat{r}$ and $\hat{s}' = \Re\hat{s}$. Therefore, by (3.3.24),

$$\tilde{q}_\ell(\hat{r}', \hat{s}') = \tilde{q}_\ell(\hat{r}, \hat{s}).$$

In other words $\tilde{q}_\ell(\hat{r}, \hat{s})$ is known if we know $\mu = \hat{r} \cdot \hat{s} = \cos\theta$. There is a function $Q_\ell(\mu)$ defined for $-1 \le \mu \le 1$ such that

$$\tilde{q}_\ell(\hat{r}, \hat{s}) = Q_\ell(\hat{r} \cdot \hat{s}). \qquad (3.3.25)$$

When we introduced the linear functional $Q_{\hat{r}}$ in subsection 3.3.3 we imagined $\hat{r}$ to be fixed; the result (3.3.25) shows that the averaging process in (3.3.17) is rotational symmetric with respect to the axis defined by the direction $\hat{r}$.

3.3.5 Properties of Q_ℓ

We gather the information of the previous section into a theorem and then derive a number of properties about Q_ℓ and bases on $\mathcal{H}_\ell$.

Theorem: For each degree $\ell \ge 0$, there is a unique scalar function $\tilde{q}_\ell(\hat{r}, \hat{s})$ with these properties:

(a) $\qquad\qquad \tilde{q}_\ell(\hat{r}, s) \in \mathcal{H}_\ell$ as a function of s.

(b) $\qquad\qquad h_\ell(\hat{r}) = \dfrac{1}{4\pi} \int_{S(1)} d^2\hat{s}\; \tilde{q}_\ell(\hat{r}, \hat{s}) h_\ell(\hat{s})$

for every $\hat{r} \in S(1)$ and every $h_\ell \in \mathcal{H}_\ell$. This function $\tilde{q}_\ell$ is of the form (3.3.25) for some real-valued function, $Q_\ell(\mu)$ on the interval $-1 \le \mu \le 1$. When $|s| \ne 1$, by (3.1.4)

$$\tilde{q}_\ell(\hat{r}, s) = s^\ell Q_\ell(\hat{r} \cdot \hat{s}). \qquad (3.3.26)$$

Result (a) shows that there is an axisymmetric homogeneous harmonic polynomial for each degree ℓ.

Corollary 1: (Spherical Harmonic Addition Theorem) Let $\beta_\ell^{-\ell}$, $\beta_\ell^{-\ell+1}, \ldots, \beta_\ell^{\ell-1}$, β_ℓ^{ℓ} be any orthonormal basis for $\mathcal{H}_\ell$. Then

$$Q_\ell(\hat{\mathbf{r}} \cdot \hat{\mathbf{s}}) = \sum_{m=-\ell}^{\ell} \beta_\ell^m(\hat{\mathbf{r}}) \, \overline{\beta_\ell^m(\hat{\mathbf{s}})} \quad \text{for all} \quad \hat{\mathbf{r}}, \hat{\mathbf{s}} \in S(1). \qquad (3.3.27)$$

Proof: By (3.3.5) and (3.3.7), if $h_\ell \in \mathcal{H}_\ell$,

$$h_\ell = \sum_{m=-\ell}^{\ell} \beta_\ell^m \langle \beta_\ell^m | h_\ell \rangle$$

$$h_\ell(\hat{\mathbf{r}}) = \sum_{m=-\ell}^{\ell} \beta_\ell^m(\hat{\mathbf{r}}) \frac{1}{4\pi} \int_{S(1)} d^2\hat{\mathbf{s}} \, \overline{\beta_\ell^m(\hat{\mathbf{s}})} \, h_\ell(\hat{\mathbf{s}})$$

$$h_\ell(\hat{\mathbf{r}}) = \frac{1}{4\pi} \int_{S(1)} d^2\hat{\mathbf{s}} \left[\sum_{m=-\ell}^{\ell} \beta_\ell^m(\hat{\mathbf{r}}) \, \overline{\beta_\ell^m(\hat{\mathbf{s}})} \right] h_\ell(\hat{\mathbf{s}}). \qquad (3.3.28)$$

Comparison with part (b) of the theorem proves (3.3.27), because $\sum_{m=-\ell}^{\ell} \beta_\ell^m(\hat{\mathbf{r}}) \, \overline{\beta_\ell^m(\hat{\mathbf{s}})}$ is clearly in $\mathcal{H}_\ell$ as a function of $\hat{\mathbf{s}}$.   ◁

As a statement of the Spherical Harmonic Addition Theorem, Corollary 1 leaves something to be desired, namely, an explicit definition of Q_ℓ as function of μ. We must wait until section 3.5 for a derivation, but for completeness we state that in fact $Q_\ell(\mu) = (2\ell + 1)P_\ell(\mu)$, where P_ℓ is the *Legendre polynomial* of degree ℓ. We can get a number of properties of the function Q_ℓ without having an explicit formula for it, for example Corollaries 2 and 3.

Corollary 2: For every orthonormal basis β_ℓ^m of $\mathcal{H}_\ell$,

$$Q_\ell(1) = \sum_{m=-\ell}^{\ell} |\beta_\ell^m(\hat{\mathbf{r}})|^2 = 2\ell + 1. \qquad (3.3.29)$$

Proof: The first part is immediate from (3.3.27) when one sets $\hat{\mathbf{r}} = \hat{\mathbf{s}}$. To find the constant we average both sides of (3.3.29) over the surface $S(1)$ and using the orthonormality property of the β_l^m we get

$$\langle Q_\ell(1) \rangle = Q_\ell(1)$$
$$= \sum_{m=-\ell}^{\ell} \langle |\beta_\ell^m(\hat{\mathbf{r}})|^2 \rangle = \sum_{m=-\ell}^{\ell} \|\beta_\ell^m\|^2$$
$$= 2\ell + 1.$$

$\triangleleft$

This corollary allows us to bound the size of a normalized element of $\mathcal{H}_\ell$ evaluated at any point: if $h_\ell \in \mathcal{H}_\ell$ and $\langle |h_\ell|^2 \rangle = 1$ then

$$|h_\ell(\hat{\mathbf{r}})| \le \sqrt{2\ell + 1} \quad \text{for all} \quad \hat{\mathbf{r}} \in S(1). \tag{3.3.30}$$

To see this note h_ℓ is a member of an orthonormal basis for $\mathcal{H}_\ell$. By (3.3.29), every such member satisfies inequality (3.3.30).

Corollary 3:

$$|Q_\ell(\mu)| \le Q_\ell(1) \quad \text{for} \; -1 \le \mu \le 1. \tag{3.3.31}$$

Proof: Choose an orthonormal basis for $\mathcal{H}_\ell$, say β_ℓ^m with $m = -\ell, \ldots, \ell$. Apply Schwarz's inequality (Debnath and Mikusiński, 1990, chap. 3),

$$\left| \sum_{m=-\ell}^{\ell} x_m \bar{y}_m \right|^2 \le \left(\sum_{m=-\ell}^{\ell} |x_m|^2 \right) \left(\sum_{m=-\ell}^{\ell} |y_m|^2 \right)$$

to (3.3.27), obtaining

$$|Q_\ell(\hat{\mathbf{r}} \cdot \hat{\mathbf{s}})|^2 \le \left(\sum_{m=-\ell}^{\ell} |\beta_\ell^m(\hat{\mathbf{r}})|^2 \right) \left(\sum_{m=-\ell}^{\ell} |\beta_\ell^m(\hat{\mathbf{s}})|^2 \right).$$

By (3.3.29) each term on the right is just $Q_\ell(1)$. Thus

$$|Q_\ell(\mu)|^2 \le |Q_\ell(1)|^2.$$

$\triangleleft$

Corollary 4: In (3.2.1), let us write

$$f_\ell(\hat{\mathbf{r}}) = \sum_{m=-\ell}^{\ell} f_\ell^m \beta_\ell^m(\hat{\mathbf{r}}). \qquad (3.3.32)$$

Then $r^\ell f_\ell(\hat{\mathbf{r}}) \in \mathcal{H}_\ell$ (obviously) and (also obviously)

$$f(\hat{\mathbf{r}}) = \sum_{\ell=0}^{\infty} f_\ell(\hat{\mathbf{r}}) \qquad \text{for all} \quad \hat{\mathbf{r}} \in S(1) \qquad (3.3.33)$$

and, finally, for all $\hat{\mathbf{r}} \in S(1)$

$$f_\ell(\hat{\mathbf{r}}) = \frac{1}{4\pi} \int_{S(1)} d^2\hat{\mathbf{s}} \; Q_\ell(\hat{\mathbf{r}} \cdot \hat{\mathbf{s}}) f(\hat{\mathbf{s}}). \qquad (3.3.34)$$

Therefore $f_\ell(\hat{\mathbf{r}})$ depends only on f and ℓ not on the basis chosen for $\mathcal{H}_\ell$. The function f_ℓ is called the *orthogonal projection* of f onto $\mathcal{H}_\ell$.
Proof: From (3.3.31), (3.2.2), and (3.3.27)

$$f_\ell(\hat{\mathbf{r}}) = \sum_{m=-\ell}^{\ell} \beta_\ell^m(\hat{\mathbf{r}}) \frac{1}{4\pi} \int_{S(1)} d^2\hat{\mathbf{s}} \; \overline{\beta_\ell^m(\hat{\mathbf{s}})} \, f(\hat{\mathbf{s}})$$

$$= \frac{1}{4\pi} \int_{S(1)} d^2\hat{\mathbf{s}} \left[\sum_{m=-\ell}^{\ell} \beta_\ell^\ell(\hat{\mathbf{r}}) \overline{\beta_\ell^m(\hat{\mathbf{s}})} \right] f(\hat{\mathbf{s}})$$

$$= \frac{1}{4\pi} \int_{S(1)} d^2\hat{\mathbf{s}} \; Q_\ell(\hat{\mathbf{r}} \cdot \hat{\mathbf{s}}) f(\hat{\mathbf{s}}). \qquad \triangleleft$$

3.4 An Orthonormal Basis for $\mathcal{H}_\ell$

It is time to develop an explicit set of functions to serve as a basis for $\mathcal{H}_\ell$. As we have seen already, a great deal can be learned without having a formula for the harmonic polynomials themselves; but when, for example, actual geomagnetic field calculations are required, implicit relations and properties of the basis are not enough. Naturally, armed with formulas for a basis, we will then be able to exhibit the kernel function Q_ℓ by substitution into (3.3.27). But how do we find an orthonormal basis? One approach might be to return to (3.1.7), the definition of a homogeneous polynomial of degree ℓ, and simply impose the condition that harmonic functions satisfy Laplace's equation. A homogeneous linear system of equations in the coefficients results, from which a basis could be constructed by finding a linearly independent subset of column vectors. These in turn could be orthogonalized via a Gram–Schmidt process. This is obviously a process more suited to a computer than a person, and it reveals nothing interesting about the basis functions. We prefer a systematic analytic development that generates the conventional basis familiar to geophysicists. We call this the *natural basis* associated with a particular Cartesian coordinate system.

3.4.1 Application of the Surface Curl, Λ

The reader may have noticed that so far, aside from algebraic manipulations, the fundamental properties of spherical harmonics have been obtained by applying linear operators, like ∇_1^2 or $\Re$, to $\mathcal{H}_\ell$. The same principle will be put to work here. Our construction is based on the operator $\Lambda = \mathbf{r} \times \nabla = \hat{\mathbf{r}} \times \nabla_1$. This is the dimensionless surface curl, discussed in detail in subsections 7.2.12 and 7.2.14. Let $\hat{\mathbf{x}}_1, \hat{\mathbf{x}}_2, \hat{\mathbf{x}}_3$ be a Cartesian basis for R^3, $\mathbf{r} = \sum r_i \hat{\mathbf{x}}_i$ and $\partial_i = \partial/\partial r_i$. The Cartesian components of Λ are $\Lambda_i = \hat{\mathbf{x}}_i^M \cdot \Lambda$; we recall from 7.2.12 that

$$\Lambda_i = \sum_{j=1}^{3} \sum_{k=1}^{3} \epsilon_{ijk} r_j^M \partial_k \tag{3.4.1}$$

where the superscript M implies the multiplication operator. From this it is clear that if $p_\ell \in \mathcal{P}_\ell$, then $\Lambda_i p_\ell \in \mathcal{P}_\ell$, or in words: acting on a homogeneous polynomial of degree ℓ with any component of the surface curl gives another such polynomial. That is,

$$\Lambda_i : \mathcal{P}_\ell \to \mathcal{P}_\ell.$$

We would like this to be true of the harmonic polynomials too, and it is. Using results from Chapter 7, we will demonstrate that the components of the operator Λ commute with both ∇_1^2 and ∇^2. Equation (7.2.120) tells us that if r, θ, λ are the spherical coordinates belonging to the Cartesian basis $\hat{\mathbf{x}}_1, \hat{\mathbf{x}}_2, \hat{\mathbf{x}}_3$ (in what follows we freely interchange this notation for our basis with $\hat{\mathbf{x}}$, $\hat{\mathbf{y}}$, $\hat{\mathbf{z}}$) then

$$\Lambda_z = \Lambda_3 = \partial_\lambda \tag{3.4.2}$$

and from (7.2.113)

$$\nabla^2 = r^{-2}[(r\partial_r)(r\partial_r + 1) + \nabla_1^2] \tag{3.4.3}$$

where by (7.2.112)

$$\nabla_1^2 = \frac{1}{\sin\theta}\,\partial_\theta(\sin\theta\partial_\theta) + \frac{1}{\sin^2\theta}\,\partial_\lambda^2. \tag{3.4.4}$$

From (3.4.4), evidently $\partial_\lambda \nabla_1^2 = \nabla_1^2 \partial_\lambda$, so $\Lambda_z \nabla_1^2 = \nabla_1^2 \Lambda_z$. But x and y are the z' axes in other Cartesian coordinate systems, so $\Lambda_x \nabla_1^2 = \nabla_1^2 \Lambda_x$ and $\Lambda_y \nabla_1^2 = \nabla_1^2 \Lambda_y$. In other words, the components of $\mathbf{\Lambda}$ commute with ∇_1^2,

$$\Lambda_i \nabla_1^2 = \nabla_1^2 \Lambda_i. \tag{3.4.5}$$

Similarly, from (3.4.3) and (3.4.4), $\partial_\lambda \nabla^2 = \nabla^2 \partial_\lambda$, so $\Lambda_z \nabla^2 = \nabla^2 \Lambda_z$. Thus

$$\Lambda_i \nabla^2 = \nabla^2 \Lambda_i. \tag{3.4.6}$$

Now suppose $h_\ell \in \mathcal{H}_\ell$. We have $\Lambda_i h_\ell \in \mathcal{P}_\ell$, and also $\nabla^2(\Lambda_i h_\ell) = (\nabla^2 \Lambda_i) h_\ell = (\Lambda_i \nabla^2) h_\ell = \Lambda_i (\nabla^2 h_\ell) = 0$. Thus $\Lambda_i h_\ell \in \mathcal{H}_\ell$, and we have the desired result

$$\Lambda_i : \mathcal{H}_\ell \to \mathcal{H}_\ell.$$

For the rest of this subsection we single out the z component of the surface curl, which is by (3.4.2) the longitudinal derivative. If f and $g \in \mathcal{H}_\ell$, then because Λ_z is a FODO, $\Lambda_z(fg) = f(\Lambda_z g) + (\Lambda_z f)g$. Averaging this equation over $S(1)$ yields

$$\langle f(\Lambda_z g)\rangle = -\langle(\Lambda_z f)g\rangle. \tag{3.4.7}$$

This follows because

$$\langle\,\Lambda_z(fg)\rangle = \langle\partial_\lambda(fg)\rangle = 0.$$

We now want to rewrite (3.4.7) in inner product form, in terms of the most important operator $-i\Lambda_z$. First we note that (3.4.7) is equivalent to

$$\langle\overline{(-i\Lambda_z f)}\,g\rangle = \langle\overline{f}(-i\Lambda_z g)\rangle.$$

Translated into inner product notation, this is

$$\langle(-i\Lambda_z)f|g\rangle = \langle f|(-i\Lambda_z)g\rangle. \qquad (3.4.8)$$

Therefore $-i\Lambda_z : \mathcal{H}_\ell \to \mathcal{H}_\ell$ is a self-adjoint operator on $\mathcal{H}_\ell$. A *self-adjoint operator* A is one such that for all x and y in an inner-product space, $\langle Ax|y\rangle = \langle x|Ay\rangle$; such an operator has only real eigenvalues (Debnath and Mikusiński, 1990, chap. 4). A linear transformation on an N-dimensional space can have no more than N eigenvalues. Another important consequence is that $\mathcal{H}_\ell$ can be written as the orthogonal direct sum of the eigenspaces of the operator (Strang, 1988). Thus, if we can find an orthonormal basis for each of these eigenspaces, we can construct a basis for $\mathcal{H}_\ell$.

In anticipation of the discovery that the eigenvalues of the operator $-i\Lambda_z$ are integers we denote them by m, although so far we know only that they are real numbers. As the reader probably already knows, m is called the *order* of the associated spherical harmonic. The eigenspaces of the operator $-i\Lambda_z$ are called $\mathcal{H}_\ell^m$. Each such $\mathcal{H}_\ell^m$ is the set of all $h_\ell \in \mathcal{H}_\ell$ such that $(-i\Lambda_z)h_\ell = mh_\ell$. Thus $h_\ell^m \in \mathcal{H}_\ell^m$ means $h_\ell^m \in \mathcal{H}_\ell$ and also $-i\Lambda_z h_\ell^m = mh_\ell^m$, or, equivalently,

$$\partial_\lambda h_\ell^m = imh_\ell^m. \qquad (3.4.9)$$

We can integrate (3.4.9) to obtain

$$h_\ell^m(r,\theta,\lambda) = h_\ell^m(r,\theta,0)e^{im\lambda}.$$

Also since $h_l^m \in \mathcal{P}_\ell$ we know its dependence on r from (3.1.4); thus

$$h_\ell^m(r,\theta,\lambda) = r^\ell h_\ell^m(1,\theta,0)e^{im\lambda}. \qquad (3.4.10)$$

It is now easy to show that if m is an eigenvalue of $-i\Lambda_z : \mathcal{H}_\ell \to \mathcal{H}_\ell$, then m must be an integer. The function $h_\ell^m(\mathbf{r})$ must be a nonzero single-valued function of $\mathbf{r}$ (it is a polynomial), and so it cannot change if the longitude λ is increased by 2π. Therefore $e^{i2\pi m} = 1$, for which the only solutions are integers.

We can also demonstrate that the eigenspaces of $-i\Lambda_z$ are necessarily orthogonal, i.e., if $m \neq m'$, $\mathcal{H}_\ell^m \perp \mathcal{H}_\ell^{m'}$. Suppose $h_\ell^m \in \mathcal{H}_\ell^m$

and $h_\ell^{m'} \in \mathcal{H}_\ell^{m'}$. Then

$$
\begin{aligned}
m\langle h_\ell^m | h_\ell^{m'} \rangle &= \langle m h_\ell^m | h_\ell^{m'} \rangle \\
&= \langle (-i\Lambda_z) h_\ell^m | h_\ell^{m'} \rangle \\
&= \langle h_\ell^m | (-i\Lambda_z) h_\ell^{m'} \rangle \\
&= \langle h_\ell^m | m' h_\ell^{m'} \rangle \\
&= m' \langle h_\ell^m | h_\ell^{m'} \rangle.
\end{aligned}
$$

Hence
$$
(m - m')\langle h_\ell^m | h_\ell^{m'} \rangle = 0.
$$

It is also straightforward to show that $\overline{\mathcal{H}_\ell^m} = \mathcal{H}_\ell^{-m}$, i.e., the complex conjugate of any member of $\mathcal{H}_\ell^m$ is a member of $\mathcal{H}_\ell^{-m}$.

Proof: If $h_\ell^m \in \mathcal{H}_\ell^m$, then $\overline{h_\ell^m} \in \mathcal{H}_\ell$. By taking complex conjugates of (3.4.9) we get
$$
\partial_\lambda \overline{h_\ell^m} = -im\overline{h_\ell^m},
$$

so $\overline{h_\ell^m} \in \mathcal{H}_\ell^{-m}$. Thus for every m, $\overline{\mathcal{H}_\ell^m}$ is a subset of $\mathcal{H}_\ell^{-m}$, i.e.,

$$
\overline{\mathcal{H}_\ell^m} \subset \mathcal{H}_\ell^{-m}.
$$

But then by the same argument we can show that

$$
\overline{\mathcal{H}_\ell^{-m}} \subset \mathcal{H}_\ell^m,
$$

and also

$$
\overline{\overline{\mathcal{H}_\ell^{-m}}} = \mathcal{H}_\ell^{-m} \subset \overline{\mathcal{H}_\ell^m}.
$$

Thus since each set is a subset of the other,

$$
\overline{\mathcal{H}_\ell^{-m}} = \mathcal{H}_\ell^{-m}.
$$

We have not yet determined how many eigenvalues the operator $-i\Lambda_z$ has. We will need $2\ell + 1$ if we are to use the eigenspaces in constructing an orthonormal basis for $\mathcal{H}_\ell$, since dim $\mathcal{H}_\ell = 2\ell + 1$. First we show that the m's are restricted to the range $-\ell \leq m \leq \ell$, then that there is an eigenvalue corresponding to every integer in this range.

Theorem: If m is an eigenvalue of $-i\Lambda_z : \mathcal{H}_\ell \to \mathcal{H}_\ell$, then $-\ell \le m \le \ell$.

Proof: From (3.4.10) we have

$$h_\ell^m(r, \theta, \lambda) = r^\ell h_\ell^m(1, \theta, 0)e^{im\lambda}.$$

Define

$$\varpi^+ = x + iy = r(\sin \theta)e^{i\lambda} \qquad \varpi^- = x - iy = r(\sin \theta)e^{-i\lambda}$$

then

$$x = \tfrac{1}{2}(\varpi^+ + \varpi^-) \qquad y = \frac{1}{2i}(\varpi^+ - \varpi^-).$$

Therefore a linear combination of ℓth degree monomials in x, y, z is also a linear combination of ℓth degree monomials in ϖ^+, ϖ^-, z. These are of the form

$$(\varpi^+)^a(\varpi^-)^b z^c = r^{a+b+c}(\sin \theta)^{a+b}(\cos \theta)^c e^{i(a-b)\lambda} \qquad (3.4.11)$$

where a, b, and c are integers and

$$a + b + c = \ell \qquad \text{and} \qquad a \ge 0, \quad b \ge 0, \quad c \ge 0.$$

If the h_ℓ^m is to be a linear combination of terms (3.4.11), those terms must all satisfy

$$a - b = m. \qquad (3.4.12)$$

But then, adding ℓ to both sides of (3.4.12) we see $\ell + m = 2a + c \ge 0$, and so $m \ge -\ell$; similarly, subtracting ℓ we find $m - \ell = -2b - c \le 0$, which gives $m \le \ell$; together these give the desired $-\ell \le m \le \ell$. ◁

Our ability to construct the natural basis for $\mathcal{H}_\ell$ generated by the Cartesian basis $\hat{\mathbf{x}}, \hat{\mathbf{y}}, \hat{\mathbf{z}}$ for R^3 will depend on proving that

$$\dim \mathcal{H}_\ell^m \ge 1 \qquad \text{if} \qquad m \in \{-\ell, -\ell+1, \ldots, 0, \ldots, \ell-1, \ell\}. \quad (3.4.13)$$

If we can prove this, then two other consequences follow immediately:

$$\dim \mathcal{H}_\ell^m = 1 \qquad (3.4.14)$$

and

$$\mathcal{H}_\ell = \mathcal{H}_\ell^{-\ell} \oplus \mathcal{H}_\ell^{-\ell+1} \oplus \cdots \oplus \mathcal{H}_\ell^0 \oplus \cdots \oplus \mathcal{H}_\ell^{\ell-1} \oplus \mathcal{H}_\ell^\ell. \qquad (3.4.15)$$

Construction of the orthonormal basis for $\mathcal{H}_\ell$ then requires only that we find one member of each of the $\mathcal{H}_\ell^m$. To see that (3.4.14, 3.4.15) follow from (3.4.13), suppose that (3.4.13) is true. Then for each $m \in \{-\ell, \ldots, \ell\}$, $\mathcal{H}_\ell^m$ contains an orthonormal basis with at least one member. The vectors in $\mathcal{H}_\ell^m$ and $\mathcal{H}_\ell^{m'}$ are orthogonal, so the collection of all the vectors in the bases for $\mathcal{H}_\ell^{-\ell}, \ldots, \mathcal{H}_\ell^\ell$ is an orthonormal subset of $\mathcal{H}_\ell$. Like any orthonormal subset, this one is linearly independent. Hence it can have at most $2\ell + 1$ members. Therefore none of the orthonormal bases of the individual $\mathcal{H}_\ell^m$'s can contain more than one member. Since (3.4.13) says that each has at least one, (3.4.14) follows. But then the basis vectors for the $\mathcal{H}_\ell^m$ taken together span $\mathcal{H}_\ell$, so $\mathcal{H}_\ell = \mathcal{H}_\ell^{-\ell} + \mathcal{H}_\ell^{-\ell+1} + \cdots + \mathcal{H}_\ell^{\ell-1} + \mathcal{H}_\ell^\ell$. Since $\mathcal{H}_\ell^m \perp \mathcal{H}_\ell^{m'}$ for $m \neq m'$, we have (3.4.15).

3.4.2 Lifting and Lowering Operators

It remains to prove (3.4.13). To do so we must construct a nonzero h_ℓ^m in each $\mathcal{H}_\ell^m$ for $m \in \{-\ell, \ldots, \ell\}$. Our approach is as follows: for $m = \pm\ell$, it is easy to construct a basis element, so we do it. Then we construct two operators: one, which we call a *lifting operator* that generates h_ℓ^{m+1} from h_ℓ^m; the other, *lowering operator* decreases m by unity. In this way we provide a chain of mappings between the $\mathcal{H}_\ell^m$, enabling us to generate nonzero h_l^m for the remaining values of m.

First we build h_ℓ^ℓ; let $m = \ell$ and test the following guess:

$$h_\ell^\ell = (x + iy)^\ell.$$

We compute the Laplacian of this function in the Cartesian basis:

$$\partial_x^2 h_\ell^\ell = \ell(\ell - 1)(x + iy)^{\ell-2}$$
$$\partial_y^2 h_\ell^\ell = -\ell(\ell - 1)(x + iy)^{\ell-2}$$
$$\partial_z^2 h_\ell^\ell = 0,$$

and by adding these equations we obtain $\nabla^2 h_\ell^\ell = 0$. Obviously $h_\ell^\ell \in \mathcal{P}_\ell$, so $h_\ell^\ell \in \mathcal{H}_\ell$. Expressed in terms of spherical polar coordinates

$$h_\ell^\ell = r^\ell \sin^\ell\theta \, e^{i\ell\lambda},$$

so

$$\partial_\lambda h_\ell^\ell = i\ell h_\ell^\ell,$$

which verifies that the guessed function is an eigenfunction of $-i\Lambda_z$ with eigenvalue ℓ; in symbols $h_\ell^\ell \in \mathcal{H}_\ell^\ell$. Since $\overline{\mathcal{H}_\ell^m} = \mathcal{H}_\ell^{-m}$, we have

$$h_\ell^{-\ell} = \overline{h_\ell^\ell} = (x - iy)^\ell \in \mathcal{H}_\ell^{-\ell},$$

a fact also obvious by a repetition of the argument just given for h_ℓ^ℓ.

The reader may wonder where the guessed functions came from. If we look at the proof of the theorem in the previous section and set $m = \ell$ in (3.4.12), we are led to the conclusion that in (3.4.11) both b and c vanish, which implies the existence of an eigenfunction exactly in the form suggested for h_ℓ^ℓ. We need hardly mention that setting $m = -\ell$ in (3.4.12) gives $h_\ell^{-\ell}$.

Having obtained two special basis functions our next task is to exhibit the promised operators. We will rely on the commutation properties of the components of surface curl derived in subsection 7.2.12 of the mathematical appendix. In particular, we need the relation (7.2.76), rewritten without the Einstein summation convention:

$$\Lambda_i \Lambda_j - \Lambda_j \Lambda_i = -\sum_{k=1}^3 \epsilon_{ijk} \Lambda_k. \qquad (3.4.16)$$

Among the nine equations (3.4.16), three are trivial and three are duplicates. The three nontrivial, nonduplicates are

$$\Lambda_y \Lambda_z - \Lambda_z \Lambda_y = -\Lambda_x$$
$$\Lambda_z \Lambda_x - \Lambda_x \Lambda_z = -\Lambda_y \qquad (3.4.17)$$
$$\Lambda_x \Lambda_y - \Lambda_y \Lambda_x = -\Lambda_z.$$

Just as $\varpi^\pm = x \pm iy$ were convenient variables to use instead of x, y in subsection 3.4.1, so we replace $\Lambda_x, \Lambda_y, \Lambda_z$ by $\Lambda^+, \Lambda^-, \Lambda_z$ where

$$\Lambda^+ = \Lambda_x + i\Lambda_y$$
$$\Lambda^- = \Lambda_x - i\Lambda_y. \qquad (3.4.18)$$

As we will soon verify, Λ^+ is the lifting operator and Λ^- the lowering operator. We can halve the computations we must do by observing that Λ^- is the complex conjugate of Λ^+. Before we do this we need to define the complex conjugate of an operator. Suppose K is an operator on complex-valued scalar fields on subset S of R^3. Then $\overline{K}$, the complex conjugate of K, is the operator on S, which assigns to any scalar field f on S the scalar field $\overline{K} f$ given by

$$\overline{K} f = \overline{(K \overline{f}\,)}. \tag{3.4.19}$$

The operator K is *real* if $K = \overline{K}$ and *imaginary* if $\overline{K} = -K$. In the sense of this definition, $\Lambda_x, \Lambda_y, \partial_x, \partial_y, \partial_z$, and r_i^M are all real operators. All the rules for complex conjugation of ordinary numbers now work for operators, so that, for example,

$$\overline{\Lambda^+} = \overline{\Lambda_x + i\Lambda_y} = \overline{\Lambda_x} + i\,\overline{\Lambda_y} = \Lambda_x - i\Lambda_y = \Lambda^-. \tag{3.4.20}$$

We have already calculated (but not named) Λ^+ and Λ^- in the mathematical appendix. From (7.2.120) we see

$$-i\,\Lambda^+ = e^{i\lambda}(\partial_\theta + i\cot\theta\partial_\lambda). \tag{3.4.21}$$

We verify that applying (3.4.20) to (3.4.21) gives the second member of (7.2.10):

$$-i\,\Lambda^- = e^{-i\lambda}(\partial_\theta - i\cot\theta\,\partial_\lambda). \tag{3.4.22}$$

To return to the chief purpose of this subsection we state the main result: if $h_\ell^m \in \mathcal{H}_\ell^m$ then $\Lambda^+ h_\ell^m \in \mathcal{H}_\ell^{m+1}$ and $\Lambda^- h_\ell^m \in \mathcal{H}_\ell^{m-1}$. That is

$$\begin{aligned}\Lambda^+ &: \mathcal{H}_\ell^m \to \mathcal{H}_\ell^{m+1} \\ \Lambda^- &: \mathcal{H}_\ell^m \to \mathcal{H}_\ell^{m-1}.\end{aligned} \tag{3.4.23}$$

Proof: By (3.4.17) and some algebra we have

$$\begin{aligned}\Lambda_z\Lambda^+ &= \Lambda_z\Lambda_x + i\Lambda_z\Lambda_y \\ &= \Lambda_x\Lambda_z - \Lambda_y + i(\Lambda_y\Lambda_z + \Lambda_x) \\ &= (\Lambda_x + i\Lambda_y)\Lambda_z - \Lambda_y + i\Lambda_x \\ &= \Lambda^+\Lambda_z + i\Lambda^+ \\ &= \Lambda^+(\Lambda_z + i^M),\end{aligned}$$

so

$$\Lambda_z \Lambda^+ = \Lambda^+ (\Lambda_z + i^M). \qquad (3.4.24)$$

By taking complex conjugates, we also have

$$\Lambda_z \Lambda^- = \Lambda^- (\Lambda_z - i^M). \qquad (3.4.25)$$

Now suppose $h_\ell^m \in \mathcal{H}_\ell^m$. Then

$$
\begin{aligned}
\Lambda_z(\Lambda^+ h_\ell^m) &= (\Lambda_z \Lambda^+) h_\ell^m \\
&= [\Lambda^+ (\Lambda_z + i^m)] h_\ell^m \\
&= \Lambda^+ [(\Lambda_z + i) h_\ell^m] \\
&= \Lambda^+ [imh_\ell^m + ih_\ell^m] \\
&= \Lambda^+ [i(m+1) h_\ell^m] \\
&= i(m+1)(\Lambda^+ h_\ell^m).
\end{aligned}
$$

Therefore $\Lambda^+ h_\ell^m \in \mathcal{H}_\ell^{m+1}$. Using (3.4.25), the same chain of argument shows $\Lambda^- h_\ell^m \in \mathcal{H}_\ell^{m-1}$. It is also possible to use the fact that $\mathcal{H}_\ell^{-m} = \overline{\mathcal{H}_\ell^m}$ and (3.4.20). ◁

We see now that Λ^+ or Λ^- can be used to generate the chain of mappings illustrated below, which successively lift or lower the order of the spherical harmonic they operate on:

$$
\mathcal{H}_\ell^{-\ell} \underset{\Lambda^-}{\overset{\Lambda^+}{\rightleftarrows}} \mathcal{H}_\ell^{\ell+1} \underset{\Lambda^-}{\overset{\Lambda^+}{\rightleftarrows}} \cdots \underset{\Lambda^-}{\overset{\Lambda^+}{\rightleftarrows}} \mathcal{H}_\ell^0 \underset{\Lambda^-}{\overset{\Lambda^+}{\rightleftarrows}} \cdots \underset{\Lambda^-}{\overset{\Lambda^+}{\rightleftarrows}} \mathcal{H}_\ell^{\ell-1} \underset{\Lambda^-}{\overset{\Lambda^+}{\rightleftarrows}} \mathcal{H}_\ell^\ell.
$$

Each end of this chain contains a nonzero vector (member of $\mathcal{H}_\ell$). The only thing to be checked is that a lifting or lowering does not result in a zero vector, which would, of course, formally satisfy the above relations but render the operation useless from the perspective of generating nontrivial elements. If we can show that Λ^+ and Λ^- in the chain never map nonzero functions into a zero function, we will have two different ways to construct nonzero members of $\mathcal{H}_\ell^m$, namely, starting at the bottom with any nonzero member of $\mathcal{H}_\ell^{-\ell}$ and working up:

$$u_\ell^m = (-i\,\Lambda^+)^{\ell+m} h_\ell^{-\ell} \qquad (3.4.26)$$

or starting at the top and working down:

$$v_\ell^m = (i\,\Lambda^-)^{\ell-m} h_\ell^\ell. \tag{3.4.27}$$

To see whether Λ^+ or Λ^- maps nonzero harmonics into zero, we use, instead of (3.4.24) and (3.4.25), formulas for $\Lambda^+\Lambda^-$ and $\Lambda^-\Lambda^+$. We have

$$\Lambda^-\Lambda^+ = (\Lambda_x - i\Lambda_y)(\Lambda_x + i\Lambda_y)$$

$$= \Lambda_x^2 + \Lambda_y^2 + i(\Lambda_x\Lambda_y - \Lambda_y\Lambda_x)$$

$$= \Lambda^2 - \Lambda_z^2 - i\Lambda_z$$

where (3.4.17) has been used to get the last line.

Recalling that $\Lambda^2 = \nabla_1^2$ we have

$$\Lambda^-\Lambda^+ = \nabla_1^2 - \Lambda_z^2 - i\Lambda_z. \tag{3.4.28}$$

Taking complex conjugates gives

$$\Lambda^+\Lambda^- = \nabla_1^2 - \Lambda_z^2 + i\Lambda_z. \tag{3.4.29}$$

Now suppose $h_\ell^m \in \mathcal{H}_\ell^m$. Then applying (3.4.28)

$$\Lambda^-\Lambda^+ h_\ell^m = \nabla_1^2 h_\ell^m - \Lambda_z^2 h_\ell^m - i\Lambda_z h_\ell^m$$

$$= -\ell(\ell+1)h_\ell^m - \partial_\lambda^2 h_\ell^m - i\partial_\lambda h_\ell^m$$

$$= -\ell(\ell+1)h_\ell^m + m^2 h_\ell^m + m h_\ell^m$$

$$= -[\ell^2 + \ell - m^2 - m]h_\ell^m.$$

Therefore

$$\Lambda^-\Lambda^+ h_\ell^m = -(\ell - m)(\ell + m + 1)h_\ell^m. \tag{3.4.30}$$

Repeating this calculation for $\Lambda^+\Lambda^-$ gives

$$\Lambda^+\Lambda^- h_\ell^m = -(\ell + m)(\ell - m + 1)h_\ell^m. \tag{3.4.31}$$

Now (3.4.30) shows that if $-\ell \leq m < \ell$ then $\Lambda^+ h_\ell^m = 0$ implies $h_\ell^m = 0$. Therefore the lifting operator applied successively to $h_\ell^{-\ell}$ never results in a zero element until the top of the chain is reached; applying it to h_ℓ^ℓ results in a vanishing function, however, which is quite consistent because we have already shown $m \leq \ell$ for all legitimate eigenfunctions. Similarly, (3.4.31) shows that if $-\ell < m \leq \ell$ then $\Lambda^- h_\ell^m = 0$ implies $h_\ell^m = 0$. Therefore applying the lowering operator successively to h_ℓ^ℓ results in nonzero functions until it is applied to $h_\ell^{-\ell}$. This proves (3.4.13) and hence (3.4.14) and (3.4.15).

3.4.3 Explicit Expressions of a Basis

Later calculations will be simplified if we scale our bases in a special way; so we define specific versions of (3.4.26, 3.4.27):

$$U_\ell^m = (-i\Lambda^+)^{l+m} U_\ell^{-\ell} \qquad (3.4.32)$$

where we take

$$U_\ell^{-\ell} = \frac{(x - iy)^\ell}{2^\ell \ell!} = \frac{r^\ell \sin{}^\ell\theta e^{-i\ell\lambda}}{2^\ell \ell!}. \qquad (3.4.33)$$

Similarly

$$W_\ell^m = (i\Lambda^-)^{l-m} W_\ell^\ell \qquad (3.4.34)$$

with

$$W_\ell^\ell = \frac{(x + iy)^\ell}{2^\ell \ell!} = \frac{r^\ell \sin{}^\ell\theta e^{i\ell\lambda}}{2^\ell \ell!}.$$

We have already noted that the chain of mappings illustrated earlier cannot be extended in either direction. For that reason, we agree to define

$$U_\ell^m = W_\ell^m = 0 \qquad \text{if} \qquad |m| > \ell, \quad m \text{ an integer.}$$

It is now possible to obtain formulas for U_ℓ^m and W_ℓ^m. We provide the details for U_ℓ^m. From (3.4.21), if $h_\ell^m \in \mathcal{H}_\ell^m$ then

$$(-i\,\Lambda^+)h_\ell^m = e^{i\lambda}(\partial_\theta - m\cot\theta)h_\ell^m.$$

And the reader may verify that this is equivalent to

$$(-i\,\Lambda^+)h_\ell^m = e^{i\lambda}(\sin\theta)^m\partial_\theta[(\sin\theta)^{-m}h_\ell^m] \in \mathcal{H}_\ell^{m+1}. \qquad (3.4.35)$$

Abbreviate $\sin\theta$ as s. Then a second application of $(-i\,\Lambda^+)$ to (3.4.35) gives

$$(-i\,\Lambda^+)^2 h_\ell^m = e^{i2\lambda}(s^{m+1}\partial_\theta s^{-m-1})(s^m\partial_\theta s^{-m})h_\ell^m \in \mathcal{H}_\ell^{m+2}$$

and p applications generate a polynomial in $\mathcal{H}_\ell^{m+p}$ from h_l^m,

$$(-i\,\Lambda^+)^p h_\ell^m = e^{ip\lambda}(s^{m+p-1}\partial_\theta s^{-mp+1})(s^{m+p-2}\partial_\theta s^{-m-p+2})\cdots$$
$$\cdots(s^{m+1}\partial_\theta s^{-m-1})(s^m\partial_\theta s^{-m})h_\ell^m .$$

Using the associative law for operators, we can rewrite this last equation as

$$e^{ip\lambda}s^{m+p}(1/s\,\partial_\theta)^p s^{-m}h_\ell^m ,$$

so

$$(-i\,\Lambda^+)^p h_\ell^m = e^{ip\lambda}(\sin\theta)^{m+p}\left(\frac{1}{\sin\theta}\partial_\theta\right)^p [(\sin\theta)^{-m}h_\ell^m] \quad \in \mathcal{H}_\ell^{m+p}.$$

$$(3.4.36)$$

If we put $m = -\ell$, $p = \ell+m$, $h_\ell^{-\ell} = U_\ell^{-\ell}$ in (3.4.36), we obtain from (3.4.33)

$$U_\ell^m = r^\ell e^{im\lambda}\frac{1}{2^\ell\ell!}(\sin\theta)^m\left(\frac{1}{\sin\theta}\partial_\theta\right)^{\ell+m}(\sin\theta)^{2\ell} \in \mathcal{H}_\ell^m. \quad (3.4.37)$$

This expression can be simplified by replacing the variable θ with μ, where $\mu = \cos\theta$, so that $\sin\theta = \sqrt{1-\mu^2}$ and

$$\partial_\mu = -\frac{1}{\sin\theta}\partial_\theta.$$

Thus we finally obtain an explicit formula for generating U_ℓ^m

$$U_\ell^m = r^\ell e^{im\lambda}\frac{(-1)^m}{2^\ell\ell!}(1-\mu^2)^{m/2}\partial_\mu^{\ell+m}(\mu^2-1)^\ell \in \mathcal{H}_\ell^m. \quad (3.4.38)$$

In an entirely similar way we can work with (3.4.34) and $(i\Lambda^-)^p h_\ell^m$ to obtain:

$$W_\ell^m = r^\ell e^{im\lambda}\frac{(-1)^m}{2^\ell\ell!}(1-\mu^2)^{-m/2}\partial_\mu^{\ell-m}(\mu^2-1)^\ell \in \mathcal{H}_\ell^m. \quad (3.4.39)$$

While the functions U_ℓ^m and W_ℓ^m have been obtained by working from opposite ends of the chain of mappings, they are not unrelated. Obviously $\overline{U_\ell^{-m}} = W_\ell^m$, a fact deducible already from (3.4.32, 3.4.33). A more interesting observation is that since $\dim \mathcal{H}_\ell^m = 1$, there must be a constant α_ℓ^m such that

$$U_\ell^m = \alpha_\ell^m W_\ell^m.$$

Therefore, although by no means obvious, it must be true that for all μ in $-1 \le \mu \le 1$,

$$(1-\mu^2)^{m/2}\partial_\mu^{\ell+m}(\mu^2-1)^\ell = \alpha_\ell^m(1-\mu^2)^{-m/2}\partial_\mu^{\ell-m}(\mu^2-1)^\ell.$$

Hence for all μ

$$(-1)^m(\mu^2-1)^m\partial_\mu^{\ell+m}(\mu^2-1)^\ell = \alpha_\ell^m\partial_\mu^{\ell-m}(\mu^2-1)^\ell. \qquad (3.4.40)$$

The two sides of this equation are polynomials of degree $\ell + m$ in μ. If they are equal, their leading terms must be equal, so

$$(-1)^m\mu^{2m}\partial_\mu^{\ell+m}\mu^{2\ell} = \alpha_\ell^m\partial_\mu^{\ell-m}\mu^{2\ell}.$$

Now

$$\partial_\mu^a\mu^b = \frac{b!}{(b-a)!}\mu^{b-a},$$

so we must have

$$(-1)^m\frac{(2\ell)!}{(\ell-m)!}\mu^{\ell+m} = \alpha_\ell^m\frac{(2\ell)!}{(\ell+m)!}.$$

Hence

$$\alpha_\ell^m = (-1)^m\frac{(\ell+m)!}{(\ell-m)!}$$

and therefore

$$U_\ell^m = (-1)^m\frac{(\ell+m)!}{(\ell-m)!}W_\ell^m$$
$$= (-1)^m\frac{(\ell+m)!}{(\ell-m)!}\overline{U_\ell^{-m}}. \qquad (3.4.41)$$

Equations (3.4.38, 3.4.39) provide us with a means for calculating two bases, U_l^m and W_l^m, for each of the $\mathcal{H}_l^m$, and (3.4.41) establishes the relationships between them. Now we relate them to a basis for the $\mathcal{H}_\ell^m$, which appears commonly in the geophysical literature. The notation in the literature is very inconsistent, particularly with regard to sign and, unfortunately, authors are not always careful to say what they mean.

The function

$$P_\ell^m(\mu) = \frac{1}{2^\ell \ell!} (1 - \mu^2)^{m/2} \partial_\mu^{\ell+m} (\mu^2 - 1)^\ell \tag{3.4.42}$$

is called the *associated Legendre function* of degree ℓ and order m; in the older literature they are sometimes referred to as *Ferrer's* associated functions. It is $(1 - \mu^2)^{m/2}$ times a polynomial of degree $\ell - m$ in μ, which involves only the powers $\mu^{\ell-m}$, $\mu^{\ell-m-2}$, $\mu^{\ell-m-4}$, $\ldots$, $\mu^{\sigma(\ell-m)}$. In terms of (3.4.42), we have

$$U_\ell^m(\mathbf{r}) = r^\ell (-1)^m e^{im\lambda} P_\ell^m(\cos \theta) \tag{3.4.43}$$

and (3.4.40) implies

$$P_\ell^m(\mu) = (-1)^m \frac{(\ell + m)!}{(\ell - m)!} P_\ell^{-m}(\mu). \tag{3.4.44}$$

3.4.4 Normalizing the Natural Basis

The functions $U_\ell^{-\ell}, U_\ell^{-\ell+1}, \ldots, U_\ell^{\ell-1}, U_\ell^\ell$ are in $\mathcal{H}_\ell$, are all nonzero, and are mutually orthogonal in the sense that

$$\langle U_\ell^m | U_\ell^{m'} \rangle = 0 \quad \text{if} \quad m \neq m'.$$

They are not an orthonormal basis for $\mathcal{H}_\ell$, because it is not true that $\langle U_\ell^m | U_\ell^m \rangle = 1$. To construct an orthonormal basis $\{Y_\ell^{-\ell}, \ldots, Y_\ell^\ell\}$ for $\mathcal{H}_\ell$, we define

$$Y_\ell^m = U_\ell^m / \langle U_\ell^m | U_\ell^m \rangle^{\frac{1}{2}}. \tag{3.4.45}$$

Then

$$\langle Y_\ell^m | Y_\ell^{m'} \rangle = \delta_{mm'}$$

and in fact

$$\langle Y_\ell^m | Y_{\ell'}^{m'} \rangle = \delta_{\ell\ell'} \delta_{mm'}. \tag{3.4.46}$$

The problem, of course, is that we don't know $\langle U_\ell^m | U_\ell^m \rangle$. To find it, recall definition of the inner product in terms of averaging on $S(1)$:

$$\langle f | g \rangle = \langle \overline{f}g \rangle = \frac{1}{4\pi} \int_{S(1)} d^2\hat{\mathbf{r}} \ \overline{f(\hat{\mathbf{r}})} \, g(\hat{\mathbf{r}}). \tag{3.4.47}$$

To prove things about U_ℓ^m, we return to (3.4.23), written here as

$$U_\ell^{m+1} = (-i\,\Lambda^+) U_\ell^m.$$

We use induction on m. We have

$$\langle U_\ell^{m+1} | U_\ell^{m+1} \rangle = \langle -i\,\Lambda^+ U_\ell^m | -i\,\Lambda^+ U_\ell^m \rangle$$
$$= \overline{(-i)}\,(-i)\langle \Lambda^+ U_\ell^m | \Lambda^+ U_\ell^m \rangle$$
$$= \langle \Lambda^+ U_\ell^m | \Lambda^+ U_\ell^m \rangle.$$

Equation (3.4.8) is equivalent to $\langle \Lambda_z f | g \rangle = -\langle f | \Lambda_z g \rangle$. Since $\hat{\mathbf{x}}$ and $\hat{\mathbf{y}}$ are $\hat{\mathbf{z}}'$ in other Cartesian coordinate systems, $\langle \Lambda_x f | g \rangle = -\langle f | \Lambda_x g \rangle$ and $\langle \Lambda_y f | g \rangle = -\langle f | \Lambda_y g \rangle$. Thus

$$\langle \Lambda^+ f | g \rangle = \langle \Lambda_x f | g \rangle + \langle i\Lambda_y f | g \rangle$$
$$= -\langle f | \Lambda_x g \rangle - i \langle \Lambda_y f | g \rangle$$
$$= -\langle f | \Lambda_x g \rangle + \langle f | i\Lambda_y g \rangle$$
$$= -\langle f | \Lambda^- g \rangle.$$

Thus

$$\langle U_\ell^{m+1} | U_\ell^{m+1} \rangle = -\langle U_\ell^m | \Lambda^- \Lambda^+ U_\ell^m \rangle.$$

Then, according to (3.4.30),

$$\langle U_\ell^{m+1} | U_\ell^{m+1} \rangle = (\ell - m)(\ell + m - 1)\langle U_\ell^m | U_\ell^m \rangle. \tag{3.4.48}$$

Then, letting $\gamma_\ell^m = \langle U_\ell^m | U_\ell^m \rangle$, we have

$$\gamma_\ell^m = (\ell - m + 1)(\ell + m)\gamma_\ell^{m-1}$$
$$= (\ell - m + 1)(\ell - m + 2)(\ell + m)(\ell + m - 1)\gamma_\ell^{m-2}$$
$$= [(\ell - m + 1)(\ell - m + 2) \cdots (\ell - m + n)]$$
$$\cdot [(\ell + m)(\ell + m - 1) \cdots (\ell + m - n + 1)]\gamma_\ell^{m-n}$$
$$= \frac{(\ell - m + n)!}{(\ell - m)!} \frac{(\ell + m)!}{(\ell + m - n)!} \gamma_\ell^{m-n}.$$

If we set $n = \ell + m$ here, we have

$$\langle U_\ell^m | U_\ell^m \rangle = \gamma_\ell^m = \frac{(2\ell)!(\ell + m)!}{(\ell - m)!} \gamma_\ell^{-\ell}. \tag{3.4.49}$$

It remains only to calculate $\gamma_\ell^{-\ell} = \langle U_\ell^{-\ell} | U_\ell^{-\ell} \rangle = \langle | U_\ell^{-\ell} |^2 \rangle$. From (3.4.33), on $S(1)$

$$|U_\ell^{-\ell}(\hat{\mathbf{r}})|^2 = \frac{\sin^{2\ell}\theta}{(2^\ell \ell!)^2},$$

so

$$\gamma_\ell^{-\ell} = \frac{1}{2^{2\ell}(\ell!)^2} \langle \sin^{2\ell}\theta \rangle. \tag{3.4.50}$$

The last step is to evaluate the integral, the average value of $\sin^{2\ell}\theta$ on the sphere:

$$\langle \sin^{2\ell}\theta \rangle = \frac{1}{4\pi} \int_0^{2\pi} d\lambda \int_0^\pi d\theta \, \sin^{2\ell+1}\theta$$
$$= \tfrac{1}{2} \int_0^\pi d\theta \, \sin^{2\ell+1}\theta = \frac{S_\ell}{2}. \tag{3.4.51}$$

To evaluate the integral we split the integrand into two factors, then integrate by parts:

$$S_\ell = \int_0^\pi d\theta \, \sin\theta \, \sin^{2\ell}\theta$$
$$= -\cos\theta \, \sin^{2\ell}\theta \Big|_0^\pi + \int_0^\pi d\theta \, 2\ell\cos^2\theta \, \sin^{2\ell-1}\theta$$
$$= 2\ell \int_0^\pi d\theta \, (1 - \sin^2\theta)\sin^{2\ell-1}\theta$$
$$= 2\ell(S_{\ell-1} - S_\ell).$$

Rearranging this leads to the relation

$$S_\ell = \frac{2\ell}{2\ell+1} S_{\ell-1}, \quad \ell = 1,2,3 \ldots .$$

It is easily seen that $S_0 = 2$ and so, by repetitive application of the relation, we obtain the formula

$$S_\ell = 2\frac{2}{3}\frac{4}{5}\frac{6}{7}\cdots\frac{2\ell}{2\ell+1} = \frac{2^{2\ell+1}(\ell!)^2}{(2\ell+1)!}.$$

We combine this with equations (3.4.50, 3.4.51):

$$\gamma_\ell^{-\ell} = \frac{1}{(2\ell+1)!}. \tag{3.4.52}$$

Therefore, (3.4.49) gives

$$\langle U_\ell^m | U_\ell^m \rangle = \frac{(\ell+m)!}{(\ell-m)!(2\ell+1)}. \tag{3.4.53}$$

Finally, therefore, we have the following orthonormal basis for $\mathcal{H}_\ell$, based on the particular Cartesian basis $\hat{\mathbf{x}}, \hat{\mathbf{y}}, \hat{\mathbf{z}}$:

$$\left.\begin{aligned}
Y_\ell^m(\mathbf{r}) &= \sqrt{2\ell+1}\,\sqrt{\frac{(\ell-m)!}{(\ell+m)!}}\, U_\ell^m \\
&= (-1)^m\,\sqrt{2\ell+1}\,\sqrt{\frac{(\ell-m)!}{(\ell+m)!}}\, r^\ell e^{im\lambda} P_\ell^m(\cos\theta)
\end{aligned}\right\} \tag{3.4.54}$$

$$P_\ell^m(\mu) = \frac{1}{2^\ell\ell!}(1-\mu^2)^{m/2}\partial_\mu^{\ell+m}(\mu^2-1)^\ell \qquad -\ell \le m \le \ell. \tag{3.4.55}$$

In view of (3.4.41)

$$Y_\ell^{-m}(\mathbf{r}) = (-1)^m\overline{Y_\ell^m(\mathbf{r})}. \tag{3.4.56}$$

3.5 Axisymmetric Spherical Harmonics

Now that we have an orthonormal basis for the space $\mathcal{H}_\ell^m$ we can use it to study various members of $\mathcal{H}_\ell$. One case of interest is that of axially symmetric spherical harmonics, corresponding to $m = 0$. The corresponding associated Legendre functions, P_ℓ^0, are known as *Legendre polynomials* (yes, they are polynomials), and are usually denoted P_ℓ. We will show that when we evaluate the kernel function $Q_\ell(\hat{\mathbf{r}} \cdot \hat{\mathbf{s}})$ with a special orientation with respect to our Cartesian basis, namely $\hat{\mathbf{s}} = \hat{\mathbf{z}}$, then the expansion (3.3.27) for Q_ℓ in terms of an orthonormal basis is axisymmetric about $\hat{\mathbf{z}}$, and contains only terms involving P_ℓ. This special feature results from the fact that $Y_\ell^m(\hat{\mathbf{z}}) = 0$, for $m \neq 0$. We will also show that the Legendre polynomials arise naturally in the expansion of the potential at $\mathbf{r}$ of a point charge located at $\mathbf{s}$. An application of the Spherical Harmonic Addition Theorem then allows us to express that expansion in terms of *any* orthonormal basis.

3.5.1 Behavior Near the z Axis

Equation (3.4.54) states that $\overline{Y_\ell^m} = (-1)^m Y_\ell^{-m}$. Thus if we examine the general structure of the polynomial $Y_\ell^m(\mathbf{r})$ for $m \geq 0$, the case $m < 0$ can be obtained from this relation. We note that λ is undefined on the z axis so we find an expression for Y_ℓ^m as a polynomial in x, y, and z. Now

$$
\begin{aligned}
r^m (1 - \mu^2)^{m/2} e^{im\lambda} &= r^m (\sin\theta)^m e^{im\lambda} \\
&= (r\sin\theta e^{i\lambda})^m \qquad (3.5.1) \\
&= (x + iy)^m .
\end{aligned}
$$

So we can write

$$
U_\ell^m(\mathbf{r}) = (-1)^m (x + iy)^m \Pi_\ell^m(\mathbf{r}) \qquad (3.5.2)
$$

where

$$
\Pi_\ell^m(\mathbf{r}) = \frac{1}{2^\ell \ell!} r^{\ell - m} \partial_\mu^{\ell + m} (\mu^2 - 1)^\ell . \qquad (3.5.3)
$$

The function $\partial_\mu^{\ell + m}(\mu^2 - 1)^\ell$ is a linear combination of $\mu^{\ell - m}$, $\mu^{\ell - m - 2}$, $\mu^{\ell - m - 4}$, ..., $\mu^{\sigma(\ell - m)}$. Therefore, since $r\mu = r\cos\theta = z$, the function $\Pi_\ell^m(\mathbf{r})$ is a linear combination of $z^{\ell - m}$, $r^2 z^{\ell - m - 2}$, $r^4 z^{\ell - m - 4}$, ..., $r^{\ell - m - \sigma(\ell - m)} z^{\sigma(\ell - m)}$. This shows three things about Π_ℓ^m:

$$
\begin{aligned}
&\Pi_\ell^m \text{ is a polynomial in } z \text{ and } r^2 \\
&\Pi_\ell^m \in \mathcal{P}_{\ell - m} \quad \text{(because } r^2 = x^2 + y^2 + z^2\text{)} \qquad (3.5.4) \\
&\partial_\lambda \Pi_\ell^m = 0.
\end{aligned}
$$

From (3.5.2) and (3.5.1), U_ℓ^m, and thus Y_ℓ^m, vanishes on the z axis like $(x+iy)^m$ if $m > 0$ and like $(x-iy)^{-m}$ if $m < 0$. In particular,

$$Y_\ell^m(\hat{\mathbf{z}}) = 0 \quad \text{if} \quad m \neq 0. \tag{3.5.5}$$

It is conceivable that Y_ℓ^m vanishes even faster than $(x+iy)^m$ as $x, y \to 0$. To check this, we must evaluate $\Pi_\ell^m(\hat{\mathbf{z}})$ in (3.5.2). From (3.5.3) this is

$$\Pi_\ell^m(\hat{\mathbf{z}}) = \frac{1}{2^\ell \ell!} \left[\partial_\mu^{\ell+m} (\mu^2 - 1)^\ell \right]_{\mu=1}.$$

Replace the variable μ with $\xi = \mu - 1$. Then $\mu^2 - 1 = \xi(2 + \xi)$ and

$$\Pi_\ell^m(\hat{\mathbf{z}}) = \frac{1}{2^\ell \ell!} \left[\partial_\xi^{\ell+m} \xi^\ell (2 + \xi)^\ell \right]_{\xi=0}. \tag{3.5.6}$$

Consider the binomial expansion of $(2+\xi)^\ell$; only the term $2^{\ell-m}\xi^m \binom{\ell}{m}$ will make a nonzero contribution to (3.5.6), so

$$\Pi_\ell^m(\hat{\mathbf{z}}) = \frac{1}{2^m m!} \frac{(\ell+m)!}{(\ell-m)!} \quad \text{if} \quad m \geq 0. \tag{3.5.7}$$

Therefore if $m \neq 0$, $Y_\ell^m(\mathbf{r})$ vanishes exactly as fast as $(x^2 + y^2)^{|m|/2}$ when $x, y \to 0$. The special case $m = 0$ gives

$$Y_\ell^0(\hat{\mathbf{r}}) = \sqrt{2\ell + 1}\, \Pi_\ell^0(\hat{\mathbf{r}}) = \sqrt{2\ell + 1}\, P_\ell(\hat{\mathbf{r}} \cdot \hat{\mathbf{z}}) \tag{3.5.8}$$

where $P_\ell(\mu)$ is the associated Legendre function with $m = 0$ (check (3.4.42)), conventionally called the *Legendre polynomial* and given explicitly by

$$P_\ell(\mu) = \frac{1}{2^\ell \ell!} \partial_\mu^\ell (\mu^2 - 1)^\ell \tag{3.5.9}$$

and, from (3.5.7),

$$P_\ell(1) = 1. \tag{3.5.10}$$

3.5.2 Calculating the Kernel Function

We return to the self-reproducing kernel, the subject of section 3.2. Now we can calculate the kernel function $Q_\ell(\mu)$, which appeared in (3.3.27).

$$Q_\ell(\hat{\mathbf{r}} \cdot \hat{\mathbf{s}}) = \sum_{m=-\ell}^{\ell} \beta_\ell^m(\hat{\mathbf{r}}) \overline{\beta_\ell^m(\hat{\mathbf{s}})} \quad \text{for all} \quad \hat{\mathbf{r}}, \hat{\mathbf{s}} \in S(1). \tag{3.5.11}$$

The formula holds for every orthonormal basis β_ℓ^m in $\mathcal{H}_\ell$ and for any $\hat{\mathbf{r}}, \hat{\mathbf{s}}, \in S(1)$. We chcose a Cartesian basis $\hat{\mathbf{x}}, \hat{\mathbf{y}}, \hat{\mathbf{z}}$, in R^3 and take the β_ℓ^m to be the harmonics Y_ℓ^m generated by the basis $\hat{\mathbf{x}}, \hat{\mathbf{y}}, \hat{\mathbf{z}}$. We set $\hat{\mathbf{s}} = \hat{\mathbf{z}}$. Then for every $\hat{\mathbf{r}} \in S(1)$, (3.5.11) gives

$$Q_\ell(\hat{\mathbf{r}} \cdot \hat{\mathbf{z}}) = \sum_{m=-\ell}^{\ell} Y_\ell^m(\hat{\mathbf{r}}) \overline{Y_\ell^m(\hat{\mathbf{z}})}.$$

But all the Y_ℓ^m for $m \neq 0$ vanish along $\hat{\mathbf{z}}$; thus from (3.5.8, 3.4.10)

$$Q_\ell(\hat{\mathbf{r}} \cdot \hat{\mathbf{z}}) = Y_\ell^0(\hat{\mathbf{r}})\overline{Y_\ell^0(\hat{\mathbf{z}})}$$
$$= \sqrt{2\ell + 1} \, P_\ell(\hat{\mathbf{r}} \cdot \hat{\mathbf{z}})\sqrt{2\ell + 1} \, P_\ell(1).$$

Thus

$$Q_\ell(\hat{\mathbf{r}} \cdot \hat{\mathbf{z}}) = (2\ell + 1)P_\ell(\hat{\mathbf{r}} \cdot \hat{\mathbf{z}}).$$

Since $\mu = \hat{\mathbf{r}} \cdot \hat{\mathbf{z}}$ can take any value from -1 to 1 if we choose $\hat{\mathbf{r}}$ suitably, we conclude that

$$Q_\ell(\mu) = (2\ell + 1)P_\ell(\mu). \tag{3.5.12}$$

We see now that the statement of the Spherical Harmonic Addition Theorem in (3.3.27) can be rewritten as

$$(2\ell + 1)P_\ell(\hat{\mathbf{r}} \cdot \hat{\mathbf{s}}) = \sum_{m=-\ell}^{\ell} \beta_\ell^m(\hat{\mathbf{r}}) \overline{\beta_\ell^m(\hat{\mathbf{s}})} \tag{3.5.13}$$

where $P_\ell(\mu)$ is the Legendre polynomial given by (3.5.9) and the set of β_ℓ^m constitutes any orthonormal basis for $\mathcal{H}_\ell$.

3.5.3 The Generating Function for Legendre Polynomials

Another expression for the Legendre polynomial $P_\ell(\mu)$ is important in solving Laplace's equation, and allows us to derive Green's function for ∇^2. For $-1 \leq \mu \leq 1$ and $|r| < \sqrt{2} - 1$, we have $|r^2 - 2\mu r| < 1$, so we can use the Binomial Theorem to write

$$(1 + r^2 - 2\mu r)^{-1/2} = \sum_{n=0}^{\infty} C_n(r^2 - 2\mu r)^n$$

$$(r - 2\mu)^n = \sum_{m=0}^{n} D_{m,n} \, r^m \mu^{n-m}$$

where

$$C_n = \frac{1}{n!}\left(-\frac{1}{2}\right)\left(-\frac{3}{2}\right)\cdots\left(\frac{1}{2}-n\right)$$

and

$$D_{m,n} = (-2)^{n-m}\binom{n}{m} = (-2)^{n-m}\frac{n!}{m!(n-m)!}.$$

Then

$$(1+r^2-2\mu r)^{-1/2} = \sum_{n=0}^{\infty}\sum_{m=0}^{n} C_n D_{m,n}\, r^{m+n}\mu^{n-m}.$$

For each integer ℓ, only finitely many terms in the sum have $m+n = \ell$, so we can collect them and write

$$(1+r^2-2\mu r)^{-1/2} = \sum_{\ell=0}^{\infty} r^\ell p_\ell(\mu) \tag{3.5.14}$$

where the functions

$$p_\ell(\mu) = \sum_{m+n=\ell} C_n\, D_{m,n}\, \mu^{n-m} = \sum_{m=0}^{[\ell/2]} C_{\ell-m}\, D_{m,\ell-m}\, \mu^{\ell-2m}. \tag{3.5.15}$$

Here $[\ell/2] = $ largest integer $\le \ell/2$. It follows that $p_\ell(\mu)$ is a linear combination of μ^ℓ, $\mu^{\ell-2}$, $\mu^{\ell-4}$, ..., $\mu^{\sigma(\ell)}$. In particular, p_ℓ is a polynomial in μ of degree ℓ. Moreover, $r^\ell p_\ell(\mu)$ is a linear combination of z^ℓ, $r^2 z^{\ell-2}$, $r^4 z^{\ell-4}$, ..., where $z = r\mu$. Thus

$$r^\ell p_\ell(\hat{\mathbf{r}}\cdot\hat{\mathbf{z}}) \in \mathcal{P}_\ell. \tag{3.5.16}$$

Now $1/|\mathbf{r}|$ is the potential of a point charge at the origin, so it is well known that $\nabla^2\frac{1}{|\mathbf{r}|} = 0$, if $|\mathbf{r}| > 0$. This may be easily seen from

$$r^2\nabla^2\frac{1}{r} = (r\partial_r^2 r + \nabla_1^2)\frac{1}{r} = 0, \qquad r > 0.$$

But $1/|\mathbf{r}-\hat{\mathbf{z}}|$ is the potential of a point charge at $\hat{\mathbf{z}}$, so we also have when $\mathbf{r} \ne \hat{\mathbf{z}}$

$$\nabla^2\frac{1}{|\mathbf{r}-\hat{\mathbf{z}}|} = 0. \tag{3.5.17}$$

If $r = |\mathbf{r}|$ and $\hat{\mathbf{r}} \cdot \hat{\mathbf{z}} = \mu$, then

$$\frac{1}{|\mathbf{r} - \hat{\mathbf{z}}|} = [(\mathbf{r} - \hat{\mathbf{z}}) \cdot (\mathbf{r} - \hat{\mathbf{z}})]^{-1/2} = [1 + r^2 - 2r\mu]^{-1/2},$$

so for $r < \sqrt{2} - 1$

$$\frac{1}{|\mathbf{r} - \hat{\mathbf{z}}|} = \sum_{\ell=0}^{\infty} r^\ell p_\ell(\hat{\mathbf{r}} \cdot \hat{\mathbf{z}}).$$

Since a power series can be differentiated term by term in its region of convergence, if $r < \sqrt{2} - 1$ we have

$$\nabla^2 \frac{1}{|\mathbf{r} - \hat{\mathbf{z}}|} = \sum_{\ell=0}^{\infty} \nabla^2 [r^\ell p_\ell(\hat{\mathbf{r}} \cdot \hat{\mathbf{z}})]. \tag{3.5.18}$$

The ℓth term in the sum in (3.5.18) is in $\mathcal{P}_{\ell-2}$, so that sum is a power series in $r^{\ell-2}$. But in order to satisfy (3.5.17) each term must vanish separately. Combining this with the result (3.5.16), we have

$$r^\ell p_\ell(\hat{\mathbf{r}} \cdot \hat{\mathbf{z}}) \in \mathcal{H}_\ell; \tag{3.5.19}$$

that is, $r^\ell p_\ell(\hat{\mathbf{r}} \cdot \hat{\mathbf{z}})$ is a harmonic polynomial of degree ℓ. Clearly, $\partial_\lambda r^\ell p_\ell(\hat{\mathbf{r}} \cdot \hat{\mathbf{z}}) = 0$, so

$$r^\ell p_\ell(\hat{\mathbf{r}} \cdot \hat{\mathbf{z}}) \in \mathcal{H}_\ell^0. \tag{3.5.20}$$

But $\dim \mathcal{H}_\ell^0 = 1$, and we know from our derivation of an orthonormal basis in the last section that $r^\ell P_\ell(\hat{\mathbf{r}} \cdot \hat{\mathbf{z}}) \in \mathcal{H}_\ell^0$. Therefore, there is a constant C_ℓ such that

$$p_\ell(\mu) = C_\ell P_\ell(\mu).$$

If we set $\mu = 1$ in (3.5.14), we get

$$\frac{1}{1 - r} = \sum_{\ell=0}^{\infty} r^\ell p_\ell(1).$$

Therefore, $p_\ell(1) = 1$. Then, by (3.5.10), $C_\ell = 1$ and

$$p_\ell(\mu) = P_\ell(\mu) = \frac{1}{2\ell + 1}\, Q_\ell(\mu). \tag{3.5.21}$$

By Corollaries 2 and 3 of section 3.3, $|Q_\ell(\mu)| \leq 2\ell + 1$, so

$$|P_\ell(\mu)| \leq 1 \qquad \text{if} \qquad -1 \leq \mu \leq 1. \tag{3.5.22}$$

Therefore, (3.5.14) converges not merely for $|r| < \sqrt{2} - 1$ but for $|r| < 1$, and we have

$$(1 + r^2 - 2\mu r)^{-1/2} = \sum_{\ell=0}^{\infty} r^\ell P_\ell(\mu) \qquad r < 1. \tag{3.5.23}$$

The term on the left side of (3.5.23) is known as the *generating function* for the Legendre polynomials.

3.5.4 Green's Function for ∇^2

The potential at $\mathbf{r}$ of a point charge at $\mathbf{s}$ is essentially $1/|\mathbf{r} - \mathbf{s}|$. We can now give the expansion of this potential in spherical harmonics relative to an origin at $\mathbf{0}$. We write

$$|\mathbf{r} - \mathbf{s}|^2 = (\mathbf{r} - \mathbf{s}) \cdot (\mathbf{r} - \mathbf{s})$$

$$= r^2 + s^2 - 2\mathbf{r} \cdot \mathbf{s}$$

$$= r^2 + s^2 - 2rs\mu$$

$$= r^2[1 + (s/r)^2 - 2(s/r)\mu].$$

We use this to write the potential in the same form as (3.5.23), allowing an expansion in Legendre polynomials,

$$\frac{1}{|\mathbf{r} - \mathbf{s}|} = \frac{1}{r}[1 + (s/r)^2 - 2(s/r)\mu]^{-1/2} = \frac{1}{r}\sum_{\ell=0}^{\infty} \left(\frac{s}{r}\right)^\ell P_\ell(\mu).$$

The series converges as long as $s < r$. It can be written

$$\frac{1}{|\mathbf{r} - \mathbf{s}|} = \frac{1}{r}\sum_{\ell=0}^{\infty} \left(\frac{s}{r}\right)^\ell P_\ell(\hat{\mathbf{s}} \cdot \hat{\mathbf{r}}), \tag{3.5.24}$$

and using the Spherical Harmonic Addition Theorem, if $\beta_\ell^{-\ell}, \ldots, \beta_\ell^\ell$ is any orthonormal basis for $\mathcal{H}_\ell$, we have

$$\frac{1}{|\mathbf{r} - \mathbf{s}|} = \frac{1}{r} \sum_{\ell=0}^{\infty} \frac{1}{(2\ell + 1)} \left(\frac{s}{r}\right)^\ell \sum_{m=-\ell}^{\ell} \beta_\ell^m(\hat{\mathbf{r}}) \overline{\beta_\ell^m(\hat{\mathbf{s}})} \quad s < r. \quad (3.5.25)$$

Equation (3.5.25) gives the potential at $\mathbf{r}$ due to a point charge at $\mathbf{s}$ if $\mathbf{r}$ is in the shell $S(s, \infty)$, i.e., $|\mathbf{r}| > |\mathbf{s}|$. Alternatively it can be interpreted as giving the potential at $\mathbf{s}$ due to a point charge at $\mathbf{r}$ when $s < r$.

3.6 The Character of the Natural Basis

It is often helpful to be able to visualize the basis functions as they vary in space, but, except for the most mathematical of readers, equations (3.4.54, 3.4.55) are not very evocative. In this section, we examine these equations and calculate some further properties that elucidate the appearance of the natural basis.

3.6.1 Nodal Lines of $\mathrm{Re}\,Y_\ell^m(\hat{\mathbf{r}})$ on $S(1)$

First we notice that it is sufficient to study the behavior of the functions on $S(1)$ because on a sphere of any other radius, say r, the functional shape remains the same, scaled by r^l. We must fix a coordinate system with its origin at the center of $S(1)$. We use geographical terms, natural in geophysics, for the angular coordinates: λ is the *longitude* and θ the *colatitude*. Also we may sometimes refer to the line $\lambda = 0$ as the Greenwich meridian. We reproduce the defining equations of the natural orthonormal basis functions here for convenience: relative to a fixed Cartesian system $\hat{\mathbf{x}}, \hat{\mathbf{y}}, \hat{\mathbf{z}}$, for $-\ell \leq m \leq \ell$:

$$Y_\ell^m(\mathbf{r}) = (-1)^m \sqrt{2\ell + 1} \sqrt{\frac{(\ell - m)!}{(\ell + m)!}} \, r^\ell e^{im\lambda} P_\ell^m(\cos\theta) \qquad (3.6.1)$$

$$P_\ell^m(\mu) = \frac{1}{2^\ell \ell!} (1 - \mu^2)^{m/2} \partial_\mu^{\ell+m} (\mu^2 - 1)^\ell. \qquad (3.6.2)$$

Notice how Y_ℓ^m is the product of a function of longitude and a function of colatitude. The longitudinal function, $e^{im\lambda}$, contains all

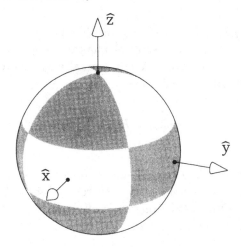

Figure 3.6.1: The function $\mathrm{Re}Y_4^2(\hat{\mathbf{r}})$ on $S(1)$. The shaded regions are those where the function is positive; the nodal lines are the boundaries of the shaded zones.

the complex-variable behavior and is particularly simple – sine or cosine variation depending upon whether the real or imaginary part is considered. For visualization purposes we describe only the real part of $Y_\ell^m(\hat{\mathbf{r}})$. The imaginary part has the same shape, but advanced in longitude by an amount $\pi/2m$. Furthermore, there is no need to worry about negative m, because we have seen in (3.4.56) that $Y_\ell^{-m} = (-1)^m \overline{Y_\ell^m}$.

In a graphical description we consider the nodal lines, those points on $S(1)$ where $\mathrm{Re}\, Y_\ell^m(\hat{\mathbf{r}})$ vanishes. The sign of this function changes each time we cross a nodal line. From (3.6.1) it is clear that $\mathrm{Re}\, Y_\ell^m(\hat{\mathbf{r}})$ has two families of nodal lines on $S(1)$: (a) those associated with the variation of $e^{im\lambda}$, which are m great circles, passing through the poles (or the $\hat{\mathbf{z}}$ axis), equally spaced in longitude; (b) small circles at the colatitudes where $P_\ell^m(\cos\theta)$ vanishes. A picture of $\mathrm{Re}\, Y_4^2(\hat{\mathbf{r}})$ is provided in Figure 3.6.1, which illustrates the general idea. While the location of the nodal lines in the longitudinal direction of travel is easily settled, at this point we have not established whether $P_\ell^m(\mu)$ has any roots for real μ, except those at $\mu = \pm 1$ when $|m| > 0$ described earlier; these correspond to zeros at the poles.

We now establish a result, important in its own right and most helpful in understanding the shape of the natural basis.

Theorem: The associated Legendre function $P_\ell^m(\mu)$ for real μ

has exactly $\ell - m$ simple zeros on the open interval $(-1,\ 1)$.

Proof: First we recall some elementary properties of polynomials of a single real variable. A polynomial with real coefficients, $p(x)$ of degree n with m real roots $x_1, x_2, x_3 \cdots x_m$, can be factored thus:

$$p(x) = P(x) \prod_{j=1}^{m} (x - x_j)^{\nu_j}$$

where $P(x)$ is a polynomial of degree $n - \sum_{j=1}^{m} \nu_j$, where the integer ν_j is the *multiplicity* of the jth root; if $\nu_j = 1$ that root is called *simple*. Obviously from the definition $0 < \nu_j \le n$. Since the polynomial is of degree n

$$\sum_{j=1}^{m} \nu_j \le n.$$

Consider the derivative of $p(x)$, a polynomial in x of degree $n - 1$:

$$\partial_x p = P(x) \sum_{j=1}^{m} \nu_j (x - x_j)^{\nu_j - 1} \prod_{k \ne j} (x - x_k)^{\nu_k}$$

$$+ \partial_x P \prod_{j=1}^{m} (x - x_j)^{\nu_j}.$$

Clearly those roots that are not simple are also roots of the derivative polynomial.

Returning to the roots of the associated Legendre function, we examine the real zeros on $(-1,\ 1)$ of the polynomial

$$q_\ell^m(\mu) = \partial_\mu^{\ell - m} (\mu^2 - 1)^\ell. \tag{3.6.3}$$

According to (3.4.40) and (3.6.2)

$$q_\ell^m(\mu) = \frac{2^\ell \ell}{\alpha_\ell^m} (1 - \mu^2)^{m/2} P_\ell^m(\mu) \tag{3.6.4}$$

where α_ℓ^m is some constant. Thus $q_\ell^m(\mu)$ and $P_\ell^m(\mu)$ share the same zeros on the interval of interest because the multiplying factor $(1 - \mu^2)^{m/2}$ is positive on it.

We begin with $m = l$ in (3.6.3). Clearly $q_\ell^\ell(\mu)$ is a polynomial of degree 2ℓ and can be written

$$q_\ell^\ell(\mu) = (\mu - 1)^\ell (\mu + 1)^\ell.$$

Thus $q_\ell^\ell(\mu)$ has two roots each of multiplicity ℓ, one at $\mu = 1$ and the other at $\mu = -1$. There are no real roots on the open interval $(-1, 1)$, and so P_ℓ^ℓ has none in that interval either. Consider the derivative $\partial_\mu q_\ell^\ell = p_\ell^{\ell-1}$: it is a polynomial of degree $2\ell - 1$ with zeros of multiplicity $\ell - 1$ at $\mu = -1$ and $\mu = +1$; therefore it has at most one other real zero. Rolle's Theorem (Thomas and Finney, 1984) states that if $a < b$ and $f(a) = f(b)$ and f is a continuously differentiable function on the closed interval $[a, b]$, the derivative $\partial_x f$ must vanish at least once in (a, b). Therefore $q_\ell^{\ell-1}(\mu)$ vanishes exactly once in $(-1, 1)$. From (3.6.4) $P_\ell^{\ell-1}(\mu)$ has exactly one zero in $(-1, 1)$.

The argument is repeated on the second derivative of $q_\ell^\ell(\mu)$, applying Rolle's Theorem to the subintervals between the roots of the first derivative. This shows $P_\ell^{\ell-2}(\mu)$ has two zeros on the open interval. The argument can be applied repeatedly until one has differentiated ℓ times. Exactly one new zero appears with each differentiation; the process can be carried out to $m = 0$ when there are ℓ zeros on the open interval. This proves the required result. ◁

The theorem shows us that Re $Y_\ell^m(\hat{\mathbf{r}})$ has $\ell - m$ zeros along lines of constant latitude. Of course it does not address the spatial distribution of the nodal lines, which can be discovered only by more detailed analysis of the function P_ℓ^m. Before proceeding to such an analysis, we provide graphs of a few low-degree spherical harmonics. In Figure 3.6.2 we show the values of

$$\text{Re}\, Y_\ell^m(1, \theta, 0),$$

which provides an idea of the oscillations of the fully normalized natural basis function as the evaluation point moves from the north pole to the south pole on the Greenwich meridian.

3.6.2 General Appearance of Y_ℓ^m for large ℓ

We are going to study the behavior of the associated Legendre functions for large degree ℓ, when the order m is not too close to

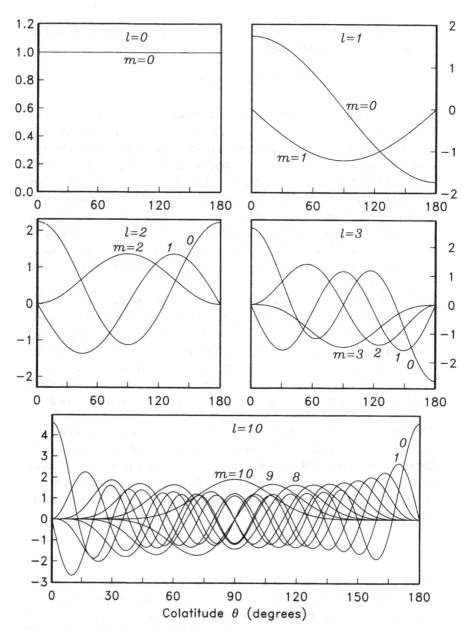

Figure 3.6.2: Behavior of $\mathrm{Re}Y_\ell^m(\hat{\mathbf{r}})$ on the Greenwich meridian, $\lambda = 0$, for several low-degree harmonics.

the degree. We will observe that the solution is oscillatory in one subinterval and exponential in another. This will give us a feeling for the general shape of the θ variation of the natural basis functions. The approach will be to work from the differential equation that these functions obey, looking for approximate solutions, valid for large ℓ.

Recall that every spherical harmonic of degree ℓ satisfies the eigenvalue equation (3.1.8) and so, in particular, the fully normalized members of the natural basis do also:

$$\nabla_1^2 Y_\ell^m = -\ell(\ell+1)Y_\ell^m.$$

We rewrite (3.6.1) thus:

$$Y_\ell^m(\mathbf{r}) = r^\ell e^{im\lambda} p(\theta) \tag{3.6.5}$$

where the latitudinal behavior is expressed directly in terms of a function of θ:

$$p(\theta) = (-1)^m \sqrt{2\ell+1} \sqrt{\frac{(\ell-m)!}{(\ell+m)!}}\; P_\ell^m(\cos\theta).$$

We have dropped the sub- and superscripts for economy's sake; also we assume $m \geq 0$ for definiteness. Expanding the surface Laplacian with (3.4.4) and performing some algebra, we find the differential equation for p:

$$\partial_\theta^2 p + \cot\theta\,\partial_\theta p + \left[\ell(\ell+1) - \frac{m^2}{\sin^2\theta}\right] p = 0. \tag{3.6.6}$$

We want to study the behavior of p as $\ell \to \infty$ while m/ℓ remains constant. It will be convenient to work with latitude γ rather than colatitude θ. The relation between these is

$$\gamma = \pi/2 - \theta.$$

We introduce some other new variables, L, γ_c, and the function $f(\gamma)$, which will allow us to simplify (3.6.6). Define

$$L = \sqrt{\ell(\ell+1)} \tag{3.6.7}$$

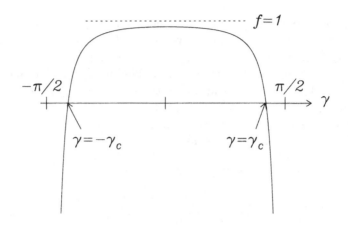

Figure 3.6.3: Sketch of the function $f(\gamma)$ given by (3.6.9).

$$\gamma_c = \cos^{-1}(m/L) \qquad (3.6.8)$$

$$f(\gamma) = 1 - \frac{\cos^2\gamma_c}{\cos^2\gamma}. \qquad (3.6.9)$$

Then (3.6.6) becomes

$$\partial_\gamma^2 p - \tan\gamma\, \partial_\gamma p + L^2 f(\gamma)p = 0. \qquad (3.6.10)$$

The function f is shown in Figure 3.6.3. Let us compare (3.6.10) with the more familiar standard differential equation

$$\partial_\gamma^2 y + \Omega^2 y = 0.$$

When Ω^2 is positive we have oscillatory solutions to the standard equation, with wavenumber Ω; when Ω^2 is negative there is exponential growth or decay. Therefore, in a band covering the equator, $|\gamma| \le \gamma_c - \epsilon$, we expect that the solutions p of (3.6.10) will be oscillatory, with wavenumbers that grow like L as $L \to \infty$. In the two zones that include the poles, $\gamma_c + \epsilon \le |\gamma| \le \pi/2 - \epsilon$, we expect solutions of (3.6.10) to grow or decay exponentially. The critical regions near $\pm\gamma_c$ will have to be examined carefully.

In the central region, $-\gamma_c + \epsilon \le \gamma \le \gamma_c - \epsilon$, we look for oscillatory solutions to (3.6.10) that have the form

$$p(\gamma) = e^{iLW(\gamma)}. \qquad (3.6.11)$$

The equation for $W(\gamma)$ equivalent to (3.6.10) is

$$f - (\partial_\gamma W)^2 = \frac{1}{(iL)} (\partial_\gamma^2 W - \tan \gamma \, \partial_\gamma W). \qquad (3.6.12)$$

The reason we expect solutions of form (3.6.11) with a well-behaved W is that if p_1, p_2 are any two linearly independent solutions of (3.6.10), which is a linear differential equation, they can never both vanish at a single γ. If they did, all solutions of (3.6.10) would vanish there, violating the fundamental existence theorem for ordinary differential equations. But then $\ln(p_1 + ip_2)$ is a well-behaved function of γ, because the argument of the logarithm never vanishes. We can take $iLW(\gamma) = \ln(p_1 + ip_2)$.

Equation (3.6.12) is a nonlinear differential equation, but it has the virtue that as L becomes large, we see a simple solution emerging. No approximation has been made yet; that is the next step. We suspect that for $L \gg 1$, (3.6.12) can be solved iteratively. That is, we can write

$$W = W_0(\gamma) + \frac{1}{(iL)} W_1(\gamma) + \frac{1}{(iL)^2} W_2(\gamma) + \cdots. \qquad (3.6.13)$$

The various W_n are found by substituting (3.6.13) into (3.6.12) and equating coefficients of $(iL)^{-n}$. The series (3.6.13) will be asymptotic rather than convergent. That is, if we terminate it at $(iL)^{-n}W_n(\gamma)$, we will obtain a function that, as $L \to \infty$, differs from a true solution of (3.6.12) by an error of order L^{-n-1}. The first three terms, corresponding to $(iL)^{-n}$, $n = 0, 1, 2$ give

$$f - (\partial_\gamma W_0)^2 = 0 \qquad (3.6.14)$$

$$-2(\partial_\gamma W_0)(\partial_\gamma W_1) = \partial_\gamma^2 W_0 - \tan \gamma \partial_\gamma W_0 \qquad (3.6.15)$$

$$-2(\partial_\gamma W_0)(\partial_\gamma W_2) - (\partial_\gamma W_1)^2 = \partial_\gamma^2 W_1 - \tan \gamma \partial_\gamma W_1. \qquad (3.6.16)$$

We choose for W_0 the particular solution, which obviously satisfies (3.6.14):

$$W_0(\gamma) = \int_0^\gamma d\phi \, f(\phi)^{1/2} = \int_0^\gamma d\phi \left[1 - \frac{\cos^2 \gamma_c}{\cos^2 \phi} \right]^{1/2}. \qquad (3.6.17)$$

Equation (3.6.15) for $W_1(\gamma)$ combined with (3.6.14) can be manipulated to read

$$-2\partial_\gamma W_1 = \partial_\gamma \ln (\cos \gamma \partial_\gamma W_0).$$

Therefore, using (3.6.17), we take as a particular solution

$$W_1(\gamma) = -\tfrac{1}{2} \ln \left[\sqrt{\cos^2 \gamma - \cos^2 \gamma_c}\right] = \ln \left[(\cos^2 \gamma - \cos^2 \gamma_c)^{-1/4}\right]. \tag{3.6.18}$$

At this point, we note that since all the W_n are real, we can write (3.6.11) as

$$p(\gamma) = A(\gamma)e^{iLV(\gamma)} \tag{3.6.19}$$

where $\ln A(\gamma) = W_1 - W_3/L^2 + W_5/L^4 - \cdots$ and $V = W_0 - W_2/L^2 + W_4/L^4 - \cdots$. Then A is a real amplitude, LV is a real phase, and with errors of order L^{-2}, $A = (\cos^2 \gamma - \cos^2 \gamma_c)^{-1/4}$, $V = W_0$. Thus

$$p(\gamma) = (\cos^2 \gamma - \cos^2 \gamma_c)^{-1/4} e^{iLW_0(\gamma)} \tag{3.6.20}$$

is an approximation to a solution of (3.6.10) with an error in amplitude of $O(L^{-2})$ and an error in the phase LW_0 of $O(L^{-1})$. If p is a solution of (3.6.10), so is its complex conjugate $\bar{p}$, and so is $Be^{-i\beta}p$, where B and β are any real constants. Hence, so is $\tfrac{1}{2}(Be^{-i\beta}p + Be^{i\beta}\bar{p}) = q$. Thus our general approximate solution for (3.6.10) is

$$q(\gamma) = \frac{B}{(\cos^2 \gamma - \cos^2 \gamma_c)^{1/4}} \cos [LW_0(\gamma) - \beta], \tag{3.6.21}$$

which contains the two arbitrary real constants, B and β. To determine the values of the constants, we match the value and the first derivative with those of the true solution at the equator $\gamma = 0$. From (3.6.21), the approximate solution gives

$$q(0) = \frac{B \cos \beta}{(\sin \gamma_c)^{1/2}} \quad \text{and} \quad \partial_\gamma q(0) = \frac{LB \sin \beta}{(\sin \gamma_c)^{1/2}}. \tag{3.6.22}$$

To find the initial values from the exact solution, we expand $(\mu^2 - 1)^\ell$ by the Binomial Theorem; then we can show from (3.6.2) and the definition of p that at $\gamma = 0$, where $\mu = \sin \gamma = 0$:

$$p(0) = \frac{(-1)^m [(2\ell + 1)(\ell + m)!(\ell - m)!]^{1/2}}{2^\ell [\tfrac{1}{2}(\ell + m)]! \, [\tfrac{1}{2}(\ell - m)]!} \cos \left(\frac{\ell - m}{2}\pi\right)$$

$$= (-1)^m \frac{2}{[\pi \sin \gamma_c]^{1/2}} \cos \left(\frac{\ell - m}{2}\pi\right) + O(\ell - m)^{-2}$$

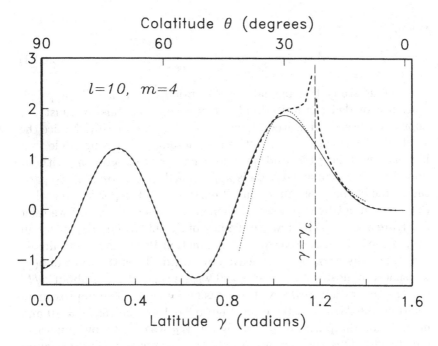

Figure 3.6.4: The function $p(\gamma)$ for $\ell = 10$, $m = 4$: solid line, exact solution of (3.6.10); heavy dashed curve, asymptotic theory from (3.6.23, 3.6.26); light dashed curve, Airy connecting solution from (3.6.31).

$$\partial_\gamma p(0) = \frac{(-1)^m [(2\ell + 1)(\ell + m)!(\ell - m)!]^{1/2}}{2^{\ell-1} [\frac{1}{2}(\ell + m - 1)]! \, [\frac{1}{2}(\ell - m - 1)]!} \sin \left(\frac{\ell - m}{2} \pi \right)$$

$$= 2L(-1)^m \left(\frac{\sin \gamma_c}{\pi} \right)^{1/2} \sin \left(\frac{\ell - m}{2} \pi \right) + O(\ell - m)^{-2}.$$

In these equations we have used the fact that, if $\ell - m$ and $\ell + m$ are both large, the factorials may be approximated with Stirling's formula, $n! = \sqrt{2\pi n} n^n e^{-n} [1 + O(1/n)]$.

These initial conditions are used to solve for the unknown constants in (3.6.22): we find $B = 2(-1)^m \pi^{-1/2}$ and $\beta = \pi(\ell - m)/2$. Also, we note that $L = \sqrt{\ell(\ell + 1)} = \ell + \frac{1}{2} + O(\ell^{-1})$, so with an error $O(L^{-2})$ in amplitude and $O(L^{-1})$ in phase we arrive at an approximate solution

$$p(\gamma) = \frac{2(-1)^m \cos \left\{ L[W_0(\gamma) - \pi \sin^2 \frac{1}{2} \gamma_c] + \frac{1}{4}\pi \right\}}{\pi^{1/2} (\cos^2 \gamma - \cos^2 \gamma_c)^{1/4}} \qquad (3.6.23)$$

where

$$W_0(\gamma) = \int_0^\gamma d\phi \left[1 - \frac{\cos^2\gamma_c}{\cos^2\phi}\right]^{1/2}. \qquad (3.6.24)$$

Let us study this partial solution briefly. It is a valid approximation provided the latitude obeys $|\gamma| \leq \gamma_c - \epsilon$; thus we must not approach the critical latitude $\gamma_c = \cos^{-1}(m/L)$ too closely. Notice the solution consists of the product of two factors: (a) an amplitude that increases with latitude, and becomes singular at γ_c, growing without bound as the singular latitude is approached; (b) an oscillatory part, the cosine of the phase function. Unfortunately, the phase integral in (3.6.24) can not be expressed in elementary functions, but if we refer to Figure 3.6.3, which shows the square of the integrand, we can get an idea of the behavior. Near $\gamma = 0$ the function $W_0(\gamma)$ increases almost proportionally to γ, but is concave downward. Thus the zeros of $p(\gamma)$ are spaced almost evenly near $\gamma = 0$ with a separation of about π/L, but the separation grows with increasing latitude. The approximate solution (3.6.23) is illustrated in Figure 3.6.4 for the case $\ell = 10$ and $m = 4$; for the purposes of the figure, W_0 was found by numerical integration. Observe how accurately the asymptotic solution agrees with the true one outside the critical zone, even when $\ell = 10$, which is not very large.

In subsection 3.6.1 we derived the result that there are always exactly $\ell - m$ zeros of $P_\ell^m(\mu)$ in the open interval $(-1, 1)$. The reader may be curious to know how many zeros are predicted by equations (3.6.23, 3.6.24). As we mentioned earlier, all the oscillations are expected to occur within the central interval $(-\gamma_c, \gamma_c)$. The zeros of cosine are spaced by π so that we see from (3.6.23, 3.6.24), and the fact that W_0 is an odd function that the number of zeros is predicted to be the largest integer below

$$Z = \frac{2L}{\pi}W_0(\gamma_c) = \frac{2L}{\pi}\int_0^{\gamma_c} d\phi \left[1 - \frac{\cos^2\gamma_c}{\cos^2\phi}\right]^{1/2}.$$

To evaluate the integral, substitute the new variable $c = \cos^2\phi$, so that $d\phi = dc/2\sqrt{c(1 - c^2)}$; then

$$Z = \frac{L}{\pi}\int_{\cos^2\gamma_c}^1 \frac{dc}{c}\left[\frac{c - \cos^2\gamma_c}{1 - c}\right]^{1/2}.$$

If we translate the integral to the complex c plane, we find branch points at the limits of integration, which we join with a branch cut. The desired integral is then half the integral around the cut; if the contour around the cut is expanded, it is easy to see that there are two contributions – one from the simple pole at $c = 0$ and the other from the integral on a very large circle. Summing these contributions gives us the answer:

$$Z = L(1 - \cos\gamma_c).$$

Finally, let us return to the original variables ℓ and m via (3.6.7, 3.6.8)

$$Z = L(1 - m/L) = L - m = \ell - m + \tfrac{1}{2} + O(\ell^{-1}).$$

Since the error of replacing L by $\ell + \tfrac{1}{2}$ is never more than 0.09 for all $\ell \geq 1$, the largest integer below Z is $\ell - m$ and the asymptotic theory always gives the exact answer for the number of oscillations, an impressive result.

The analysis we have carried out is the initial phase of the *WKBJ approximation* for studying second-order ordinary differential equations. The completion of the analysis of (3.6.10), which we only sketch, means that we must next venture into the region $|\gamma| > \gamma_c$, where the coefficient f in (3.6.10) is negative, so the solutions grow or decay exponentially instead of oscillating. Following the same kind of treatment as before, we can see that the general solution is

$$(\cos^2\gamma_c - \cos^2\gamma)^{-1/4}[Ce^{-LU_0(\gamma)} + De^{LU_0(\gamma)}]$$

where

$$U_0(\gamma) = \int_{\gamma_c}^{\gamma} d\phi \left[\frac{\cos^2\gamma_c}{\cos^2\phi} - 1\right]^{1/2}. \qquad (3.6.25)$$

For definiteness we look at the solution when $\gamma > 0$, the northern hemisphere. We know that as $L \to \infty$, $(2\ell+1)^{-1/2}p$ remains bounded, so we must take $D = 0$ to eliminate the growing component, and then in $\gamma_c + \epsilon \leq \gamma \leq \pi/2$

$$p(\gamma) = \frac{C}{(\cos^2\gamma_c - \cos^2\gamma)^{1/4}} e^{-LU_0(\gamma)}. \qquad (3.6.26)$$

To find the unknown constant C, we must connect (3.6.23) with (3.6.26) through the critical point γ_c where $f(\gamma_c) = 0$ and where the solutions switch from oscillatory behavior to exponential. If $|\gamma - \gamma_c| \ll$ 1, we can approximate $f(\gamma)$ by $f'(\gamma_c)(\gamma - \gamma_c)$, so (3.6.10) becomes approximately

$$\partial_\gamma^2 p - \tan \gamma \, \partial_\gamma p + L^2 f'(\gamma_c)(\gamma - \gamma_c)p = 0. \qquad (3.6.27)$$

It is the vanishing of the coefficient f that allows a singular solution into the equation, but this solution belongs only to the approximate solution, not the original. So the strategy now is to solve an approximation in the neighborhood of the singularity and suppress the singular component by appropriate choice of coefficients. The model equation is Airy's differential equation:

$$\partial_x^2 p - xp = 0. \qquad (3.6.28)$$

To mimic this equation we choose in (3.6.27) a new stretched independent variable x, given by

$$x = -L^{2/3} f'(\gamma_c)^{1/3}(\gamma - \gamma_c) = L^{1/3}(2 \tan \gamma_c)^{1/3}(\gamma - \gamma_c). \qquad (3.6.29)$$

Then (3.6.27) becomes

$$\partial_x^2 p - xp = O(L^{-2/3}). \qquad (3.6.30)$$

If we ignore the error term when $L \gg 1$ we obtain (3.6.28), whose general solution is $EAi(x) + FBi(x)$ where Ai is the *Airy function* and Bi is its associated function (Abramowitz and Stegun, 1965, p. 476).

As $x \to \infty$, $Bi(x) \sim x^{-1/4}\exp\left(\frac{2}{3} x^{3/2}\right)$, so we must take $F = 0$. In the critical region near γ_c, $p(\gamma) = EAi(x)$ where x is given by (3.6.29). This approximation is valid as long as $|\gamma - \gamma_c| \ll 1$, even if L and $x \to \infty$. The constant E is determined by requiring that as $x \to -\infty$, $EAi(x)$ fits smoothly onto (3.6.23) as $\gamma \to \gamma_c$. This matching of the two asymptotic expansions determines E and gives

$$p(\gamma) = (-1)^m \left[\frac{L}{2 \sin \gamma_c \cos^2\gamma_c} \right]^{1/6} 2Ai(x) \qquad (3.6.31)$$

where x is from (3.6.29). The approximation (3.6.31) is valid as $L \to \infty$ in the range $\gamma_c - \eta L^{-2/3} < \gamma < \gamma_c + \eta L^{-2/3}$, where η is any constant. This solution is also illustrated in Figure 3.6.4.

Now we can determine C in (3.6.26) by demanding that (3.6.31) with $x \to \infty$ fits smoothly onto (3.6.26) with $\gamma \to \gamma_c$. The result is

$$p(\gamma) = \frac{(-1)^m}{\pi^{1/2}(\cos^2\gamma_c - \cos^2\gamma)^{1/4}} e^{-LU_0(\gamma)} \tag{3.6.32}$$

where $U_0(\gamma)$ is given by (3.6.25). The approximation (3.6.30) is valid in the range $\gamma_c + \eta L^{-2/3} \le \gamma \le \pi/2$ for any constant η, as $L \to \infty$. For a fuller treatment of this sort of asymptotic analysis, the reader should consult Bender and Orszag (1978).

Later on we will need to understand how U_0 increases with γ, and so we perform an approximate analysis. We introduce the factor $\sin\phi$ into (3.6.25) to make the integral elementary: then after some straightforward calculus

$$U_0(\gamma) > \int_{\gamma_c}^{\gamma} d\phi \, \sin\phi \left[\frac{\cos^2\gamma_c}{\cos^2\phi} - 1 \right]^{1/2} \tag{3.6.33}$$
$$= \cos\gamma_c \, g(\cos\gamma_c/\cos\gamma)$$

where the function $g(\rho)$ is given by

$$g(\rho) = -\ln\left(\rho - \sqrt{\rho^2 - 1}\right) - \sqrt{1 - 1/\rho^2} \tag{3.6.34}$$

and $g(\rho)$ is a positive, monotone increasing function for $\rho > 1$; indeed, g tends to $\ln\rho$ as ρ tends to infinity.

3.6.3 Horizontal Wavelength of Y_ℓ^m

It is quite natural to ask, What is the horizontal wavelength of $Y_\ell^m(\hat{\mathbf{r}})$? We might expect this question to have a simple answer and it does. For this to make much sense, the scale of variation of the function should be short compared with the radius of the sphere, and so we call on our high-degree asymptotic expansion. From the asymptotic form (3.6.23) and definition (3.6.5) we see that provided the latitude γ obeys $|\gamma| < \gamma_c$ we can approximate Y_ℓ^m by

$$Y_\ell^m(\hat{\mathbf{r}}) = A \cos(n\gamma + \phi)e^{im\lambda}$$

where A, n, and ϕ are slowly varying functions of γ. We will calculate what n is in a moment, but the other two functions are not important in this discussion. Rewriting in complex form we see

$$Y_\ell^m(\hat{\mathbf{r}}) = \tfrac{1}{2} A \left(e^{i(m\lambda+n\gamma+\phi)} + e^{i(m\lambda-n\gamma-\phi)} \right)$$
$$= A_+ e^{i(m\lambda+n\gamma)} + A_- e^{i(m\lambda-n\gamma)}.$$

This looks like the sum of two complex traveling waves. In the local horizontal coordinate system the eastward wavenumber of both waves on the unit sphere is

$$k_E = m/\cos\gamma$$

and the northward wavenumbers are

$$k_N^+ = n \quad \text{and} \quad k_N^- = -n,$$

so that for both waves the wavenumber magnitude is

$$k = \sqrt{n^2 + m^2/\cos^2\gamma} \qquad\qquad (3.6.35)$$

and of course the corresponding wavelength is $2\pi/k$.

To find n at a particular latitude we return to (3.6.23, 3.6.24). In the abbreviated notation of this section, (3.6.23) is

$$p(\gamma) = A \cos\left(L W_0(\gamma) + \phi_1\right) \qquad\qquad (3.6.36)$$

and (3.6.24) states

$$W_0(\gamma) = \int_0^\gamma d\phi \, [1 - \cos^2\gamma_c/\cos^2\gamma]^{1/2}. \qquad\qquad (3.6.37)$$

Let us take the integrand of (3.6.37) to be roughly constant as the cosine in (3.6.36) goes through one period, which is equivalent to replacing the curve W, locally by its tangent. Then we find

$$n = L[1 - \cos^2\gamma_c/\cos^2\gamma]^{1/2}.$$

Recall from (3.6.8) that

$$\cos\gamma_c = m/L,$$

so we have

$$n = [L^2 - m^2/\cos^2\gamma]^{1/2}.$$

Substituting this into (3.6.35), we discover that the wavenumber k of the two traveling waves is just $L = \sqrt{\ell(\ell+1)}$, which is very accurately approximated by $\ell + \frac{1}{2}$. Hence the gratifying result that independent of the order m and the latitude γ, the wavelength on $S(1)$ is approximately $2\pi/(\ell + \frac{1}{2})$.

The fact that this result has been shown with the natural basis attached to a particular coordinate system does not restrict it to these particular functions. Clearly any spherical harmonic can be composed of linear combinations of the Y_ℓ^m and therefore on a small scale will appear to comprise superpositions of plane waves with wavelength $2\pi/(\ell + \frac{1}{2})$ with wavevectors in different directions. The only gap in this argument is the fact that we had to restrict ourselves to latitudes below the critical γ_c and m not too near ℓ, the conditions of validity of the asymptotic formulas. The constraint to low latitudes is easily lifted by rotating the coordinate system, so that the local region lies on the equator of a new frame in which the basis functions can be defined. Now the spherical harmonic referred to the original frame is a linear combination of natural basis functions evaluated at a point that must obey the latitude condition. It is possible to develop an asymptotic theory for the case in which m is close to ℓ, but we will not spend any time on it. We merely report that the same result regarding a wavelength can be obtained. The extreme case of $m = \ell$ is easy to analyze and we leave it as an exercise for the reader to show that once again the approximate wavelength at the equator is just $2\pi/(\ell + \frac{1}{2})$.

To isolate a single traveling wave turns out to be impossible within the space $\mathcal{H}_\ell$. Instead one must introduce the singular solutions of the differential equation (3.6.6), so-called Legendre functions of the second kind, Q_ℓ^m. A typical traveling wave would be

$$W_\ell^m = [P_\ell^m(\cos\theta) + (2i/\pi)Q_\ell^m(\cos\theta)]e^{im\lambda}.$$

These functions are singular at the poles. Physically, the singularity can be thought of as the place where the spherical surface has brought the wave to a focus. Mathematically, the singularity can be regarded as an artifact of the spherical polar coordinate system.

3.7 Numerical Calculations and the Like

As we will see in later chapters, the natural basis is used in the
description of the geomagnetic field as well as in a variety of theoretical
settings. One often needs to compute numerical values for a range of
degrees and orders at a large number of points. Naturally computers
are employed for this kind of work. Perhaps surprisingly, efficient
computational methods for finding spherical harmonic expansions are
not based on explicit expressions for the homogeneous polynomials,
but rely upon relationships between the spherical harmonics of
differing degree and order at a fixed point; these are called *recurrence
relationships*, and they have applications in theoretical problems as
well. But first, for the sake of completeness, we exhibit primitive
expressions for the functions comprising the first few members of the
natural basis.

3.7.1 Explicit Formulas for $P_\ell^m(\mu)$

Recall once more that our version of the natural orthonormal basis is
written

$$Y_\ell^m(\mathbf{r}) = (-1)^m \sqrt{2\ell+1} \sqrt{\frac{(\ell-m)!}{(\ell+m)!}} \, r^\ell e^{im\lambda} P_\ell^m(\cos\theta) \qquad (3.7.1)$$

$$P_\ell^m(\mu) = \frac{1}{2^\ell \ell!} (1-\mu^2)^{m/2} \partial_\mu^{\ell+m} (\mu^2-1)^\ell. \qquad (3.7.2)$$

From (3.4.44) we can write an alternative to (3.7.2) that will prove
useful:

$$P_\ell^m(\mu) = \frac{(-1)^m}{2^\ell \ell!} \frac{(\ell+m)!}{(\ell-m)!} (1-\mu^2)^{-m/2} \partial_\mu^{\ell-m} (\mu^2-1)^\ell. \qquad (3.7.3)$$

Also recall that the functions $P_\ell^m(\mu)$ are the associated Legendre
functions and that the special case of order zero, $m = 0$, is written
$P_\ell(\mu)$ and is called a Legendre polynomial, which is consistent,
because it is a polynomial in μ. As we noted in subsection 3.4.3,
not everyone accepts this definition. Our convention follows that of
the classical literature, for example, Chapter XV of Whittaker and
Watson (1927). The widely cited reference work of Abramowitz and
Stegun (1965) promotes an alternative definition with a multiplying

factor $(-1)^m$. Most authors in geomagnetism abide by our definition of $P_\ell^m(\mu)$, although there is little agreement about the proper form of the normalization in (3.7.1) or its equivalent. It is obvious that (3.7.2) is a general algorithm for generating the functions; we perform the differentiations and present a short list of the low-degree functions. These formulas are useful for doing short calculations by hand; we explain how large-scale computations are done in subsection 3.7.3.

In what follows $\mu = \cos\theta$ and $s = \sin\theta$, where θ is the colatitude, the angle measured from the fixed vector $\hat{z}$. We list only the expressions for positive m in view of (3.4.44) and (3.4.56).

$$P_0(\mu) = 1$$

$$P_1(\mu) = \mu \qquad P_1^1(\mu) = s$$
$$P_2(\mu) = (3\mu^2 - 1)/2 \quad P_2^1(\mu) = 3\mu s \quad P_2^2(\mu) = 3s^2$$

$$P_3(\mu) = \mu(5\mu^2 - 3)/2$$
$$P_3^1(\mu) = 3s(5\mu^2 - 1)/2$$
$$P_3^2(\mu) = 15s^2\mu \qquad P_3^3(\mu) = 15s^3$$

$$P_4(\mu) = (35\mu^4 - 30\mu^2 + 3)/8$$
$$P_4^1(\mu) = 5s\mu(7\mu^2 - 3)/2$$
$$P_4^2(\mu) = 15s^2(7\mu^2 - 1)/2$$
$$P_4^3(\mu) = 105s^3\mu \qquad P_4^4(\mu) = 105s^4$$

$$P_5(\mu) = \mu(63\mu^4 - 70\mu^2 + 15)/8$$
$$P_5^1(\mu) = 15s(21\mu^4 - 14\mu^2 + 1)/8$$
$$P_5^2(\mu) = 105s^2\mu(3\mu^2 - 1)/2$$
$$P_5^3(\mu) = 105s^3(9\mu^2 - 1)/2$$
$$P_5^4(\mu) = 945s^4\mu \qquad P_5^5(\mu) = 945s^5$$

The corresponding normalization constants from (3.7.1) for these functions are so easily computed, they will not be tabulated here. We briefly note that the geophysical literature contains many different normalizations and scalings of the natural basis. The most popular is the *Schmidt quasi-normalized* basis. These basis functions are taken to be purely real, with a sine and a cosine member for each order (except $m = 0$ of course) rather than complex as in our treatment; they are not of unit norm, but instead satisfy $\langle \beta_\ell^m | \beta_\ell^m \rangle = 1/(2\ell + 1)$.

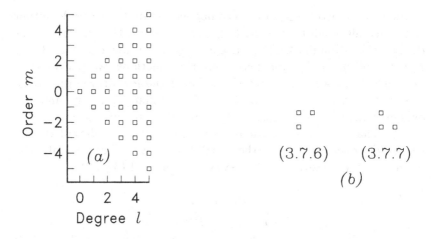

Figure 3.7.1: (a) A small portion of the lattice, showing sites where nontrivial values of Y_ℓ^m or P_ℓ^m occur. (b) Spatial relation on the lattice of the three-term connections listed.

We find later in the next subsection a version of Schmidt normalization convenient for displaying certain recurrence relations. For a further discussion and application see subsection 4.4.2 and Langel (1987). Notice, that while Langel adopts the same definition as we do for the associated Legendre functions, he denotes them $P_{n,m}(\cos\theta)$. In this he is following the notation of Chapman and Bartels (1962), the definitive text for an earlier generation of geomagnetists.

3.7.2 Three-Term Recurrence Relationships

Up to this point it has been natural to consider the basis functions as complex fields with the domain $S(1)$. Now we adopt a different perspective, motivated by expansions like (3.2.1), where the basis β_ℓ^m is taken to be the natural orthonormal basis:

$$f(\hat{\mathbf{r}}) = \sum_{\ell=0}^{\infty} \sum_{m=-\ell}^{\ell} f_\ell^m Y_\ell^m(\hat{\mathbf{r}}).$$

When we focus on the evaluation of this expression for a particular point we need to know the values of the complex numbers $Y_\ell^m(\hat{\mathbf{r}})$ for a fixed vector $\hat{\mathbf{r}}$ and all integer pairs (ℓ, m) satisfying $0 \le \ell$, $-\ell \le m \le \ell$. We can imagine a two-dimensional lattice, a coordinate plane in which only the integers are significant; if ℓ corresponds to the horizontal coordinate, we study the values of the complex-valued function $Y_\ell^m(\hat{\mathbf{r}})$ in a wedge as shown in Figure 3.7.1(a). By convention we simply define $Y_\ell^m = 0$ if $|m| > \ell$. It is convenient to drop the argument $\hat{\mathbf{r}}$, since it never varies.

Our interest will center on the connection between the value of Y_ℓ^m or P_ℓ^m and those at the neighbors in the lattice. We begin by developing relationships for P_ℓ^m between values at a point (ℓ, m) and those at two immediate neighbors: these are often called *three-term recurrence relations* for obvious reasons. We need the famous Leibnitz rule (Thomas and Finney, 1984): let f, g both be n-times continuously differentiable functions of μ; then

$$\partial_\mu^n(fg) = f\partial_\mu^n g + \frac{n}{1!}(\partial_\mu f)\partial_\mu^{n-1}g + \frac{n(n-1)}{2!}(\partial_\mu^2 f)\partial_\mu^{n-2}g$$
$$+ \cdots + \binom{n}{m}(\partial_\mu^m f)\partial_\mu^{n-m}g + \cdots + (\partial_\mu^n f)g. \qquad (3.7.4)$$

First we rearrange the defining equation (3.7.2) thus:

$$2^\ell \ell!(1 - \mu^2)^{-m/2}P_\ell^m(\mu) = \partial_\mu^{\ell+m}(\mu^2 - 1)^\ell. \qquad (3.7.5)$$

The Leibnitz rule can be applied in several ways to (3.7.5). First we write $\partial_\mu^{\ell+m} = \partial_\mu^{\ell+m-1}\partial_\mu$ and then we apply the associative property of the differential operator:

$$\partial_\mu^{\ell+m}(\mu^2 - 1)^\ell = \partial_\mu^{\ell+m-1}[\partial_\mu(\mu^2 - 1)^\ell]$$
$$= \partial_\mu^{\ell+m-1}2\ell\mu(\mu^2 - 1)^{\ell-1}$$
$$= 2\ell\partial_\mu^{\ell+m-1}[\mu(\mu^2 - 1)^{\ell-1}].$$

At this point we invoke (3.7.4) with $n = \ell + m - 1$, $f = \mu$, and $g = (\mu^2 - 1)^{\ell-1}$; then we identify the terms that result from (3.7.5):

$$\partial_\mu^{\ell+m}(\mu^2 - 1)^\ell = 2\ell\mu\partial_\mu^{\ell+m-1}(\mu^2 - 1)^{\ell-1}$$
$$+ 2\ell(\ell + m - 1)\partial_\mu^{\ell+m-2}(\mu^2 - 1)^{\ell-1}$$
$$= 2\ell\mu\frac{2^{\ell-1}(\ell-1)!P_{\ell-1}^m}{(1 - \mu^2)^{m/2}}$$
$$+ 2\ell(\ell + m - 1)\frac{2^{\ell-1}(\ell-1)!P_{\ell-1}^{m-1}}{(1 - \mu^2)^{(m-1)/2}}.$$

After tidying this up we discover a three-term recurrence relationship:

$$P_\ell^m = \mu P_{\ell-1}^m + (1 - \mu^2)^{1/2}(\ell + m - 1)P_{\ell-1}^{m-1}.$$

We defer the question of the applicability of this formula; it is clearly permissible to increase ℓ by one and to rearrange slightly, introducing s for $(1 - \mu^2)^{1/2} = \sin\theta$:

$$\mu P_\ell^m = P_{\ell+1}^m - (\ell + m)sP_\ell^{m-1}. \tag{3.7.6}$$

In terms of the lattice picture, we see that (3.7.6) is a linear relationship between values at a point, the value immediately to the right, and the one directly below the point; see Figure 3.7.1(b). Thus it is a connection between nearest neighbors in the lattice with a simple geometrical relationship. Our goal is to find all such nearest-neighbor connections.

Exactly the same factorization of the differential operator can be applied to the alternative definition (3.7.3). We state the result without giving the straightforward algebra:

$$(\ell + m + 1)\mu P_\ell^m = sP_\ell^{m+1} + (\ell - m + 1)P_{\ell+1}^m. \tag{3.7.7}$$

The geometrical configuration is shown in Figure 3.7.1(b)

We can obtain other relationships by different sets of operations. For example, the differential operator can be factored like this: $\partial_\mu^{\ell+m} = \partial_\mu^{\ell+m-2}\partial_\mu^2$. But it turns out that the two equations (3.7.6, 3.7.7) are all we require to generate all the contiguous, three-term relationships, as we now show.

It is intuitively appealing to think of moving the geometrical figures around the lattice, vertically by adding or subtracting from the order m, horizontally by adding or subtracting from the degree ℓ. When a particular point in the lattice is covered by elements in the terms of two relationships, it can be eliminated and a new connection built. Thus, for example, we see that without the need for any displacement, (3.7.6) and (3.7.7) overlap at P_ℓ^m and $P_{\ell+1}^m$, either of which can be eliminated. In this way we can get the connections between three values in a row vertically:

$$2m\mu P_\ell^m = sP_\ell^{m+1} - (\ell - m + 1)(\ell + m)sP_\ell^{m-1}$$

or, after a shift of the geometric figure to the left by one unit, between values on the top and bottom left diagonals with the center:

$$2mP_\ell^m = (\ell + m)(\ell + m - 1)sP_{\ell-1}^{m-1} + sP_{\ell-1}^{m+1}.$$

Suppose that we add one to m in (3.7.6); this moves the geometrical object on the lattice up by one unit. Algebraically,

$$\mu P_\ell^{m+1} = P_{\ell+1}^{m+1} - (\ell + m + 1)sP_\ell^m. \qquad (3.7.8)$$

This equation and (3.7.7) connect the four values P_ℓ^m, $P_{\ell+1}^m$, P_ℓ^{m+1}, and $P_{\ell+1}^{m+1}$, or the four values on the corners of a unit cell in the lattice. By adding the equations together with suitable weighting factors, any selected value can be eliminated and a three-term recurrence relation results between the remaining three associated Legendre functions. Now we have the relationships for all the "L"-shaped figures with unit arms in the lattice. This process can be continued in a perfectly systematic way to develop the list of recurrences below.

We list all such relationships for a central point and its nearest neighbors, and the point and its neighbors on the diagonals, but not any that mix these two.

$$\mu P_\ell^m = P_{\ell+1}^m - (\ell + m)sP_\ell^{m-1} \qquad (3.7.9)$$

$$(\ell + m + 1)\mu P_\ell^m = sP_\ell^{m+1} + (\ell - m + 1)P_{\ell+1}^m \qquad (3.7.10)$$

$$(\ell - m)\mu P_\ell^m = (\ell + m)P_{\ell-1}^m - sP_\ell^{m+1} \qquad (3.7.11)$$

$$\mu P_\ell^m = P_{\ell-1}^m + (\ell - m + 1)sP_\ell^{m-1} \qquad (3.7.12)$$

$$2m\mu P_\ell^m = sP_\ell^{m+1} + (\ell - m + 1)(\ell + m)sP_\ell^{m-1} \qquad (3.7.13)$$

$$(2\ell + 1)\mu P_\ell^m = (\ell - m + 1)P_{\ell+1}^m + (\ell + m)P_{\ell-1}^m \quad (3.7.14)$$

$$(2m - (2\ell + 1)s^2)P_\ell^m = sP_{\ell-1}^{m+1}$$
$$+ (\ell - m + 2)(\ell - m + 1)sP_{\ell+1}^{m-1} \qquad (3.7.15)$$

$$(2m + (2\ell + 1)s)P_\ell^m = sP_{\ell+1}^{m+1}$$
$$+ (\ell + m)(\ell + m - 1)sP_{\ell-1}^{m-1} \qquad (3.7.16)$$

$$2mP_\ell^m = (\ell + m)(\ell + m - 1)sP_{\ell-1}^{m-1} + sP_{\ell-1}^{m+1} \qquad (3.7.17)$$

$$(2\ell + 1)sP_\ell^m = (\ell + m)(\ell + m - 1)P_{\ell-1}^{m-1}$$

$$- (\ell - m + 1)(\ell - m + 2)P_{\ell+1}^{m-1} \qquad (3.7.18)$$

$$2mP_\ell^m = sP_{\ell+1}^{m+1} + (\ell - m + 2)(\ell - m + 1)sP_{\ell+1}^{m-1} \qquad (3.7.19)$$

$$(2\ell + 1)sP_\ell^m = P_{\ell+1}^{m+1} - P_{\ell-1}^{m+1} \qquad (3.7.20)$$

For both theoretical and computational work we would like fully normalized versions of these results. From the perspective of economy of notation, a version of the Schmidt quasi-normalized functions mentioned earlier has much to recommend it in this regard. Define

$$P_{\ell,m}(\mu) = (-1)^m \sqrt{\frac{(\ell - m)!}{(\ell + m)!}} P_\ell^m(\mu), \qquad (3.7.21)$$

so that

$$Y_\ell^m(\hat{\mathbf{r}}) = \sqrt{2\ell + 1}\, e^{im\lambda} P_{\ell,m}(\cos\theta). \qquad (3.7.22)$$

Thus when the order m varies and the degree ℓ is fixed, the relative sizes of these new functions and the fully normalized basis remains the same. Furthermore, when ℓ is very large, the unnormalized associated Legendre functions suffer a great variation in relative size as m sweeps from $-\ell$ to ℓ. Such a large change in size as the un-normalized functions exhibit is a matter of concern when practical computations are to be done, and this problem is ameliorated with the fully or quasi-normalized functions; we address this question in the next subsection. To list the recurrence relations for the quasi-normalized functions in a compact form, we introduce the following convenient abbreviations: let

$$\binom{+}{p} = \sqrt{\ell + m + p}$$

$$\binom{-}{p} = \sqrt{\ell - m + p}$$

and similarly

$$\binom{+\ -}{p\ q} = \sqrt{(\ell + m + p)(\ell - m + q)}$$

$$\binom{-\ -}{p\ q} = \sqrt{(\ell - m + p)(\ell - m + q)}$$

and so on, where the sign above the integer indicates the sign of m within the corresponding factor on the right side. With this notation in place, only algebra is needed to obtain the following results from equations (3.7.21) and (3.7.9–3.7.20):

$$\left(\begin{smallmatrix} - \\ 1 \end{smallmatrix}\right)\mu P_{\ell,m} = \left(\begin{smallmatrix} + \\ 1 \end{smallmatrix}\right)P_{\ell+1,m} + \left(\begin{smallmatrix} + \\ 0 \end{smallmatrix}\right)sP_{\ell,m-1} \tag{3.7.23}$$

$$\left(\begin{smallmatrix} + \\ 1 \end{smallmatrix}\right)\mu P_{\ell,m} = \left(\begin{smallmatrix} - \\ 1 \end{smallmatrix}\right)P_{\ell+1,m} - \left(\begin{smallmatrix} - \\ 0 \end{smallmatrix}\right)sP_{\ell,m+1} \tag{3.7.24}$$

$$\left(\begin{smallmatrix} - \\ 0 \end{smallmatrix}\right)\mu P_{\ell,m} = \left(\begin{smallmatrix} + \\ 0 \end{smallmatrix}\right)P_{\ell-1,m} + \left(\begin{smallmatrix} + \\ 1 \end{smallmatrix}\right)sP_{\ell,m+1} \tag{3.7.25}$$

$$\left(\begin{smallmatrix} + \\ 0 \end{smallmatrix}\right)\mu P_{\ell,m} = \left(\begin{smallmatrix} - \\ 0 \end{smallmatrix}\right)P_{\ell-1,m} - \left(\begin{smallmatrix} - \\ 1 \end{smallmatrix}\right)sP_{\ell,m-1} \tag{3.7.26}$$

$$2m\mu P_{\ell,m} = -\left(\begin{smallmatrix} + & - \\ 1 & 0 \end{smallmatrix}\right)sP_{\ell,m+1} - \left(\begin{smallmatrix} + & - \\ 0 & 1 \end{smallmatrix}\right)sP_{\ell,m-1} \tag{3.7.27}$$

$$(2\ell+1)\mu P_{\ell,m} = \left(\begin{smallmatrix} + & - \\ 1 & 1 \end{smallmatrix}\right)P_{\ell+1,m} + \left(\begin{smallmatrix} + & - \\ 0 & 0 \end{smallmatrix}\right)P_{\ell-1,m} \tag{3.7.28}$$

$$(2m-(2\ell+1)s^2)P_{\ell,m} = -\left(\begin{smallmatrix} - & - \\ 0 & -1 \end{smallmatrix}\right)sP_{\ell-1,m+1} - \left(\begin{smallmatrix} - & - \\ 2 & 1 \end{smallmatrix}\right)sP_{\ell+1,m-1} \tag{3.7.29}$$

$$(2m+(2\ell+1)s^2)P_{\ell,m} = -\left(\begin{smallmatrix} + & + \\ 2 & 1 \end{smallmatrix}\right)sP_{\ell+1,m+1} - \left(\begin{smallmatrix} + & + \\ 0 & -1 \end{smallmatrix}\right)sP_{\ell-1,m-1} \tag{3.7.30}$$

$$2mP_{\ell,m} = -\left(\begin{smallmatrix} + & + \\ 0 & -1 \end{smallmatrix}\right)sP_{\ell-1,m-1} - \left(\begin{smallmatrix} - & - \\ 0 & -1 \end{smallmatrix}\right)sP_{\ell-1,m+1} \tag{3.7.31}$$

$$(2\ell+1)sP_{\ell,m} = \left(\begin{smallmatrix} - & - \\ 2 & 1 \end{smallmatrix}\right)P_{\ell+1,m-1} - \left(\begin{smallmatrix} + & + \\ 0 & -1 \end{smallmatrix}\right)P_{\ell-1,m-1} \tag{3.7.32}$$

$$2mP_{\ell,m} = -\left(\begin{smallmatrix} + & + \\ 2 & 1 \end{smallmatrix}\right)sP_{\ell+1,m+1} - \left(\begin{smallmatrix} - & - \\ 2 & 1 \end{smallmatrix}\right)sP_{\ell+1,m-1} \tag{3.7.33}$$

$$(2\ell+1)sP_{\ell,m} = -\left(\begin{smallmatrix} + & + \\ 2 & 1 \end{smallmatrix}\right)P_{\ell+1,m+1} + \left(\begin{smallmatrix} - & - \\ -1 & 0 \end{smallmatrix}\right)P_{\ell-1,m+1} \tag{3.7.34}$$

Finally, we often need to express the gradient of a spherical harmonic in terms of other harmonics. Recall that our method of generating a set of basis functions for $\mathcal{H}_\ell$ relied upon repeated differentiation and so we know, for example, from (3.4.32), that

$$U_\ell^{m+1} = -i\Lambda^+ U_\ell^m \tag{3.7.35}$$

and by (3.4.21)

$$-i\Lambda^+ = e^{i\lambda}(\partial_\theta + i\cot\theta\,\partial_\lambda).$$

When the expression for U_ℓ^m as an associated Legendre function is recovered from (3.4.43) and we perform a few algebraic rearrangements, we find that (3.7.35) is equivalent to

$$P_\ell^{m+1} = s\partial_\mu P_\ell^m + \frac{m\mu}{s}P_\ell^m,$$

which we can choose to interpret as

$$s^2\partial_\mu P_\ell^m = sP_\ell^{m+1} - m\mu P_\ell^m. \tag{3.7.36}$$

In terms of the quasi-normalized functions this is

$$
\begin{aligned}
s^2 \partial_\mu P_{\ell,m} &= -(\genfrac{}{}{0pt}{}{+}{1}\genfrac{}{}{0pt}{}{-}{0}) s P_{\ell,m+1} - m\mu P_{\ell,m} \\
&= -\sqrt{(\ell+m+1)(\ell-m)}\, s P_{\ell,m+1} - m\mu P_{\ell,m}.
\end{aligned}
\tag{3.7.37}
$$

Obviously one may use the three-term relationships to obtain the derivatives in terms of other harmonics: for example, combining (3.7.10) with (3.7.36) to eliminate P_ℓ^{m+1} (by simply subtracting the two equations) yields

$$
s^2 \partial_\mu P_\ell^m = (\ell+1)\mu P_\ell^m - (\ell-m+1) P_{\ell+1}^m.
\tag{3.7.38}
$$

3.7.3 Numerical Calculations

As we remarked earlier, it is unusual in applications of spherical harmonics to require the evaluation of $Y_\ell^m(\mathbf{r})$ for a single degree and order. Instead the most common situation arises in the expansion of a function where all the degree and order harmonics up to some limit are needed for a fixed point $\mathbf{r}$. In the computation of the functions $Y_\ell^m(\mathbf{r})$ and its gradient, only the factor associated with $P_\ell^m(\cos\theta)$ presents any difficulty numerically. Then it is computationally more efficient to make use of the recurrence relations described in the previous section. As computers become more powerful, questions of computational speed that were seemingly important at one time soon lose their significance, but in this case, if high-degree functions are desired, evaluation of the explicit polynomial expressions should be avoided if reasonable accuracy is to be maintained. Before discussing this question, we should, in a numerical setting, explore what a high-degree expansion is. In geomagnetism, spherical harmonic expansions for the main geomagnetic field are limited usually to a maximum degree and order of around 13; inclusion of terms to account for crustal components of the field have led to some models extending to degree 63 (Cain et al., 1989a). As an aside, we note that spherical harmonic expansions for the earth's gravity have been taken to degree 100. We will require computational methods to function properly to this degree and order.

We have seen in (3.5.10) that $P_\ell(1) = P_\ell^0(1) = 1$ for any degree ℓ and indeed from (3.5.22) $|P_\ell(\mu)| \le 1$ when $|\mu| \le 1$. Suppose for some high degree one employed the explicit polynomial expression to calculate $P_\ell(1)$: this requires the summation of about $\ell/2$ numerical terms. We will see that the terms are large in magnitude, and they obviously cancel one another if unity is to be the final answer. In computers calculations are done to a fixed number of significant figures at each stage, so that if two terms of order 1000 are subtracted to yield an answer of unity, three significant figures of accuracy are necessarily lost. To get an idea of how large the coefficients are in $P_\ell(\mu)$, we examine just the leading term. From (3.7.2)

$$
\begin{aligned}
P_\ell(\mu) &= \frac{1}{2^\ell \ell!}\partial_\mu^\ell(\mu^{2\ell}) \; + \; O\mu^{\ell-2} \\
&= \frac{(2\ell)!}{2^\ell(\ell!)^2}\mu^\ell \; + \; \cdots \\
&\simeq \frac{2^\ell \mu^\ell}{\sqrt{2\pi}} \; + \; \cdots
\end{aligned}
$$

where in the last line we have invoked Stirling's approximation for the factorial. Even for a degree-13 expansion, the leading coefficient is over 3000, and thus represents more than three lost figures when μ is near 1; for degree 100 the loss is nearly 30 significant figures. This is too great for most computers, which typically carry 14 significant decimal digits in their most accurate mode of arithmetic.

The principle of the recurrence approach is to compute P_ℓ^m or $P_{\ell,m}$ for a degree-order pair and for a neighbor in the lattice. There are several lines in the lattice where this can be done conveniently, notably $m = \ell$. Then a three-term relation can be used to march across the plane, filling in numerical values. Not every relationship may be suitable, as we will see. But first we should decide which functions to compute.

If efficiency were the only consideration, the relations for the unnormalized associated Legendre functions would appear preferable, because we see from (3.7.23–3.7.34) that the fully or quasi-normalized function requires calculation of square roots, which are many times more time-consuming computationally than multiplications and additions, and square roots do not appear in (3.7.9–3.7.20). It is already noticeable from the degree-5 functions given in subsection 3.7.1 that, as ℓ grows, $P_\ell^m(\mu)$ can attain large values. In fact, focusing on $P_\ell^\ell(\mu)$, we have already seen from (3.4.33, 3.4.41, 3.4.43) that

$$P_\ell^\ell(\cos\theta) = \frac{(2\ell)!}{2^\ell \ell!} \sin{}^\ell\theta, \qquad (3.7.39)$$

and with the aid of Stirling's formula again, we find the multiplying factor is approximately $\sqrt{2}(2\ell/e)^\ell$. At $\ell = 13$ the factor is about 10^{13}, which is quite manageable, but when $\ell = 100$ it has reached nearly 10^{187}. In all computers there is a limit to the largest magnitude number that can be stored in a single computer word; commonly it is around 10^{39}. While it is possible, even in computers with small magnitude limits, to make programming arrangements to save such large numbers in more than one word, this is a dreadful inconvenience. The size of the fully or quasi-normalized functions for all arguments is very modest, as we proved in subsection 3.3.5: equation (3.3.30) states that

$$|Y_\ell^m(\hat{\mathbf{r}})| \le \sqrt{2\ell+1},$$

and this means from (3.7.22) that $|P_{\ell,m}| \le 1$.

We discuss one solution among many possible approaches to the computational problem. In our scheme, we compute all the different order functions $P_{\ell,m}$ for a fixed degree ℓ, that is, we recur vertically in the lattice. For this purpose we rearrange (3.7.27):

$$P_{\ell,m} = -\frac{2(m+1)(\mu/s)P_{\ell,m+1} + \sqrt{(\ell+m+2)(\ell-m-1)}P_{\ell,m+2}}{\sqrt{(\ell+m+1)(\ell-m)}}.$$
$$(3.7.40)$$

The solution can be started at the point $m = \ell - 1$ because the equivalent to (3.7.39) for the quasi-normalized functions is

$$P_{\ell,\ell}(\mu) = \frac{(-1)^\ell\sqrt{(2\ell)!}}{2^\ell \ell!}\, s^\ell = (-1)^\ell\sqrt{\frac{1}{2} \cdot \frac{3}{4} \cdot \frac{5}{6} \cdots \frac{2\ell-1}{2\ell}}\, s^\ell \quad (3.7.41)$$

where the second form of the expression must be used to keep the numbers within the computer representation. We might provide $P_{\ell,\ell-1}$, but it is instructive to notice that the recurrence scheme works when we set $P_{\ell,\ell+1} = 0$. Thus the relation (3.7.40) is applied repeatedly for $m = \ell - 1, \ell - 2, \ldots 1, 0$. The apparent penalty in speed incurred by the square roots in (3.7.40) can easily be avoided if one recognizes that only $2\ell + 1$ different values are used; if the square

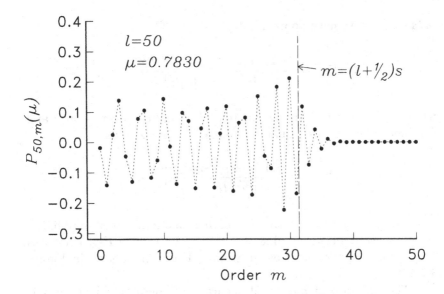

Figure 3.7.2: The values of $P_{\ell,m}(\mu)$ for fixed $\mu = 0.7830$ and $\ell = 50$ and nonnegative orders as they are found by the relationship (3.7.40), starting on the right and working to the left.

roots of the integers from 0 to 2ℓ for the largest degree anticipated are stored, no square roots need be computed in the application of the recurrence. Then the arithmetic work for each associated Legendre function can be reduced to five multiplications, one division, and one addition of floating-point numbers, ignoring operations on indexes and the store and fetch operations. It follows that the calculation of the functions P_{lm} for all orders m at a fixed ℓ is on average only a few times more expensive than the explicit evaluation of the function for one m.

Let us examine the general behavior of the solution at fixed degree and argument: Figure 3.7.2 provides an illustration. Notice how the values generally oscillate with slowly growing amplitude and then decay swiftly to small values.

This is easily understood by means of the asymptotic solution obtained in the previous section. Recall the definition $L = \sqrt{\ell(\ell+1)}$; in section 3.6.2 we discovered that when $m/L < \cos\gamma = s$ the solution is an oscillatory function of μ, and switches to exponential decline above the critical argument. The same kind of behavior is obtained when we focus on m as the independent variable: from (3.6.23) when

$m/s < L - \epsilon$ we have approximately

$$P_{\ell,m} = \frac{(-1)^m c_1 \cos\left(\frac{1}{2}\pi m + \phi_1(m)\right)}{[s^2 - m^2/L^2]^{1/4}}$$

where ϕ_1 is a slowly varying function of m and c_1 depends on μ and ℓ. When $m/s > L + \epsilon$, equations (3.6.32–3.6.34) give us

$$P_{\ell,m} = \frac{(-1)^m c_2 \exp\left(-m\phi_2(m/sL)\right)}{[m^2/L^2 - s^2]^{1/4}}$$

where $\phi_2(\rho) > g(\rho)$, a positive, increasing function defined in (3.6.34). Hence above the critical value of m, the magnitude of $P_{\ell,m}$ decreases faster than exponentially as m grows. This can be seen in Figure 3.7.3, for example.

The question of numerical accuracy of complex calculations is generally a difficult one if precise bounds on error are desired. We can give a brief sketch of one aspect. We rewrite (3.7.40) as an example of a general three-term relation

$$p_{n+1} = a_n p_n + b_n p_{n-1} \qquad (3.7.42)$$

in which the sequences $a_1, a_2, a_3 \ldots$ and $b_1, b_2, b_3, \ldots$ are fixed. Then it is obvious that the whole sequence $p_0, p_1, p_2, p_3, \ldots$ is determined by the first two elements, p_0 and p_1. Further, if p_n and q_n are two sequences satisfying (3.7.42), so is the sequence r_n given by $r_n = \alpha p_n + \beta q_n$, where α and β are arbitrary constants. Since every sequence is dictated by its first two members, it follows that any sequence can be written as a linear combination of two linearly independent sequences. We may identify another solution to the recurrence scheme by a round-about approach: the differential equation (3.6.6) used in the asymptotic analysis has a solution $P_\ell^m(\mu)$, which we have seen is analytic at $\mu = 1$. But this is a *regular singular point* of the differential equation and other solutions exist that are unbounded in its neighborhood – the Legendre functions of the second kind, written Q_ℓ^m, which we met briefly at the end of section 3.6.3. It can be shown that this set of functions satisfies the same recurrence relations as the familiar P_ℓ^m; see Whittaker and Watson (1927, chap. XX). Thus after appropriate scaling these provide the second type of solution to

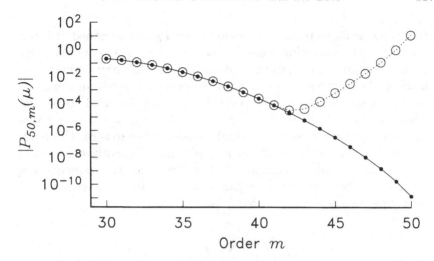

Figure 3.7.3: Approximate values of $|P_{\ell,m}(\mu)|$ for fixed $\mu = 0.7830$ and $\ell = 50$ found by recursion: solid dots, by downward recursion from $m = \ell$; open circles by upward recursion starting at $m = 0$. Calculations were carried out in arithmetic with about 7 decimal places of accuracy.

the recurrence scheme. By an asymptotic analysis identical to the one we carried out in 3.6.2, we can show that the $Q_{\ell,m}$ (written this way to reflect the normalization) behave very similarly to the $P_{\ell,m}$ for fixed μ and varying m when $m/s < L$, but above the critical order the values increase rapidly rather than decrease. The significance of this observation for the recursive computation is that the second kind of solution decreases relative to the desired one in the range above s/L when (3.7.40) is employed as shown, with order decreasing. However, if the recursion were applied in the other direction, now with m increasing, relative growth of the second solution would be expected, and severe loss of accuracy is predicted. This phenomenon is illustrated in Figure 3.7.3, where we compare the magnitudes of the high-order functions computed with finite arithmetic precision by recursion with m both decreasing and increasing.

The figure illustrates one final matter that needs attention when large-degree functions are computed. If the colatitude is small and the degree high, (3.7.39) shows that $P_{\ell,\ell}$ can become very small in magnitude. Going to a more extreme example than the one in the figure, we note that for $\ell = 100$ and $\theta < 21°$, the magnitude of $P_{\ell,\ell} < 10^{-45}$, which is too small to be distinguished from zero on many computers. While it is certainly harmless to substitute zero for the true value as far as providing an answer for $P_{\ell,\ell}$ is concerned, it is not satisfactory as a starting point for the recursion. There are two ways to solve the problem: as mentioned earlier, one can carry a scale factor along in the recurrence scheme until the values come within computer range. Alternatively, the asymptotic form can be used to predict where the size of the solution becomes tractable, and the recursion can start from there, setting all higher-order terms to zero. Of course the asymptotic solution will not normally be sufficiently accurate to provide the size of the solution, and so the computed sequence will differ from the true one by a constant factor to be determined by appealing to this identity, derived from (3.3.29):

$$(P_{\ell,0})^2 + 2 \sum_{m=1}^{\ell} (P_{\ell,m})^2 = 1.$$

This normalization fixes the magnitude of the sequence, and the sign can be obtained from the fact that when the latitude is beyond the critical value the sign of $P_{\ell,m}$ is by (3.6.32) that of $(-1)^m$.

Exercises

1. Let $\mathbf{u}$, $\mathbf{v}$, $\mathbf{w}$ be fixed real vectors in R^3. Define real scalar fields f, g, h on R^3 by requiring for each $\mathbf{r} \in R^3$ that $f(\mathbf{r}) = \mathbf{u} \cdot \mathbf{r}$, $g(\mathbf{r}) = \mathbf{v} \cdot \mathbf{r}$, $h(\mathbf{r}) = \mathbf{w} \cdot \mathbf{r}$.
   (a) Show that f, g, $h \in \mathcal{H}_1$.
   (b) Show that $\{f, g, h\}$ is a basis for $\mathcal{H}_1$ if and only if $\{\mathbf{u}, \mathbf{v}, \mathbf{w}\}$ are linearly independent.
   (c) What properties must the set $\{\mathbf{u}, \mathbf{v}, \mathbf{w}\}$ have if $\{f, g, h\}$ is to be an orthonormal basis for $\mathcal{H}_1$? Prove your answer. [Hint: Use the components of $\mathbf{u}, \mathbf{v}, \mathbf{w}$ relative to some Cartesian basis $\hat{\mathbf{x}}_1, \hat{\mathbf{x}}_2, \hat{\mathbf{x}}_3$ to calculate $\langle f^2 \rangle, \langle fg \rangle$, etc.]

2. (a) Express $Y_5^2(\mathbf{r})$ and $Y_5^{-2}(\mathbf{r})$ as polynomials in x, y, z. Leave the result in factored form.
   (b) Evaluate $Y_\ell^m(\hat{\mathbf{r}})$ numerically for the positions $\hat{\mathbf{r}} = \hat{\mathbf{z}}$ and $\hat{\mathbf{r}} = \hat{\mathbf{x}}$ and degree $\ell = 100$, for orders $m = 0, \pm1, \pm99, \pm100$. Use asymptotic approximations if necessary.

4

GAUSS' THEORY OF THE MAIN FIELD

In this chapter we apply the theory of spherical harmonics to the magnetic field of the earth. As we discussed at the end of section 2.4, the magnetic field immediately surrounding the earth in the atmosphere can be well approximated by the gradient of a harmonic function. There are sources of magnetic field above the atmosphere in the form of currents, in the ionosphere and the radiation belts of the magnetosphere; below the earth's surface in the crust and in the core there are further sources. Our aim is to give a description of the magnetic field in the source-free cavity, which we take to be a spherical shell. The form of the description will be that of a spherical harmonic expansion. We discuss how the expansion might be found in principle and how it is estimated in practice. We also discuss the relationships between the sources of the magnetic field and the expansion coefficients. A modern field model derived from satellite observations is used to illustrate some of these ideas.

4.1 Finding All the Harmonics in a Shell

Suppose that $0 \le a < c \le \infty$. We want to find all scalar fields ψ on the spherical shell $S(a, c)$ such that

$$\nabla^2 \psi = 0 \quad \text{in} \quad S(a, c). \quad (4.1.1)$$

The scalar field ψ will be the magnetic scalar potential representing the magnetic field in the shell; recall equations (2.4.34, 2.4.35). We begin by choosing for each $\mathcal{H}_\ell$ an orthonormal basis $\{\beta_\ell^{-\ell}, \ldots, \beta_\ell^{\ell}\}$.

These bases are arbitrary, but fixed throughout the following discussion. Thus we have

$$\nabla_1^2 \beta_\ell^m(\hat{\mathbf{r}}) = -\ell(\ell+1)\beta_\ell^m(\hat{\mathbf{r}}) \quad \text{for all} \quad \hat{\mathbf{r}} \in S(1) \tag{4.1.2}$$

$$\langle \overline{\beta_\ell^m} \, \beta_{\ell'}^{m'} \rangle = \delta_{\ell\ell'} \, \delta_{mm'} = \frac{1}{4\pi} \int_{S(1)} d^2\hat{\mathbf{r}} \; \overline{\beta_\ell^m}(\hat{\mathbf{r}}) \beta_{\ell'}^{m'}(\hat{\mathbf{r}}). \tag{4.1.3}$$

If $a < r < c$, then every function on the spherical surface $S(r)$ can be thought of as a function on $S(1)$, and thus expanded in spherical harmonics. In particular,

$$\psi(\mathbf{r}) = \psi(r\hat{\mathbf{r}}) = \sum_{\ell=0}^{\infty} \sum_{m=-\ell}^{\ell} \psi_\ell^m(r)\beta_\ell^m(\hat{\mathbf{r}}). \tag{4.1.4}$$

The coefficients in the expansion, ψ_ℓ^m, will, of course, depend on which r is chosen, so they are functions of r, as shown. For each r they can be calculated as $\langle f(r\hat{\mathbf{s}})\rangle_{\hat{\mathbf{s}}}$ which means average over $\hat{\mathbf{s}}$ on $S(1)$:

$$\psi_\ell^m(r) = \langle \overline{\beta_\ell^m(\hat{\mathbf{s}})} \, \psi(r\hat{\mathbf{s}})\rangle_{\hat{\mathbf{s}}} = \frac{1}{4\pi} \int_{S(1)} d^2\hat{\mathbf{s}} \; \overline{\beta_\ell^m(\hat{\mathbf{s}})} \, \psi(r\hat{\mathbf{s}}). \tag{4.1.5}$$

The coefficients $\psi_\ell^m(r)$ will also depend on which orthonormal basis is chosen for $\mathcal{H}_\ell$. However, we will see that the function

$$\psi_\ell(r\hat{\mathbf{r}}) = \sum_{m=-\ell}^{\ell} \psi_\ell^m(r)\beta_\ell^m(\hat{\mathbf{r}}) \tag{4.1.6}$$

does not depend on the basis. It is a property of ψ alone, and is called the ℓth *degree part of* ψ. To see that it does not depend on the basis, use (4.1.5, 4.1.6) to write it:

$$\psi_\ell(r\hat{\mathbf{r}}) = \frac{1}{4\pi} \int_{S(1)} d^2\hat{\mathbf{s}} \left[\sum_{m=-\ell}^{\ell} \overline{\beta_\ell^m(\hat{\mathbf{s}})} \, \beta_\ell^m(\hat{\mathbf{r}}) \right] \psi(r\hat{\mathbf{s}})$$

or by the Spherical Harmonic Addition Theorem (3.5.13):

$$\psi_\ell(r\hat{\mathbf{r}}) = \frac{(2\ell+1)}{4\pi} \int_{S(1)} d^2\hat{\mathbf{s}} \; P_\ell(\hat{\mathbf{r}} \cdot \hat{\mathbf{s}})\psi(r\hat{\mathbf{s}}). \tag{4.1.7}$$

Since the basis does not appear in (4.1.7), we have proved the independence of ψ_ℓ from any particular choice. Notice that ψ_ℓ is analogous to the orthogonal projection of a function on $S(1)$ onto $\mathcal{H}_\ell$ discussed in subsection 3.3.5.

Now we discover the form of ψ_ℓ^m that results when we impose the condition that ψ is harmonic. We write (4.1.1) as $r^2 \nabla^2 \psi = 0$, or

$$[r\partial_r(r\partial_r + 1) + \nabla_1^2]\psi = 0.$$

For a fixed r, multiply this equation by $\overline{\beta_\ell^m}(\hat{s})$ and average over $S(1)$. The result is

$$\frac{1}{4\pi}\int_{S(1)} d^2\hat{s}\ \overline{\beta_\ell^m(\hat{s})}\ r\partial_r(r\partial_r + 1)\psi(r\hat{s}) + \langle\overline{\beta_\ell^m}(\hat{s})\nabla_1^2\psi(r\hat{s})\rangle_{S(1)} = 0.$$

Taking the derivatives outside the integral sign and using the self-adjointness of ∇_1^2 on $S(1)$, a special case of (7.4.22), we find

$$(r\partial_r)(r\partial_r + 1)\langle\overline{\beta_\ell^m}(\hat{s})\,\psi(r\hat{s})\rangle_{S(1)} + \langle\overline{\nabla_1^2\beta_\ell^m}(\hat{s})\,\psi(r\hat{s})\rangle_{S(1)} = 0.$$

Finally, by employing (4.1.2) and (4.1.5) we see

$$r\partial_r(r\partial_r + 1)\psi_\ell^m(r) - \ell(\ell + 1)\psi_\ell^m(r) = 0. \tag{4.1.8}$$

The general solution of this ordinary second-order differential equation can easily be found by application of the standard method of power series development (Birkhoff and Rota, 1989); it is

$$Ar^\ell + \frac{B}{r^{\ell+1}}.$$

In order to have coefficients with the same physical dimensions as $\nabla\psi$ (i.e., tesla, gauss, or nT in the magnetic case), we create dimensionless quantities by scaling r by a factor a, the radius of the inner shell; we write the general solution of (4.1.8) as

$$\psi_\ell^m(r) = ag_\ell^m\left(\frac{a}{r}\right)^{\ell+1} + ak_\ell^m\left(\frac{r}{a}\right)^\ell. \tag{4.1.9}$$

Thus we have shown that every solution of (4.1.1) can be written in the form

$$\psi(\mathbf{r}) = \psi(r\hat{\mathbf{r}}) = a \sum_{\ell=0}^{\infty} \sum_{m=-\ell}^{\ell} \left[g_\ell^m \left(\frac{a}{r}\right)^{\ell+1} + k_\ell^m \left(\frac{r}{a}\right)^\ell \right] \beta_\ell^m(\hat{\mathbf{r}}) \quad (4.1.10)$$

where g_ℓ^m and k_ℓ^m are scalar constants.

We report without proof some facts about the series (4.1.10). (Proofs are in Kellogg, 1953.) The series converges absolutely and uniformly to ψ in $S(a + \epsilon, c - \epsilon)$ for every $\epsilon > 0$. Moreover, it can be differentiated term by term any number of times there. This fact, applied to (4.1.4), is another way to obtain (4.1.8). The coefficients g_ℓ^m and k_ℓ^m can be chosen arbitrarily except for a convergence condition, and the resulting ψ obtained from (4.1.10) will be a solution of (4.1.1). If one wants the associated magnetic field $-\nabla\psi$ to be square-integrable on the inner and outer surfaces of the shell, that is on $S(a)$ and $S(c)$, which is the same as $\langle|\nabla\psi|^2\rangle_{S(r)} < \infty$ for $r = a, c$, the convergence conditions on the coefficients are

$$\sum_{\ell=0}^{\infty} \sum_{m=-\ell}^{\ell} (\ell + 1)(2\ell + 1)|g_\ell^m|^2 < \infty \quad (4.1.11)$$

and

$$\sum_{\ell=0}^{\infty} \sum_{m=-\ell}^{\ell} \ell(2\ell + 1)|k_\ell^m|^2 \left(\frac{a}{c}\right)^{2\ell} < \infty. \quad (4.1.12)$$

We are interested in the magnetic field $\mathbf{B}$ in the atmosphere $S(a, c)$, where

$$\mathbf{B} = -\nabla\psi \quad (4.1.13)$$

and ψ satisfies (4.1.1). To calculate $\mathbf{B}$ from (4.1.10) we write

$$\mathbf{B} = \hat{\mathbf{r}}B_r + \mathbf{B}_s = -\nabla\psi = -(\hat{\mathbf{r}}\partial_r + r^{-1}\nabla_1)\psi,$$

so

$$B_r = \partial_r\psi \quad \text{and} \quad \mathbf{B}_s = -r^{-1}\nabla_1\psi.$$

Then from (4.1.10) and the ability to differentiate under the sum:

$$B_r(r\hat{\mathbf{r}}) = \sum_{\ell=0}^{\infty} \sum_{m=-\ell}^{\ell} \left[(\ell + 1) \left(\frac{a}{r}\right)^{\ell+2} g_\ell^m - \ell \left(\frac{r}{a}\right)^{\ell-1} k_\ell^m \right] \beta_\ell^m(\hat{\mathbf{r}})$$

$$(4.1.14)$$

$$\mathbf{B}_s(r\hat{\mathbf{r}}) = \sum_{\ell=0}^{\infty} \sum_{m=-\ell}^{\ell} \left[\left(\frac{a}{r}\right)^{\ell+2} g_\ell^m + \left(\frac{r}{a}\right)^{l-1} k_\ell^m \right] \boldsymbol{\nabla}_1 \beta_\ell^m(\hat{\mathbf{r}}). \quad (4.1.15)$$

4.2 Uniqueness of the Coefficients

If the scalar potential ψ satisfies (4.1.1), how many different sets of coefficients g_ℓ^m, k_ℓ^m can be used to describe ψ via (4.1.10)? The function $\psi_\ell^m(r)$ is uniquely determined by ψ, and if ψ_ℓ^m is known for all r in $a < r < c$, then (4.1.9) determines g_ℓ^m and k_ℓ^m uniquely. Thus ψ has exactly one set of coefficients g_ℓ^m, k_ℓ^m such that (4.1.10) works. These coefficients are called the scalar potential's *Gauss coefficients* relative to the basis $\{\beta_\ell^m\}$.

How much do we have to know about ψ in $S(a,c)$ besides (4.1.1) in order to find its Gauss coefficients? If we know $\psi_\ell^m(r)$ for two different values of r, at r_1 and r_2, then (4.1.9) gives two equations with nonzero determinant for g_ℓ^m and k_ℓ^m, so they can be calculated. And $\psi_\ell^m(r)$ is calculable via (4.1.5) if ψ is known on $S(r)$. Thus if we know that ψ satisfies (4.1.1), and we know ψ on two spherical surfaces, $S(r_1)$ and $S(r_2)$, we can find ψ's Gauss coefficients and hence find ψ in $S(a,c)$.

In the geomagnetic problem it is not ψ the potential that is observable, but the magnetic field, $\mathbf{B}$, given by

$$\mathbf{B} = -\boldsymbol{\nabla}\psi. \quad (4.2.1)$$

Moreover, until satellite data were available, $\mathbf{B}$ could be measured on only one (approximately) spherical surface, the surface of the earth. In fact, these data suffice to determine the Gauss coefficients (except for k_0^0; see later) of ψ, and hence to determine $\mathbf{B}$ throughout $S(a,c)$.

We will show that if $\mathbf{B}$ is known on one $S(b)$ with $a < b < c$ then all the g_ℓ^m and k_ℓ^m are determined, except for k_0^0. Then ψ is determined in $S(a,c)$ to within an additive constant, and $\mathbf{B}$ is uniquely determined in $S(a,c)$. As a first step in the proof, we multiply (4.1.14) by $\overline{\beta_{\ell'}^{m'}(\hat{\mathbf{r}})}$, set $r = b$, and average over $S(1)$. Because of (4.1.3) the result is

$$(\ell'+1)\left(\frac{a}{b}\right)^{\ell'+2} g_{\ell'}^{m'} - \ell'\left(\frac{b}{a}\right)^{\ell'-1} k_{\ell'}^{m'} = \langle \overline{\beta_{\ell'}^{m'}(\hat{\mathbf{s}})} B_r(b\hat{\mathbf{s}})\rangle.$$

For aesthetic reasons, we replace ℓ', m' by ℓ, m. Then this equation is

$$(\ell + 1) \left(\frac{a}{b}\right)^{\ell+2} g_\ell^m - \ell \left(\frac{b}{a}\right)^{\ell-1} k_\ell^m = \frac{1}{4\pi} \int_{S(1)} d^2\hat{\mathbf{s}} \; \overline{\beta_\ell^m(\hat{\mathbf{s}})} \, B_r(b\hat{\mathbf{s}}).$$
(4.2.2)

To obtain another equation for g_ℓ^m and k_ℓ^m from $\mathbf{B}$ on $S(b)$ we must use (4.1.15). To do so, we need the identity

$$\langle \boldsymbol{\nabla}_1 f \cdot \boldsymbol{\nabla}_1 g \rangle = -\langle f \nabla_1^2 g \rangle.$$
(4.2.3)

To prove this we note that $\boldsymbol{\nabla}_1$ is a vector FODO and therefore (7.2.65) gives

$$\boldsymbol{\nabla}_1 \cdot (f \boldsymbol{\nabla}_1 g) = \boldsymbol{\nabla}_1 f \cdot \boldsymbol{\nabla}_1 g + f \nabla_1^2 g.$$

Applying (7.4.21) to this equation gives the identity (4.2.3). It follows from (4.2.3) that for any ℓ, m, ℓ', m',

$$\langle \boldsymbol{\nabla}_1 \overline{\beta_{\ell'}^{m'}} \cdot \boldsymbol{\nabla}_1 \beta_\ell^m \rangle = -\langle \overline{\beta_{\ell'}^{m'}} \nabla_1^2 \beta_\ell^m \rangle = \ell(\ell+1)\langle \overline{\beta_{\ell'}^{m'}} \beta_\ell^m \rangle,$$

so

$$\langle (\boldsymbol{\nabla}_1 \overline{\beta_{\ell'}^{m'}}) \cdot (\boldsymbol{\nabla}_1 \beta_\ell^m) \rangle = \ell(\ell+1)\delta_{\ell\ell'}\delta_{mm'}.$$

If we dot $\boldsymbol{\nabla}_1 \overline{\beta_{\ell'}^{m'}}$ into (4.1.15), average over $S(1)$, set $r = b$, and replace ℓ', m' by ℓ, m, we get

$$\left(\frac{a}{b}\right)^{\ell+2} g_\ell^m + \left(\frac{b}{a}\right)^{\ell-1} k_\ell^m = -\frac{1}{\ell(\ell+1)} \frac{1}{4\pi} \int_{S(1)} d^2\hat{\mathbf{s}} \; \boldsymbol{\nabla}_1 \overline{\beta_\ell^m(\hat{\mathbf{s}})} \cdot \mathbf{B}_s(b\hat{\mathbf{s}}).$$
(4.2.4)

Solving (4.2.2) and (4.2.4) for g_ℓ^m and k_ℓ^m gives

$$(2\ell + 1) \left(\frac{a}{b}\right)^{\ell+2} g_\ell^m = \langle \overline{\beta_\ell^m(\hat{\mathbf{s}})} \, B_r(b\hat{\mathbf{s}}) \rangle_{\hat{\mathbf{s}}} - \frac{\langle (\boldsymbol{\nabla}_1 \overline{\beta_\ell^m(\hat{\mathbf{s}})}) \cdot \mathbf{B}_s(b\hat{\mathbf{s}}) \rangle_{\hat{\mathbf{s}}}}{(\ell+1)}$$
(4.2.5)

$$(2\ell + 1) \left(\frac{b}{a}\right)^{\ell-2} k_\ell^m = \langle \overline{\beta_\ell^m(\hat{\mathbf{s}})} \, B_r(b\hat{\mathbf{s}}) \rangle_{\hat{\mathbf{s}}} - \frac{\langle (\boldsymbol{\nabla}_1 \overline{\beta_\ell^m(\hat{\mathbf{s}})}) \cdot \mathbf{B}_s(b\hat{\mathbf{s}}) \rangle_{\hat{\mathbf{s}}}}{\ell}.$$
(4.2.6)

Thus the expansion coefficients g_ℓ^m and k_ℓ^m can be explicitly calculated from suitable averages of the field $\mathbf{B}$ on a single spherical surface within $S(a, c)$.

Note that (4.2.6) fails for $\ell = 0$. The reason is apparent from (4.2.1) and (4.1.10). In (4.1.10), the term involving k_0^0 is a constant, independent of r or $\hat{\mathbf{r}}$. But if $\mathbf{B}$ is what we measure, (4.2.1) shows that ψ is determined only up to an arbitrary additive constant. Therefore, given (4.1.1) and (4.2.1), if we know $\mathbf{B}$ in $S(a, c)$, we are free to choose k_0^0 in (4.1.10) in any way we please. For convenience we choose

$$k_0^0 = 0. \tag{4.2.7}$$

One other result of our use of (4.2.1), (4.1.10), (4.1.1) for magnetic fields comes from using (4.2.5) to calculate g_0^0. The result is

$$(a/b)^2 g_0^0 = \overline{\beta_0^0} \, \langle B_r(b\hat{\mathbf{s}}) \rangle_{\hat{\mathbf{s}}}$$

because β_0^0 is constant and $\nabla_1 \beta_0^0 = \mathbf{0}$. But then we note that

$$\langle B_r(b\hat{\mathbf{s}}) \rangle_{\hat{\mathbf{s}}} = \frac{1}{4\pi} \int_{S(1)} d^2\hat{\mathbf{s}} \; B_r(b\hat{\mathbf{s}}) = \frac{1}{4\pi b^2} \int_{S(b)} d^2\mathbf{s} \; B_r(\mathbf{s})$$

$$= \frac{1}{4\pi b^2} \int_{\partial B(b)} d^2\mathbf{s} \; \hat{\mathbf{s}} \cdot \mathbf{B}(\mathbf{s}) = \frac{1}{4\pi b^2} \int_{B(b)} d^3\mathbf{s} \; (\nabla \cdot \mathbf{B}) = 0$$

where $B(b)$ is the solid ball of radius b centered on $\mathbf{0}$. Therefore

$$g_0^0 = 0. \tag{4.2.8}$$

Equation (4.2.8) is forced on us by the Maxwell equation $\nabla \cdot \mathbf{B} = 0$, i.e., by the nonexistence of magnetic monopoles. Equation (4.2.7), on the other hand, is an arbitrary choice, made for convenience. Together, these two equations enable us to rewrite (4.1.10), (4.1.14), (4.1.15) as

$$\psi(r\hat{\mathbf{r}}) = a \sum_{\ell=1}^{\infty} \sum_{m=-\ell}^{\ell} \left[g_\ell^m \left(\frac{a}{r} \right)^{\ell+1} + k_\ell^m \left(\frac{r}{a} \right)^\ell \right] \beta_\ell^m(\hat{\mathbf{r}}) \tag{4.2.9}$$

$$B_r(r\hat{\mathbf{r}}) = \sum_{\ell=1}^{\infty} \sum_{m=-\ell}^{\ell} \left[(\ell+1) \left(\frac{a}{r} \right)^{\ell+2} g_\ell^m - \ell \left(\frac{r}{a} \right)^{\ell-1} k_\ell^m \right] \beta_\ell^m(\hat{\mathbf{r}})$$

$$\tag{4.2.10}$$

$$\mathbf{B}_s(r\hat{\mathbf{r}}) = -\sum_{\ell=1}^{\infty} \sum_{m=-\ell}^{\ell} \left[\left(\frac{a}{r} \right)^{\ell+2} g_\ell^m + \left(\frac{r}{a} \right)^{\ell-1} k_\ell^m \right] \nabla_1 \beta_\ell^m(\hat{\mathbf{r}}). \tag{4.2.11}$$

We have seen that the Gauss coefficients g_ℓ^m, k_ℓ^m are uniquely determined if $\mathbf{B}$ is known on one spherical surface $S(b)$, so that knowledge determines ψ and $\mathbf{B}$ uniquely throughout $S(a,c)$.

We note that now the zeroth degree part of the potential, $\psi_0(r) = 0$, and so by (4.1.5) $\langle \beta_0^0(\hat{s})\psi(r\hat{s})\rangle_{\hat{s}} = 0$. But β_0^0 is a constant, so $\langle \psi(r\hat{s})\rangle_{\hat{s}} = 0$ for every r in $a < r < c$. Thus

$$\langle \psi \rangle_{S(r)} = 0 \quad \text{for all } r \text{ in } a < r < c. \tag{4.2.12}$$

We have shown that a magnetic field $\mathbf{B}$ with no sources in $S(a,c)$ has a representation (4.2.1) with a ψ satisfying (4.2.12), and that this ψ is uniquely determined by $\mathbf{B}$. The condition (4.2.12) eliminates the arbitrary additive constant.

Two degenerate cases will be of interest. First, suppose $c = \infty$, so that (4.2.9–4.2.11) work in the whole space outside $S(a)$. Suppose also that $|\mathbf{B}| \to 0$ as $r \to \infty$. Then as $b \to \infty$, the right side of (4.2.6) $\to \infty$. The same must be true of the left side, so

$$k_\ell^m = 0 \quad \text{for all} \quad \ell, m \tag{4.2.13}$$

if $\mathbf{B}$ has no sources outside $S(a)$. That is, a magnetic field $\mathbf{B}$, all of whose sources are inside $S(a)$, has no k_ℓ^m terms. One might feel that this argument is unnecessarily elaborate. Why invoke (4.2.6)? Is it not obvious from (4.1.10) or (4.1.14) that if some $k_\ell^m \neq 0$ then $|\mathbf{B}|$ cannot $\to 0$ as $r \to \infty$? The answer is no. Consider the example

$$\psi = \sum_{\ell=0}^{\infty} \frac{(-1)^\ell}{\ell!} r^\ell = e^{-r}$$

$$B_r = -\sum_{\ell=0}^{\infty} \frac{(-1)^\ell}{\ell!} r^\ell = -e^{-r} = -\partial_r \psi$$

$$\mathbf{B}_s = 0 = r^{-1} \nabla_1 \psi.$$

It is essential to proving (4.2.13) that the separate terms in (4.1.10) be orthogonal on $S(1)$.

Next, suppose $a = 0$ so that (4.2.9–4.2.11) work in the ball $B(c)$, except possibly at $\mathbf{r} = 0$. Suppose also that $\mathbf{B}$ remains bounded as $r \to 0$. (This would happen, for example, if $\nabla \cdot \mathbf{B} = 0$ and $\nabla \times \mathbf{B} = \mathbf{0}$ in the ball $B(c)$.) Then (4.2.5) shows that

$$g_\ell^m = 0 \quad \text{for all} \quad \ell, m. \tag{4.2.14}$$

That is, a magnetic field $\mathbf{B}$ all of whose sources are outside $S(c)$ has no g_ℓ^m terms. Because of this situation, the coefficients k_ℓ^m in (4.2.9) are called the *external* Gauss coefficients, and the g_ℓ^m are called the *internal* Gauss coefficients of $\mathbf{B}$.

Now a subtle uniqueness question arises. Suppose $\mathbf{B}$ has sources inside $S(a)$ and sources outside $S(c)$. As we have seen in section 2.5, we have in all of R^3

$$\mathbf{B} = \mathbf{B}^{(E)} + \mathbf{B}^{(I)} \qquad (4.2.15)$$

where $\mathbf{B}^{(E)}$ is the magnetic field produced by the sources outside $S(c)$ (the external sources), and $\mathbf{B}^{(I)}$ is the magnetic field produced by the sources inside $S(a)$ (the internal sources). Inside $S(c)$ we have

$$\mathbf{B}^{(E)} = -\boldsymbol{\nabla}\psi^{(E)} \quad \text{with} \quad \nabla^2\psi^{(E)} = 0,$$

so there are external Gauss coefficients $k_{\ell m}^{(E)}$ such that in $B(c)$

$$\psi^{(E)}(r\hat{\mathbf{r}}) = a\sum_{\ell=1}^{\infty}\sum_{m=-\ell}^{\ell}\left(\frac{r}{a}\right)^\ell k_{\ell m}^{(E)}\beta_\ell^m(\hat{\mathbf{r}})$$

if the reader will pardon the relocation of the superscript m. Outside $S(a)$, we have

$$\mathbf{B}^{(I)} = -\boldsymbol{\nabla}\psi^{(I)} \quad \text{with} \quad \nabla^2\psi^{(I)} = 0,$$

so there are internal Gauss coefficients $g_{\ell m}^{(I)}$ such that in $S(a,\infty)$

$$\psi^{(I)}(r\hat{\mathbf{r}}) = a\sum_{\ell=1}^{\infty}\sum_{m=-\ell}^{\ell}\left(\frac{a}{r}\right)^{\ell+1} g_{\ell m}^{(I)}\beta_\ell^m(\hat{\mathbf{r}}).$$

In $S(a,c)$, all these equations are valid, so in $S(a,c)$

$$\mathbf{B} = -\boldsymbol{\nabla}\psi \quad \text{with} \quad \nabla^2\psi = 0 \quad \text{and} \quad \psi = \psi^{(I)} + \psi^{(E)}.$$

Then

$$\psi(r\hat{\mathbf{r}}) = a\sum_{\ell=1}^{\infty}\sum_{m=-\ell}^{\ell}\left[\left(\frac{a}{r}\right)^{\ell+1} g_{\ell m}^{(I)} + \left(\frac{r}{a}\right)^\ell k_{\ell m}^{(E)}\right]\beta_\ell^m(\hat{\mathbf{r}}). \qquad (4.2.16)$$

But ψ has only one expansion (4.2.9) in $S(a,c)$. Therefore, the coefficients in (4.2.16) and (4.2.9) must be the same. That is

$$g_{\ell m}^{(I)} = g_\ell^m \quad \text{and} \quad k_{\ell m}^{(E)} = k_\ell^m. \qquad (4.2.17)$$

In other words, the Gauss coefficients of the internal sources are the internal Gauss coefficients of $\mathbf{B}$ and the Gauss coefficients of the external sources are the external Gauss coefficients of $\mathbf{B}$. By observing $\mathbf{B}$ on one spherical surface $S(b)$ in $S(a,c)$, we can find the Gauss coefficients of $\mathbf{B}^{(E)}$ and $\mathbf{B}^{(I)}$. We can separately calculate throughout $S(a,c)$ the magnetic fields $\mathbf{B}^{(I)}$ due to sources inside $S(a)$ and the magnetic field $\mathbf{B}^{(E)}$ due to sources outside $S(c)$.

4.3 Observing the Sources in Principle

Observations of $\mathbf{B}$ on $S(a)$, the earth's surface, enable us to find g_ℓ^m and k_ℓ^m in (4.2.9). We know therefore that the magnetic field outside $S(a)$ produced by the source inside $S(a)$ is $\mathbf{B}^{(I)} = -\nabla \psi^{(I)}$ with

$$\psi^{(I)}(r\hat{\mathbf{r}}) = a \sum_{\ell=1}^{\infty} \left(\frac{a}{r}\right)^{\ell+1} \sum_{m=-\ell}^{\ell} g_\ell^m \, \beta_\ell^m(\hat{\mathbf{r}}) \qquad r > a \qquad (4.3.1)$$

and the magnetic field inside $S(c)$ produced by the sources outside $S(c)$ is $\mathbf{B}^{(E)} = -\nabla \psi^{(E)}$ with

$$\psi^{(E)}(r\hat{\mathbf{r}}) = a \sum_{\ell=1}^{\infty} \left(\frac{r}{a}\right)^{\ell} \sum_{m=-\ell}^{\ell} k_\ell^m \, \beta_\ell^m(\hat{\mathbf{r}}) \qquad r < c.$$

Thus the coefficients g_ℓ^m describe some property of the sources inside $S(a)$, and the coefficients k_ℓ^m concern the sources outside $S(c)$. What can we learn about the sources themselves from these coefficients?

First, consider the magnetization $\mathbf{M}$ inside the earth. Some is permanent and some is induced by the core field's effects on the crust. $\mathbf{M}$ is the total. Suppose there are no free currents, so $\mathbf{B}$ and $\mathbf{H}$ are produced by $\mathbf{M}$ alone. Then $\nabla \times \mathbf{H} = 0$ everywhere, while $\nabla \times \mathbf{B} = 0$ only outside $S(a)$. Therefore we introduce a magnetic scalar potential ψ defined everywhere in R^3 by

$$\mu_0 \mathbf{H} = -\nabla \psi.$$

Then $\mathbf{B} = -\nabla\psi$ outside $S(a)$. The magnetic dipole moment of a small volume $d^3\mathbf{s}$ at positions $\mathbf{s}$ is $\mathbf{M}(\mathbf{s})d^3\mathbf{s}$. At $\mathbf{r}$, this moment produces the magnetic potential

$$d\psi(\mathbf{r}) = -\frac{\mu_0}{4\pi}\mathbf{M}(\mathbf{s}) \cdot \nabla_\mathbf{r} \frac{1}{|\mathbf{r}-\mathbf{s}|} d^3\mathbf{s} \qquad (4.3.2)$$

where $\nabla_\mathbf{r}$ is the gradient with respect to $\mathbf{r}$. We need to prove this formula. From (2.2.24), for an infinitesimal region with dipole moment $\mathbf{m}$,

$$\mathbf{A} = -\frac{\mu_0}{4\pi}\,\mathbf{m}\times\nabla\frac{1}{r} \quad \text{and} \quad \mathbf{B} = \nabla\times\mathbf{A} \quad \text{so}$$

$$\mathbf{B} = -\frac{\mu_0}{4\pi}\,\nabla\times\left(\mathbf{m}\times\nabla\frac{1}{r}\right)$$

$$= -\frac{\mu_0}{4\pi}\left\{\left(\nabla^2\frac{1}{r}+\nabla\frac{1}{r}\cdot\nabla\right)\mathbf{m}-(\nabla\cdot\mathbf{m}+\mathbf{m}\cdot\nabla)\frac{1}{r}\right\}.$$

Since $\mathbf{m}$ does not depend on position, and $\nabla^2\frac{1}{r}=0$, we have

$$\mathbf{B} = \frac{\mu_0}{4\pi}\,(\mathbf{m}\cdot\nabla)\nabla\frac{1}{r}.$$

Relative to a Cartesian basis,

$$B_i = \frac{\mu_0}{4\pi}\,m_j\,\partial_j\,\partial_i\frac{1}{r} = \partial_i\frac{\mu_0}{4\pi}\,m_j\,\partial_j\frac{1}{r},$$

since $\partial_i\,m_j = 0$ and $\partial_j\,\partial_i = \partial_i\,\partial_j$. Thus

$$\mathbf{B} = \nabla\left[\frac{\mu_0}{4\pi}\,(\mathbf{m}\cdot\nabla)\frac{1}{r}\right] = -\nabla\psi$$

where

$$\psi = -\frac{\mu_0}{4\pi}\,(\mathbf{m}\cdot\nabla)\frac{1}{r},$$

which establishes (4.3.2) as required.

Returning to (4.3.2) we note that, if $\hat{\mathbf{x}}_1, \hat{\mathbf{x}}_2, \hat{\mathbf{x}}_3$ is a Cartesian basis,

$$\nabla_{\mathbf{r}} = \hat{\mathbf{x}}_i \frac{\partial}{\partial r_i}, \qquad \nabla_{\mathbf{s}} = \hat{\mathbf{x}}_i \frac{\partial}{\partial s_i}$$

then

$$\nabla_{\mathbf{r}} \frac{1}{|\mathbf{r} - \mathbf{s}|} = -\nabla_{\mathbf{s}} \frac{1}{|\mathbf{r} - \mathbf{s}|}$$

and therefore

$$d\psi(\mathbf{r}) = \frac{\mu_0}{4\pi} \mathbf{M} \cdot \nabla_{\mathbf{s}} \frac{1}{|\mathbf{r} - \mathbf{s}|} d^3\mathbf{s}.$$

Integrating this over $B(a)$, the spherical ball of radius a centered at $\mathbf{0}$, gives

$$\psi(\mathbf{r}) = \frac{\mu_0}{4\pi} \int_{B(a)} d^3\mathbf{s}\, \mathbf{M}(\mathbf{s}) \cdot \nabla_{\mathbf{s}} \frac{1}{|\mathbf{r} - \mathbf{s}|}. \qquad (4.3.3)$$

Now (3.5.25) can be written

$$\frac{1}{|\mathbf{r} - \mathbf{s}|} = \sum_{\ell=0}^{\infty} \frac{1}{(2\ell + 1)} \frac{1}{r^{\ell+1}} \sum_{m=-\ell}^{\ell} \beta_{\ell}^m(\hat{\mathbf{r}})\, \overline{\beta_{\ell}^m(\mathbf{s})} \qquad s < r. \quad (4.3.4)$$

We are really interested in (4.3.3) where $r > a$. Since the integral (4.3.3) is over a region with $s < a$, we will always have $s < r$, and we can substitute (4.3.4) in (4.3.3). Interchanging summation and integration, a step justified by the uniform convergence of (4.3.4), we obtain

$$\psi(\mathbf{r}) = \frac{\mu_0}{4\pi} \sum_{\ell=0}^{\infty} \frac{1}{(2\ell + 1)} \frac{1}{r^{\ell+1}} \sum_{m=-\ell}^{\ell} \beta_{\ell}^m(\hat{\mathbf{r}}) \int_{B(a)} d^3\mathbf{s}\, \mathbf{M}(\mathbf{s}) \cdot \nabla_{\mathbf{s}}\, \overline{\beta_{\ell}^m(\mathbf{s})}.$$

$$(4.3.5)$$

Comparing this with (4.3.1), and using the uniqueness of the g_{ℓ}^m, we conclude that

$$g_{\ell}^m = \frac{\mu_0}{4\pi} \frac{1}{(2\ell + 1)a^{\ell+2}} \int_{B(a)} d^3\mathbf{s}\, \mathbf{M}(\mathbf{s}) \cdot \nabla_{\mathbf{s}} \beta_{\ell}^m(\mathbf{s}). \qquad (4.3.6)$$

Thus the coefficients g_{ℓ}^m in this model can be computed directly from the magnetization distribution. As we will soon see, however, knowledge of g_{ℓ}^m gives very incomplete information about $\mathbf{M}$.

Next, suppose $\mathbf{M} = \mathbf{0}$ in the earth, and $\mathbf{B}$ and $\mathbf{H}$ are produced entirely by free currents $\mathbf{J}^{(F)}$ in the earth. While not valid for surface rocks, this condition is believed to hold below 20 km or so, because the Curie temperature of almost all naturally occurring materials is exceeded below this depth. Then $\mathbf{B} = \mu_0\mathbf{H}$ everywhere, but $\nabla \times \mathbf{H} \neq \mathbf{0}$ in the earth, so no scalar magnetic potential is defined there. We must use the magnetic vector potential $\mathbf{A}$, given by

$$\mathbf{A}(\mathbf{r}) = \frac{\mu_0}{4\pi} \int_{B(a)} d^3\mathbf{s} \, \frac{\mathbf{J}^{(F)}(\mathbf{s})}{|\mathbf{r} - \mathbf{s}|}.$$

Then $\mathbf{B}(\mathbf{r}) = \nabla \times \mathbf{A}(\mathbf{r})$. Carrying the $\nabla_{\mathbf{r}}\times$ under the integral sign gives Ampère's law,

$$\mathbf{B}(\mathbf{r}) = \frac{\mu_0}{4\pi} \int_{B(a)} d^3\mathbf{s} \, \left(\nabla_{\mathbf{r}} \frac{1}{|\mathbf{r} - \mathbf{s}|} \right) \times \mathbf{J}^{(F)}(\mathbf{s})$$

$$= \frac{\mu_0}{4\pi} \int_{B(a)} d^3\mathbf{s} \, \frac{(\mathbf{s} - \mathbf{r}) \times \mathbf{J}^{(F)}(\mathbf{s})}{|\mathbf{r} - \mathbf{s}|^3}.$$

We will use this formula to calculate B_r in $r > a$, and compare the coefficients with those in (4.2.10). We have

$$rB_r(\mathbf{r}) = \mathbf{r} \cdot \mathbf{B}(\mathbf{r}) = \frac{\mu_0}{4\pi} \int_{B(a)} d^3\mathbf{s} \, \frac{\mathbf{r} \cdot [(\mathbf{s} - \mathbf{r}) \times \mathbf{J}^{(F)}(\mathbf{s})]}{|\mathbf{r} - \mathbf{s}|^3}.$$

But

$$\mathbf{r} \cdot [(\mathbf{s} - \mathbf{r}) \times \mathbf{J}^{(F)}(\mathbf{s})] = \mathbf{r} \cdot [\mathbf{s} \times \mathbf{J}^{(F)}(\mathbf{s})] = (\mathbf{r} - \mathbf{s}) \cdot [\mathbf{s} \times \mathbf{J}^{(F)}(\mathbf{s})],$$

so

$$rB_r(\mathbf{r}) = \frac{\mu_0}{4\pi} \int_{B(a)} d^3\mathbf{s} \, \frac{(\mathbf{r} - \mathbf{s})}{|\mathbf{r} - \mathbf{s}|^3} \cdot \mathbf{s} \times \mathbf{J}^{(F)}(\mathbf{s})$$

$$= \frac{\mu_0}{4\pi} \int_{B(a)} d^3\mathbf{s} \, \left(\nabla_{\mathbf{s}} \frac{1}{|\mathbf{r} - \mathbf{s}|} \right) \cdot [\mathbf{s} \times \mathbf{J}^{(F)}(\mathbf{s})].$$

Now $(\nabla f) \cdot \mathbf{v} = \nabla \cdot (f\mathbf{v}) - f(\nabla \cdot \mathbf{v})$ so, by Gauss' Theorem

$$rB_r(\mathbf{r}) = \frac{\mu_0}{4\pi} \int_{\partial B(a)} d^2\mathbf{s}\,\hat{\mathbf{s}} \cdot \left[\frac{\mathbf{s} \times \mathbf{J}^{(F)}(\mathbf{s})}{|\mathbf{r} - \mathbf{s}|} \right]$$

$$- \frac{\mu_0}{4\pi} \int_{B(a)} d^3\mathbf{s} \, \frac{1}{|\mathbf{r} - \mathbf{s}|} \nabla_{\mathbf{s}} \cdot (\mathbf{s} \times \mathbf{J}^{(F)}(\mathbf{s})).$$

The surface integrand vanishes and hence the surface integral is zero. Also we may introduce the $\mathbf{\Lambda}$ operator of section 8.2 through $\nabla_{\mathbf{s}} \cdot (\mathbf{s} \times \mathbf{v}) = -\mathbf{\Lambda} \cdot \mathbf{v}(\mathbf{s})$: so

$$rB_r(\mathbf{r}) = \frac{\mu_0}{4\pi} \int_{B(a)} d^3 \mathbf{s} \, \frac{1}{|\mathbf{r} - \mathbf{s}|} \mathbf{\Lambda} \cdot \mathbf{J}^{(F)}(\mathbf{s}). \qquad (4.3.7)$$

Now we use (3.5.25) only for $r > a$, so we can insert (4.3.4). Interchanging the orders of summation and integration, we obtain

$$rB_r(\mathbf{r}) = \frac{\mu_0}{4\pi} \sum_{\ell=0}^{\infty} \frac{1}{(2\ell+1)r^{\ell+1}} \sum_{m=-\ell}^{\ell} \beta_\ell^m(\mathbf{r}) \int_{B(a)} d^3 \mathbf{s} \, \overline{\beta_\ell^m(\mathbf{s})} \, \mathbf{\Lambda} \cdot \mathbf{J}^{(F)}(\mathbf{s}).$$

Comparing this with (4.2.10) gives

$$g_\ell^m = \frac{\mu_0}{4\pi} \frac{1}{(\ell+1)(2\ell+1)a^{\ell+2}} \int_{B(a)} d^3 \mathbf{s} \, \overline{\beta_\ell^m(\mathbf{s})} \, [\mathbf{\Lambda} \cdot \mathbf{J}^{(F)}(\mathbf{s})]. \quad (4.3.8)$$

Since Maxwell's equations are linear in the sources, the total field is the sum of the fields produced by the separate sources. Therefore, if both $\mathbf{J}^{(F)}$ and $\mathbf{M}$ are present in the earth, g_ℓ^m is the sum of (4.3.6) and (4.3.8); that is,

$$g_\ell^m = \frac{\mu_0}{4\pi} \frac{1}{(2\ell+1)a^{\ell+2}} \int_{B(a)} d^3 \mathbf{s}$$
$$\left\{ \frac{[\mathbf{\Lambda} \cdot \mathbf{J}^{(F)}(\mathbf{s})] \, \overline{\beta_\ell^m(\mathbf{s})}}{(\ell+1)} + \mathbf{M}(\mathbf{s}) \cdot \nabla \overline{\beta_\ell^m(\mathbf{s})} \right\}. \qquad (4.3.9)$$

If there are surface currents $\mathbf{J}_s^{(F)}$, then the volume integral is understood to include a surface integral. We will have to consider only spherical surfaces $S(b)$, and on these it is easy to calculate $\mathbf{\Lambda} \cdot \mathbf{J}_s^{(F)}$, and to include its integral over $S(b)$ in (4.3.9).

To find k_ℓ^m, use (4.3.7) with the region of integration being $R^3 \backslash B(c)$, which means all of space minus the ball of radius c. We interchange $\mathbf{r}$ and $\mathbf{s}$ in (3.5.25) so that it is valid for $r < s$. Then an argument like the foregoing gives

$$k_\ell^m = -\frac{\mu_0}{4\pi} \frac{a^{\ell-1}}{\ell(2\ell+1)} \int_{R^3 \backslash B(c)} d^3 \mathbf{s} \, \frac{\overline{\beta_\ell^m(\hat{\mathbf{s}})}}{s^{\ell+1}} [\mathbf{\Lambda} \cdot \mathbf{J}^{(F)}(\mathbf{s})]. \qquad (4.3.10)$$

Naturally, we ignore permanent and induced magnetization above the atmosphere.

Knowing the coefficients g_ℓ^m tells us remarkably little about the sources inside the earth, $\mathbf{J}^{(F)}$ and $\mathbf{M}$. Even neglecting $\mathbf{M}$ (a good approximation, as we will see, when $\ell \leq 10$), we learn very little about $\mathbf{J}^{(F)}$ as we now illustrate. In $0 \leq r \leq a$ we can always find unique functions $f_\ell^m(r)$ such that

$$\mathbf{\Lambda} \cdot \mathbf{J}^{(F)}(\mathbf{r}) = \sum_{\ell=0}^{\infty} \sum_{m=-\ell}^{\ell} f_\ell^m(r)\beta_\ell^m(\hat{\mathbf{r}}) \qquad 0 \leq r \leq a$$

and thus the functions $f_\ell^m(r)$ provide a complete description of $\mathbf{\Lambda} \cdot \mathbf{J}^{(F)}(\mathbf{r})$. (Actually, the sum starts at $\ell = 1$ because $\langle \mathbf{\Lambda} \cdot \mathbf{v} \rangle = 0$, for any vector field $\mathbf{v}$.) We use (4.3.9) to tell us what we can learn about $f_\ell^m(r)$ from g_ℓ^m. Taking $\mathbf{M} = 0$ in (4.3.9) gives

$$\frac{4\pi}{\mu_0}(\ell+1)(2\ell+1)a^{\ell+2}g_\ell^m = \sum_{\ell'=0}^{\infty} \sum_{m'=-\ell'}^{\ell'} \int_{B(a)} d^3\mathbf{r} \, \overline{\beta_\ell^m(\mathbf{r})} \, f_{\ell'}^{m'}(r)\beta_{\ell'}^{m'}(\hat{\mathbf{r}}).$$

But

$$\int_{B(a)} d^3\mathbf{r} \, \overline{\beta_\ell^m(\mathbf{r})} \, \beta_{\ell'}^{m'}(\hat{\mathbf{r}})f_{\ell'}^{m'}(r) = \int_{B(a)} d^3\mathbf{r} \, r^\ell f_{\ell'}^{m'}(r) \, \overline{\beta_\ell^m(\mathbf{r})} \, \beta_{\ell'}^{m'}(\hat{\mathbf{r}})$$

$$= \int_0^a dr \, r^2 \, r^\ell f_{\ell'}^{m'}(r) \int_{S(1)} d^2\hat{\mathbf{r}} \, \overline{\beta_\ell^m(\hat{\mathbf{r}})} \, \beta_{\ell'}^{m'}(\hat{\mathbf{r}})$$

$$= 4\pi\delta_{\ell\ell'} \, \delta_{mm'} \int_0^a dr \, r^{\ell+2} f_\ell^m(r).$$

Thus

$$\frac{1}{\mu_0}(\ell+1)(2\ell+1)a^{\ell+2}g_\ell^m = \int_0^a dr \, r^{\ell+2} f_\ell^m(r). \qquad (4.3.11)$$

The functions $f_\ell^m(r)$ can be anything. All we know about each of them from $\mathbf{B}$ measured outside the earth is the single integral (4.3.11).

As another illustration of how little (4.3.9) tells us about sources, consider a spherical planet of radius a with a "core" K of radius c

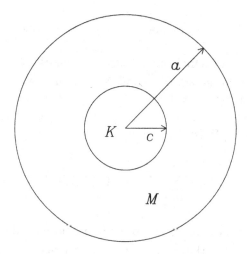

Figure 4.3.1: The planet with mantle M and core K as described in the text.

and a "mantle" M. Suppose K contains magnetic sources whose nature we do not examine. They may be $\mathbf{M}$ or $\mathbf{J}^{(F)}$ or both. Suppose the mantle M is an insulator and has no permanent magnetization, but does have a small magnetic susceptibility γ that depends only on radius r, not on θ or λ. Let $\mathbf{H}^K$ and $\mathbf{B}^K$ be the magnetic displacement and magnetic field produced by the sources in K. They magnetize the mantle, and its intensity of magnetization is $\mathbf{M} = \gamma\mathbf{H}$. If $\gamma \ll 1$, then $\mathbf{H} = \mathbf{H}^K +$ terms of order γ, so to first order in γ,

$$\mathbf{M} = \gamma\mathbf{H}^K \qquad \text{in the mantle } M. \qquad (4.3.12)$$

We ask, What is the magnetic field produced outside $S(a)$ by the mantle magnetization $\mathbf{M}$? From (4.3.9), the Gauss coefficients g_ℓ^m for this mantle-produced field are

$$g_\ell^m = \frac{1}{4\pi a^{\ell+2}(2\ell+1)} \int_M d^3\mathbf{r}\ \gamma(r)\mathbf{B}^K(\mathbf{r}) \cdot \boldsymbol{\nabla}\ \overline{\beta_\ell^m(\mathbf{r})} \qquad (4.3.13)$$

where we have used the fact that $\mathbf{B}^K = \mu_0\mathbf{H}^K$ in M.

But in M, $\mathbf{B}^K$ is a linear combination of fields

$$\boldsymbol{\nabla}\ \frac{\beta_{\ell'}^{m'}(\hat{\mathbf{r}})}{r^{\ell'+1}}$$

due to individual spherical harmonics. We will show that for each of these fields, the integral in (4.3.13) vanishes, so that the whole integral vanishes, and $g_\ell^m = 0$. In other words, to first-order in γ the magnetization of the mantle by the core is invisible outside the mantle. This result depends on the following theorem:

Theorem: Suppose $H_\ell \in \mathcal{H}_\ell$ and $\overline{H_n} \in \mathcal{H}_n$. Suppose $\mathbf{E}_\ell(\mathbf{r}) = \boldsymbol{\nabla} H_\ell(\mathbf{r})$ and $\mathbf{I}_n(\mathbf{r}) = \boldsymbol{\nabla}[\overline{H_n(\hat{\mathbf{r}})}r^{-n-1}]$. Obviously $\mathbf{E}_\ell$ is in the form of an exterior magnetic field associated with a single harmonic of degree ℓ, while $\mathbf{I}_n$ is a magnetic field of interior origin of degree n. Then for any $c > 0$,

$$\langle \mathbf{E}_\ell(c\hat{\mathbf{r}}) \cdot \mathbf{I}_n(c\hat{\mathbf{r}}) \rangle_{\hat{\mathbf{r}}} = \frac{1}{4\pi} \int_{S(1)} d^2\hat{\mathbf{r}}\ \mathbf{E}_\ell(c\hat{\mathbf{r}}) \cdot \mathbf{I}_n(c\hat{\mathbf{r}}) = 0. \qquad (4.3.14)$$

Proof:

$$\mathbf{E}_\ell = \left[\hat{\mathbf{r}}\partial_r + \frac{1}{r}\boldsymbol{\nabla}_1 \right] H_\ell(\mathbf{r}) = \frac{1}{r}[\hat{\mathbf{r}}\,r\partial_r + \boldsymbol{\nabla}_1]H_\ell(\mathbf{r})$$

$$= \frac{1}{r}[\ell\hat{\mathbf{r}} H_\ell(\mathbf{r}) + \boldsymbol{\nabla}_1 H_\ell(\mathbf{r})] = \frac{r^\ell}{r}\,[\ell\hat{\mathbf{r}} H_\ell(\hat{\mathbf{r}}) + \boldsymbol{\nabla}_1 H_\ell(\hat{\mathbf{r}})]$$

$$\mathbf{I}_n = \left[\hat{\mathbf{r}}\partial_r + \frac{1}{r}\boldsymbol{\nabla}_1 \right] \frac{\overline{H_n(\hat{\mathbf{r}})}}{r^{n+1}} = \frac{1}{r}[\hat{\mathbf{r}}\,r\partial_r + \boldsymbol{\nabla}_1]r^{-n-1}\overline{H_n(\hat{\mathbf{r}})}$$

$$= \frac{1}{r}[-(n+1)\hat{\mathbf{r}}\overline{H_n(\hat{\mathbf{r}})} + \boldsymbol{\nabla}_1\overline{H_n(\hat{\mathbf{r}})}]r^{-n-1}.$$

Let us dot these two fields together, then average over the unit sphere:

$$\mathbf{E}_\ell(c\hat{\mathbf{r}})\cdot\mathbf{I}_n(c\hat{\mathbf{r}})$$

$$= c^{\ell-n-3}[-\ell(n+1)H_\ell(\hat{\mathbf{r}})\overline{H_n(\hat{\mathbf{r}})} + \boldsymbol{\nabla}_1 H_\ell(\hat{\mathbf{r}}) \cdot \boldsymbol{\nabla}_1\overline{H_n(\hat{\mathbf{r}})}]$$

$$c^{n+3-\ell}\langle \mathbf{E}_\ell(c\hat{\mathbf{r}}) \cdot \mathbf{I}_n(c\hat{\mathbf{r}}) \rangle_{\hat{\mathbf{r}}} = -\ell(n+1)\langle H_\ell\overline{H_n}\rangle + \langle \boldsymbol{\nabla}_1 H_\ell \cdot \boldsymbol{\nabla}_1\overline{H_n}\rangle.$$

But recall from (4.2.3) that $\langle \boldsymbol{\nabla}_1 f \cdot \boldsymbol{\nabla}_1 g \rangle = -\langle f\nabla_1^2 g\rangle$; so

$$\langle \boldsymbol{\nabla}_1 H_\ell \cdot \boldsymbol{\nabla}_1\overline{H_n}\rangle = -\langle H_\ell\nabla_1^2\overline{H_n}\rangle = n(n+1)\langle H_\ell\overline{H_n}\rangle.$$

Thus

$$c^{n-\ell+3}\langle \mathbf{E}_\ell(c\hat{\mathbf{r}}) \cdot \mathbf{I}_n(c\hat{\mathbf{r}})\rangle_{\hat{\mathbf{r}}} = (n+1)(n-\ell)\langle H_\ell \overline{H_n}\rangle. \qquad (4.3.15)$$

If $\ell = n$, obviously (4.3.15) gives 0. If $\ell \neq n$, then $\langle H_\ell \overline{H_n}\rangle = 0$, so (4.3.15) again gives 0. ◁

Corollary: For every $H_n \in \mathcal{H}_n$:

$$\int_M d^3\mathbf{r} \; \gamma(r)\mathbf{I}_n(\mathbf{r}) \cdot \nabla \; \overline{\beta_\ell^m(\mathbf{r})} = 0.$$

Proof: The integral is

$$\int_c^a dr \; r^2\gamma(r) \int_{S(1)} d^2\hat{\mathbf{r}} \; \mathbf{I}_n(r\,\hat{\mathbf{r}}) \cdot \nabla \; \overline{\beta_\ell^m(r\hat{\mathbf{r}})}.$$

Applying the previous theorem the surface integral vanishes for every r. ◁

Two interesting conclusions follow from this calculation. Runcorn (1975) notes that $\mathbf{B}$ near the surface of the moon seems to be less than 0.05 nT but that many samples of lunar rock have $\mathbf{M} \approx 10^{-1}$ A/m. If a spherical shell were uniformly magnetized to this amount, and produced such a small dipole field, its thickness could be no more than about 300 meters. Runcorn suggests that the moon once had a molten core that ran a dynamo, and that this magnetized the "mantle" of the moon as it cooled through its Curie point. As we have just seen, the mantle's magnetization will produce no external field, but moon rocks can be appreciably magnetized.

A second interesting conclusion is that if a magnetic field is generated in the core of the earth, its value outside the earth will be unaffected by the magnetic polarization it induces in the mantle, as long as γ, the magnetic susceptibility of the mantle, is much less than unity, and depends only on r. Most rocks have $\gamma < 0.05$ (Clark, 1966, p. 548), so γ is detectable in surface measurements of $\mathbf{B}$ only because $\nabla_1\gamma \neq \mathbf{0}$.

4.4 Measuring the Gauss Coefficients

In practice, $\mathbf{B}$ is not measured *everywhere* on $S(a)$, the surface of the earth, but at fewer than 200 magnetic observatories. Therefore the idealized scheme for obtaining the Gauss coefficients, described on the preceding pages, is unattainable in practice. Until very recently, those coefficients have been obtained by truncating the three series (4.2.9, 4.2.10, 4.2.11) at some largest degree ℓ^* and then choosing the g_ℓ^m and k_ℓ^m by a least squares fit of (4.2.10, 4.2.11) to the observatory data. Gauss originated this scheme, and with the small number of observatories available to him in the mid-nineteenth century, he took $\ell^* = 4$. He found that, within the accuracy of his data and his fit, $k_\ell^m = 0$. In other words, he quantitatively confirmed Gilbert's conclusion of 1600 that the sources of the earth's magnetic field were inside the earth. Modern measurements indicate that over most of the earth, except during magnetic storms, the annual mean field is internally produced except for perhaps 0.1%, which comes from magnetospheric currents. For the moment, we will ignore $\mathbf{B}^{(E)}$, and assume that $k_\ell^m = 0$. Therefore we assume $\mathbf{B} = -\nabla \psi$ if $r > a$, where

$$\psi(r\hat{\mathbf{r}}) = a \sum_{\ell=1}^{\infty} \left(\frac{a}{r} \right)^{\ell+1} \sum_{m=-\ell}^{\ell} g_\ell^m \beta_\ell^m(\hat{\mathbf{r}}). \qquad (4.4.1)$$

The magnetic observatories are few, and are very unevenly distributed. (See Langel, 1987, for a table of magnetic observatories and a historical overview.) There is a "hole" 10^4 km in diameter in the South Pacific where no data are available. Moreover, local crustal fields are often several percent of the total field, and can be much larger. These fields cannot be adequately treated with the values of ℓ^* used to reduce observatory data. Typically, no improvements in the fit to the observatories are obtained if ℓ^* increases beyond about 15, and $\ell^* = 10$ is often used. This corresponds to a minimum horizontal wavelength of $2\pi/10.5$ radians on the earth's surface $S(a)$. With $a = 6371$ km, that wavelength is about 3800 km, very much longer than some of the strongest sources of crustal magnetization. These short-wavelength signals are "modeled" by adjusting the g_ℓ^m to produce a best fit to the observatory data. In reality the short wavelengths are aliased into the long ones, which means the amplitudes of the low-degree spherical harmonics are inflated.

Both the problem of poor distribution of observatories and the problem that observatories sit on local magnetic anomalies (i.e., their measured $\mathbf{B}$ has an appreciable contribution from $\ell \gg \ell^*$) are solved by measuring $\mathbf{B}$ in a satellite in low-altitude polar orbit. The satellite covers the whole earth, and at high altitudes the effects of high ℓ are much attenuated, as seen from (4.4.1). To measure a **B** of 30,000 nT accurate to 6 nT in each component (MAGSAT's estimated accuracy) requires that the orientation of the satellite be known to 2×10^{-4} radians, or 40 seconds of arc. If orientation error is not to be a major error source, this should be reduced by a factor of 10. Such accurate orientation of a satellite is very expensive, so it was originally hoped (in the POGO and OGO satellites of $\sim$ 1970) that one could measure $|\mathbf{B}|$, which is a much easier measurement. The very high density of satellite measurements would then amount to knowing $|\mathbf{B}|$ on a spherical surface $S(b)$, where b is the radius of the satellite's (circular) orbit. Indeed, the satellite period in low orbit is about 90 minutes, and the earth rotates under the satellite, so after D days one has measured $|\mathbf{B}|$ along roughly equally spaced tracks $360/(16D)$ degrees apart. To achieve $1°$ spacing requires about 22 days.

4.4.1 Nonuniqueness of Fields Based on Total Field Observations

Unfortunately, it turns out that even if $|\mathbf{B}|$ is known exactly everywhere on $S(b)$, the Gauss coefficients in (4.4.1) may not be obtainable from those measurements. There exist pairs of magnetic fields $\mathbf{B}_1$ and $\mathbf{B}_2$ which satisfy $\boldsymbol{\nabla} \cdot \mathbf{B} = 0$ and $\boldsymbol{\nabla} \times \mathbf{B} = 0$ outside $S(a)$ and such that $|\mathbf{B}_1| = |\mathbf{B}_2|$ everywhere on $S(b)$, but $\mathbf{B}_1 \neq \pm\mathbf{B}_2$. These fields would have different g_ℓ^m but would give exactly the same measurement of total intensity in a satellite on $S(b)$. This problem led to the decision to measure MAGSAT's orientation and thus to measure **B**, not just $|\mathbf{B}|$, from the satellite. The details of the nonuniqueness proof for $|\mathbf{B}|$ are given by Backus (1970). He proves only that there are some fields which have the nonuniqueness "disease." This does not prove that all fields have this disease, or even that the earth's field does. Empirically, however, it has been observed that fitting MAGSAT intensity data $|\mathbf{B}|$ produces models whose **B** can differ from the MAGSAT-measured **B** by 1000 nT near the equator.

Table 4.4.1 shows the recent estimates of the Gauss coefficients from MAGSAT with $\ell^* = 13$. The mean radius of the earth is assumed to be 6371.2 km; mean epoch is 1979.85. The longitude $\lambda = 0$ is the Greenwich meridian. Recall 1 nT = 10^{-9} tesla = 10^{-5} gauss = 1 gamma; Table 4.4.2 describes how well the model fits the data.

4.4.2 The Spectrum

In order to use Table 4.4.1. one must know the conventions of notation

Table 4.4.1. The MGST(3/80) Field Model

ℓ	m	g_ℓ^m, nT	h_ℓ^m, nT	ℓ	m	g_ℓ^m, nT	h_ℓ^m, nT
1	0	-29990.1	0.0	9	8	1.5	-6.7
1	1	-1958.1	5609.5	9	9	-4.5	3.5
2	0	-1993.2	0.0	10	0	-3.1	0.0
2	1	3027.7	-2129.5	10	1	-3.6	1.3
2	2	1662.4	-192.3	10	2	2.4	0.6
3	0	1269.6	0.0	10	3	-5.2	2.5
3	1	-2180.1	-331.4	10	4	-2.0	5.5
3	2	1251.7	270.6	10	5	4.6	-4.3
3	3	833.3	-250.6	10	6	3.2	-0.1
4	0	937.2	0.0	10	7	0.6	-1.4
4	1	782.7	211.9	10	8	2.1	3.5
4	2	399.2	-257.6	10	9	3.5	-0.4
4	3	-419.8	51.3	10	10	-1.0	-6.5
4	4	198.7	-297.6	11	0	2.2	0.0
5	0	-213.7	0.0	11	1	-1.3	0.7
5	1	357.1	42.9	11	2	-1.8	1.9
5	2	261.0	148.8	11	3	2.2	-1.1
5	3	-73.3	-150.5	11	4	0.2	-2.7
5	4	-162.8	-78.8	11	5	-0.5	0.7
5	5	-46.5	90.6	11	6	-0.3	-0.1
6	0	48.9	0.0	11	7	1.9	-2.3
6	1	65.3	-14.4	11	8	1.9	0.0
6	2	41.1	93.8	11	9	-0.3	-1.6
6	3	-192.2	70.9	11	10	2.1	-0.8
6	4	3.9	-43.2	11	11	2.4	0.8
6	5	14.1	-2.7	12	0	-1.7	0.0
6	6	-107.8	17.2	12	1	0.6	0.9
7	0	70.6	0.0	12	2	-0.1	0.3
7	1	-58.6	-81.2	12	3	-0.1	2.2
7	2	1.9	-27.2	12	4	0.3	-1.5
7	3	20.5	-5.1	12	5	0.6	0.6
7	4	-12.3	16.4	12	6	-0.6	0.3
7	5	-0.2	18.6	12	7	-0.4	-0.3
7	6	10.9	-22.9	12	8	-0.0	0.3
7	7	-2.1	-9.2	12	9	-0.7	-0.3
8	8	18.1	0.0	12	10	0.0	-1.5
8	1	6.7	6.7	12	11	0.1	-0.5
8	2	-0.2	-17.8	12	12	1.4	-0.3
8	3	-10.8	4.2	13	0	0.1	0.0
8	4	-7.1	-22.1	13	1	-0.4	-0.2
8	5	4.4	9.3	13	2	0.3	0.4
8	6	2.4	16.0	13	3	-0.5	1.7
8	7	6.0	-13.1	13	4	-0.1	0.1
8	8	-1.8	-15.8	13	5	1.2	-0.6
9	0	5.7	0.0	13	6	-0.5	-0.1
9	1	10.1	-21.3	13	7	0.1	1.0
9	2	1.0	15.0	13	8	-0.4	0.1
9	3	-12.8	9.0	13	9	-0.0	0.9
9	4	9.3	-4.9	13	10	0.0	0.4
9	5	-3.1	-6.7	13	11	-0.1	-0.4
9	6	-1.3	8.7	13	12	0.7	-1.1
9	7	6.6	9.5	13	13	0.4	0.5

Table 4.4.2. Residuals of Selected MAGSAT Data to Some Published
Field Models. Units are nT

	IGRF 1975	AWC/75	IGS/75	POGO (2/(72)	MGST (3/80)
Scalar:					
mean deviation	-90	61	23	9	0
standard deviation	125	127	120	107	8
B_r:					
mean deviation	29	46	40	25	21
standard deviation	204	153	137	211	44
B_θ:					
mean deviation	44	-10	8	12	12
standard deviation	146	115	114	145	107
B_ϕ:					
mean deviation	62	62	61	61	62
standard deviation	181	157	155	208	129

Initial geomagnetic field model from Langel, et al. (1980).

in the modern geomagnetic community. The spherical harmonics in (4.4.1) are not normalized so that $\langle|\beta_\ell^m|^2\rangle = 1$, but so that $\langle|\beta_\ell^m|^2\rangle = 1/(2\ell + 1)$. Such harmonics are called Schmidt quasi-normalized or simply Schmidt-normalized; we encountered them in section 3.7 during the discussion of recurrence relationships. In this chapter they are written with a breve (˘), as will their corresponding Gauss coefficients. The Schmidt-normalized harmonics corresponding to fully normalized β_ℓ^m in (4.4.1) are written $\breve{\beta}_\ell^m$. They are

$$\breve{\beta}_\ell^m(\mathbf{r}) = (2\ell + 1)^{-\frac{1}{2}} \beta_\ell^m(\mathbf{r}), \qquad (4.4.2)$$

and the Gauss coefficients for them are

$$\breve{g}_\ell^m = (2\ell + 1)^{1/2} g_\ell^m, \qquad (4.4.3)$$

so that (4.4.1) becomes

$$\psi(r\hat{\mathbf{r}}) = a \sum_{\ell=1}^{\infty} \left(\frac{a}{r}\right)^{\ell+1} \sum_{m=-\ell}^{\ell} \breve{g}_\ell^m \breve{\beta}_\ell^m(\hat{\mathbf{r}}). \qquad (4.4.4)$$

The orthogonality conditions are now

$$\langle \breve{\beta}_\ell^m \breve{\beta}_{\ell'}^{m'} \rangle = \delta_{\ell\ell'}\delta_{mm'}/(2\ell+1). \tag{4.4.5}$$

The corresponding Spherical Harmonic Addition Theorem becomes

$$P_\ell(\hat{\mathbf{r}} \cdot \hat{\mathbf{s}}) = \sum_{m=-\ell}^{\ell} \breve{\beta}_\ell^m(\hat{\mathbf{r}})\overline{\breve{\beta}_\ell^m(\hat{\mathbf{r}})} \tag{4.4.6}$$

and (3.5.25), Green's function for ∇^2, is

$$\frac{1}{|\mathbf{r}-\mathbf{s}|} = \frac{1}{r}\sum_{\ell=0}^{\infty}\left(\frac{s}{r}\right)^\ell \sum_{m=-\ell}^{\ell} \breve{\beta}_\ell^m(\hat{\mathbf{r}})\overline{\breve{\beta}_\ell^m(\hat{\mathbf{s}})} \qquad s < r. \tag{4.4.7}$$

We will call $\{\breve{\beta}_\ell^{-\ell}, \ldots, \breve{\beta}_\ell^\ell\}$ a Schmidt quasi-normalized basis for $\mathcal{H}_\ell$.

The form of the particular Schmidt basis used in geomagnetism has $\hat{\mathbf{z}}$ as the north polar axis of rotation of the earth, and $\hat{\mathbf{x}}$ is on the Greenwich meridian, so λ is east longitude. The conventional basis consists of real functions only, which we may identify with the real and imaginary parts of $\breve{Y}_\ell^m$, where

$$\breve{Y}_\ell^m = Y_\ell^m/\sqrt{2\ell+1} \tag{4.4.8}$$

where the natural, fully normalized basis functions $Y_\ell^m(\hat{\mathbf{r}})$ are the familiar functions discussed at great length in Chapter 3, and defined, for example, in (3.4.54, 3.4.55). Specifically, the conventional Schmidt basis, $\{\breve{U}_\ell^{-\ell}, \ldots, \breve{U}_\ell^\ell\}$, is defined by

$$\breve{U}_\ell^m(\mathbf{r}) = r^\ell(\cos m\lambda)\breve{P}_\ell^m(\cos\theta) \quad \text{if} \quad 1 \le m \le \ell$$

$$\breve{U}_\ell^m(\mathbf{r}) = r^\ell \breve{P}_\ell^0(\cos\theta) \qquad\qquad \text{if} \quad m = 0$$

$$\breve{U}_\ell^{-m}(\mathbf{r}) = r^\ell(\sin m\lambda)\breve{P}_\ell^m(\cos\theta) \quad \text{if} \quad 1 \le m \le \ell$$

where

$$
\left.
\begin{aligned}
\breve{P}_\ell^m(\mu) &= \sqrt{2}\sqrt{\frac{(\ell-m)!}{(\ell+m)!}}\, P_\ell^m(\mu) \quad \text{if} \quad 1 \le m \le \ell \\[2ex]
&= P_\ell(\mu) \qquad\qquad\qquad\qquad \text{if} \quad m = 0
\end{aligned}
\right\}. \qquad (4.4.9)
$$

The Gauss coefficients of the geomagnetic field $\mathbf{B}$ relative to the Schmidt-normalized basis are $\breve{g}_\ell^m$, so $\mathbf{B} = -\boldsymbol{\nabla}\psi$ where

$$
\psi(r\hat{\mathbf{r}}) = a \sum_{\ell=1}^{\infty} \left(\frac{a}{r}\right)^{\ell+1} \sum_{m=-\ell}^{\ell} \breve{g}_\ell^m \breve{U}_\ell^m(\hat{\mathbf{r}}). \qquad (4.4.10)
$$

In the geomagnetic literature, our $\breve{g}_\ell^m$ is written g_ℓ^m if $m \le 0$, and our $\breve{g}_\ell^{-m}$ is written h_ℓ^m if $m > 0$. Thus (4.4.10) is usually written (without the breve on the associated Legendre function!):

$$
\psi(r\hat{\mathbf{r}}) = a \sum_{\ell=1}^{\infty} \left(\frac{a}{r}\right)^{\ell+1} \sum_{m=0}^{\ell} [g_\ell^m \cos m\lambda + h_\ell^m \sin m\lambda] \breve{P}_\ell^m(\cos\theta). \qquad (4.4.11)
$$

For $m = 0$, the $\sin m\lambda$ term vanishes, so h_ℓ^0 is irrelevant. We call it zero.

The authors find it difficult to do theory with (4.4.11), so we usually convert it back to the form (4.4.10) using

$$
\left.
\begin{aligned}
\breve{g}_\ell^m &= \text{table entry for } g_\ell^m \text{ if } m \ge 0 \\[2ex]
\breve{g}_\ell^{-m} &= \text{table entry for } h_\ell^m \text{ if } m > 0
\end{aligned}
\right\}. \qquad (4.4.12)
$$

The table of Gauss coefficients (Table 4.4.1) can be put to a number of interesting uses. To discuss them, it is helpful to recall from section 4.1 the ℓth degree part of a harmonic function in $S(a,c)$. For any fixed

r, $\psi(r\hat{\mathbf{r}})$ describes a function on $S(1)$, and its ℓth degree part, $\psi_\ell(r\hat{\mathbf{r}})$, is given by

$$\psi_\ell(r\hat{\mathbf{r}}) = \sum_{m=-\ell}^{\ell} \psi_\ell^m(r)\beta_\ell^m(\hat{\mathbf{r}}) \qquad (4.4.13)$$

where $\{\beta_\ell^{-\ell},\ldots,\beta_\ell^\ell\}$ is any orthonormal basis for $\mathcal{H}_\ell$ and $\psi_\ell^m(r) = \langle\psi(r\hat{\mathbf{r}})\beta_\ell^m(\hat{\mathbf{r}})\rangle_{\hat{\mathbf{r}}}$. Recall from (4.1.7) that in fact the ℓth degree part does not depend on the choice of basis. From (4.4.10) we see

$$\psi_\ell(r\hat{\mathbf{r}}) = a\left(\frac{a}{r}\right)^{\ell+1} \sum_{m=-\ell}^{\ell} \breve{g}_\ell^m \breve{U}_\ell^m(\hat{\mathbf{r}}) \qquad (4.4.14)$$

$$= a\left(\frac{a}{r}\right)^{\ell+1} \sum_{m=0}^{\ell} [g_\ell^m \cos m\lambda + h_\ell^m \sin m\lambda]\breve{P}_\ell^m(\cos\theta). \quad (4.4.15)$$

Then clearly from (4.4.10, 4.4.14):

$$\psi = \sum_{\ell=1}^{\infty} \psi_\ell. \qquad (4.4.16)$$

As an application of Table 4.4.1, we can find $\mathbf{m}$, the vector dipole moment of the earth. We saw in section 4.3 that the scalar magnetic potential of a dipole with moment $\mathbf{m}$ is

$$\psi = -\frac{\mu_0}{4\pi}\mathbf{m}\cdot\boldsymbol{\nabla}\frac{1}{r}.$$

Now

$$\boldsymbol{\nabla}r = \hat{\mathbf{r}} \quad \text{so} \quad \boldsymbol{\nabla}\frac{1}{r} = -\frac{\hat{\mathbf{r}}}{r^2},$$

and so

$$\psi = \frac{\mu_0}{4\pi}\frac{\mathbf{m}\cdot\hat{\mathbf{r}}}{r^2} = \frac{\mu_0}{4\pi}\frac{\mathbf{m}\cdot\mathbf{r}}{r^3}.$$

Thus the potential from a point dipole falls off like $1/r^2$. Since from (4.4.14, 4.4.16) ψ_ℓ is proportional to $r^{-\ell-1}$, we see that

$$\psi_1 = -\frac{\mu_0}{4\pi}\mathbf{m} \cdot \boldsymbol{\nabla}\frac{1}{r}.$$

But by (4.4.15)

$$\psi_1 = \frac{a^3}{r^3}r[g_1^0\breve{P}_1^0(\cos\theta) + (g_1^1\cos\lambda + h_1^1\sin\lambda)\breve{P}_1^1(\cos\theta)]$$

and by (4.4.9)

$$\breve{P}_1^0(\cos\theta) = \cos\theta \quad \text{and} \quad \breve{P}_1^1(\cos\theta) = \sin\theta.$$

Therefore, identifying x, y, and z from their spherical polar equivalents

$$\psi_1 = \frac{a^3}{r^3}[g_1^0 z + g_1^1 x + h_1^1 y],$$

so

$$\frac{\mu_0}{4\pi}\mathbf{m} \cdot \mathbf{r} = a^3[g_1^1\hat{\mathbf{x}} + h_1^1\hat{\mathbf{y}} + g_1^0\hat{\mathbf{z}}] \cdot \mathbf{r}.$$

Therefore

$$\mathbf{m} = \frac{4\pi a^3}{\mu_0}[g_1^1\hat{\mathbf{x}} + h_1^1\hat{\mathbf{y}} + g_1^0\hat{\mathbf{z}}]. \tag{4.4.17}$$

Let us now refer to the table for numerical values of the degree-1 Gauss coefficients. Since $g_1^0 < 0$ but $|g_1^0| \gg |h_1^1|$ and $|g_1^1|$, $\mathbf{m}$ points nearly along the earth's axis of rotation, but from north to south. The dipole magnitude is $|\mathbf{m}| = 7.906 \times 10^{22}$ amp meter^2, and the dipole axis passing through $\mathbf{0}$ intersects the earth's surface at colatitude $11°.207$, west longitude $70°.76$.

It is difficult to digest all the information contained in Table 4.4.1. One partial summary describes how the field is apportioned among the various horizontal wavelengths. Recall from section 3.6.3 that surface harmonics of degree ℓ have horizontal wavelength of roughly

$2\pi a/(\ell+\frac{1}{2})$, so that this amounts to examining how much of the field belongs to each ℓ. Since fields of high ℓ decrease very rapidly with r, the answer will depend on which $S(r)$ we examine. One numerical measure of how much of the field on $S(r)$ belongs to harmonics of degree ℓ is simply the average over the sphere $S(r)$ of the square of ψ_ℓ:

$$\langle|\psi_\ell(r\hat{\mathbf{r}})|^2\rangle_{\hat{\mathbf{r}}} = a^2 \left(\frac{a}{r}\right)^{2\ell+2} \left\langle \left|\sum_{m=-\ell}^{\ell} \breve{g}_\ell^m \breve{U}_\ell^m(\hat{\mathbf{r}})\right|^2 \right\rangle$$

$$= a^2 \left(\frac{a}{r}\right)^{2\ell+2} \frac{1}{(2\ell+1)} \sum_{m=-\ell}^{\ell} |\breve{g}_\ell^m|^2. \qquad (4.4.18)$$

Equation (4.4.18) shows that $\sum_{m=-\ell}^{\ell} |\breve{g}_\ell^m|^2$ is independent of the Cartesian basis used to compute it, but otherwise (4.4.18) is of little interest because $\langle|\psi_\ell|^2\rangle_{S(r)}$ has no easy physical interpretation. A more promising quantity to study is $\langle|\mathbf{B}_\ell|^2\rangle_{S(r)}$, where we define the ℓth degree part of $\mathbf{B}$ by

$$\mathbf{B}_\ell = -\boldsymbol{\nabla}\psi_\ell = -\left(\hat{\mathbf{r}}\partial_r\psi_\ell + \frac{1}{r}\boldsymbol{\nabla}_1\psi_\ell\right)$$

$$= \frac{1}{r}[\hat{\mathbf{r}}(\ell+1)\psi_\ell - \boldsymbol{\nabla}_1\psi_\ell]. \qquad (4.4.19)$$

First we show that two different degree parts of a field are in a sense orthogonal with respect to averaging over spheres of constant radius. We note that if $\mathbf{B}$ is real, so is $\mathbf{B}_\ell$, because ψ is real, and then ψ_ℓ is real because of (4.4.13). Therefore we can compute $\langle\mathbf{B}_\ell\cdot\mathbf{B}_n\rangle_{S(r)}$ as $\langle\overline{\mathbf{B}_\ell}\cdot\mathbf{B}_n\rangle_{S(r)}$. From (4.4.19)

$$\overline{\mathbf{B}_\ell}\cdot\mathbf{B}_n = \frac{1}{r^2}(\ell+1)(n+1)\overline{\psi_\ell}\psi_n + \frac{1}{r^2}\boldsymbol{\nabla}_1\overline{\psi_\ell}\cdot\boldsymbol{\nabla}_1\psi_n,$$

so, averaging over $S(r)$,

$$\langle \overline{\mathbf{B}_\ell \cdot \mathbf{B}_n} \rangle_{S(r)} = \frac{1}{r^2}(\ell + 1)(n + 1)\langle \overline{\psi_\ell(r\hat{\mathbf{r}})}\psi_n(r\hat{\mathbf{r}})\rangle_{\hat{\mathbf{r}}}$$

$$+ \frac{1}{r^2}\langle \boldsymbol{\nabla}_1\overline{\psi_\ell(r\hat{\mathbf{r}})} \cdot \boldsymbol{\nabla}_1\psi_n(r\hat{\mathbf{r}})\rangle_{\hat{\mathbf{r}}}.$$

But

$$\langle \boldsymbol{\nabla}_1\overline{\psi_\ell} \cdot \boldsymbol{\nabla}_1\psi_n \rangle_{(S(r))} = -\langle \overline{\psi_\ell}\nabla_1^2\psi_n \rangle_{S(r)} \quad \text{and} \quad \nabla_1^2\psi_n = -n(n+1)\psi_n.$$

Thus, finally,

$$\langle \overline{\mathbf{B}_\ell \cdot \mathbf{B}_n} \rangle_{S(r)} = \frac{1}{r^2}(n+1)(\ell+n+1)\langle \overline{\psi_\ell}\psi_n \rangle_{S(r)}. \tag{4.4.20}$$

From (4.4.14), clearly $\langle \overline{\psi_\ell}\psi_n \rangle_{S(r)} = 0$ if $\ell \neq n$, so from (4.4.19),

$$\langle \overline{\mathbf{B}_\ell \cdot \mathbf{B}_n} \rangle_{S(r)} = \left(\frac{a}{r}\right)^{2\ell+4}(\ell+1)\left(\sum_{m=-\ell}^{\ell}|\breve{g}_\ell^m|^2\right)\delta_{\ell n}. \tag{4.4.21}$$

Thus the average over $S(r)$ of the dot product of two different harmonic degree parts of the same magnetic field vanishes.

We have already identified ψ_1 with the potential of a dipole at the center of the earth, and for obvious reasons the corresponding magnetic field $\mathbf{B}_1$ is called the *dipole part* of the geomagnetic field. Furthermore, the potential $\psi - \psi_1$ and its magnetic field are referred to as the *nondipole* parts of the geomagnetic potential and field. The radial component of this partial field is shown for the model of Table 4.4.1 in Figure 1.3.2.

The quantity

$$R_\ell(r) = \langle |\mathbf{B}_\ell|^2\rangle_{S(r)} = \left(\frac{a}{r}\right)^{2\ell+4}(\ell+1)\sum_{m=0}^{\ell}[(g_\ell^m)^2 + (h_\ell^m)^2] \tag{4.4.22}$$

was introduced by Mauersberger, 147 (1956) and independently by Luecke (1957); an important study by Lowes (1974) brought R_ℓ into prominence. The set of values of $R_\ell(r)$ for $\ell = 1, 2, 3, \ldots$ at a fixed radius r is sometimes called the *Mauersberger–Lowes spectrum* on $S(r)$. From (4.4.16) it is obvious that

$$\mathbf{B} = \sum_{\ell=1}^{\infty} \mathbf{B}_\ell. \tag{4.4.23}$$

If we average the squared magnitude of the magnetic field over $S(r)$, we find in view of the orthogonality implied by (4.4.21) that

$$\langle |\mathbf{B}|^2 \rangle_{S(r)} = \sum_{\ell=1}^{\infty} \langle |\mathbf{B}_\ell|^2 \rangle_{S(r)} = \sum_{\ell=1}^{\infty} R_\ell(r). \tag{4.4.24}$$

Thus the average squared field over any sphere is the sum over the spectrum.

Figure 4.4.1 shows a plot of the spectrum at the earth's surface taken from a fit to the MAGSAT data with $\ell^* = 23$ given by Langel and Estes (1982). The usual interpretation of the spectrum is that $\mathbf{B}_\ell(a\hat{\mathbf{r}})$ comes from the core if $1 \le \ell \le 12$, from the crust if $16 \le \ell$, and from both if $\ell = 13, 14, 15$. The argument for this interpretation is as follows:

If the sources of $\mathbf{B}$ all lie inside $S(r)$, then (4.4.1) converges and, from (4.4.22),

$$R_\ell(r) = R_\ell(a) \left(\frac{a}{r}\right)^{2\ell+4}. \tag{4.4.25}$$

From (4.4.24),

$$\langle |\mathbf{B}|^2 \rangle_{S(r)} = \left(\frac{a}{r}\right)^4 \sum_{\ell=1}^{\infty} R_\ell(a) \left(\frac{a}{r}\right)^{2\ell}. \tag{4.4.26}$$

The series on the right of (4.4.26) converges as long as all the sources lie inside $S(r)$. The individual terms in the series are then certainly

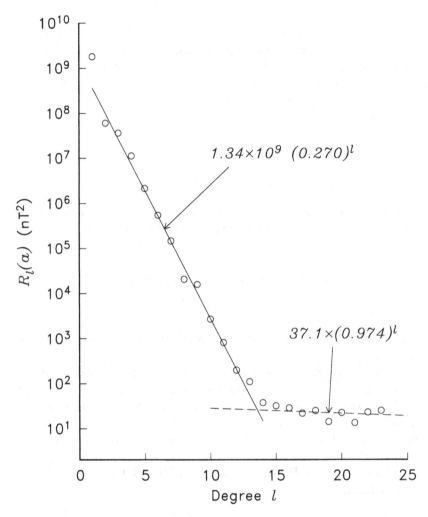

Figure 4.4.1: Mauersberger–Lowes spectrum on $S(a)$ after Langel and Estes (1982).

bounded, so there is a K such that $R_\ell(a)(a/r)^{2\ell} \le K$, or

$$R_\ell(a) \le K(r/a)^{2\ell}. \qquad (4.4.27)$$

If we could actually observe the $R_\ell(a)$ for all ℓ, we could use (4.4.27) to argue that if the sources are inside $S(r)$, there is a constant K such that

$$\ln R_\ell(a) + 2\ell\ln(a/r) \le K \qquad (4.4.28)$$

for all ℓ. Therefore, if we choose an r so small that the left side of (4.4.28) $\to \infty$ as $\ell \to \infty$, we can be sure that there are sources outside $S(r)$.

This is the only rigorous conclusion contained in (4.4.27). The converse is false. It is perfectly possible to have (4.4.28) valid for all ℓ, and yet have sources outside $S(r)$. For example, a uniformly magnetized sphere could have any radius $< a$, and (4.4.28) would be valid for all $r > 0$ because $R_\ell(a) = 0$ if $\ell > 1$.

And, of course, we cannot observe $R_\ell(a)$ for all ℓ. Figure 4.4.1, in fact, goes only to $\ell^* = 23$. Nevertheless, the linear dependence of $\ln R_\ell(a)$ on ℓ in that figure may be more than a coincidence. The simplest interpretation of Figure 4.4.1 is that if $S(r)$ is the smallest sphere containing all the sources, then for that value of r, the $R_\ell(r) = \langle |\mathbf{B}_\ell|^2 \rangle_{S(r)}$ are not very dependent on ℓ (the "white noise source hypothesis"). If we take $R_\ell(r) =$ constant then we have

$$R_\ell(a) = (r/a)^4 R_\ell(r)(r/a)^{2\ell}. \tag{4.4.29}$$

In Figure 4.4.1, the curve drawn for the surface of the earth can be written

$$R_\ell(a) = \left(\frac{r_1}{a}\right)^4 R_\ell^{(1)}(r_1) \left(\frac{r_1}{a}\right)^{2\ell} + \left(\frac{r_2}{a}\right)^4 R_\ell^{(2)}(r_2) \left(\frac{r_2}{a}\right)^{2\ell}. \tag{4.4.30}$$

This suggests that there are two statistically independent sources, one at radius r_1 and one at radius r_2. For $\ell \leq 12$, the source inside $S(r_1)$ dominates the data, while for $\ell \geq 15$ the source, which extends up to $S(r_2)$, dominates the data. Using the parameters of best fit from Figure 4.4.1, we obtain

$$r_1 = 3310 \text{ km}$$

$$\langle |\mathbf{B}_\ell^{(1)}|^2 \rangle_{S(r_1)}^{1/2} = R_\ell^{(1)}(r_1)^{1/2} = 136 \ \mu\text{T}$$

$$r_2 = 6288 \text{ km} \ (= 6371 - 83)$$

$$\langle |\mathbf{B}_\ell^{(2)}|^2 \rangle_{S(r_2)}^{1/2} = R_\ell^{(2)}(r_2)^{1/2} = 6.3 \text{ nT}.$$

The seismically observed core radius is 3486 km. Clearly we cannot use magnetic data to measure this radius accurately, but r_1 is close enough to 3486 km to suggest that the core contributes the $\mathbf{B}_\ell$ with $1 \leq \ell \leq 12$.

A glance at Figure 4.4.1 shows that the straight-line fit above $\ell = 15$ is not too good. The curve by Cain et al. (1989b) is better determined because it has more data, but it appears to be swamped by noises of measurement (satellite position and orientation errors, instrument errors) above about $\ell = 45$. For $45 \leq \ell \leq 66$, $R_\ell(r_3)$ is nearly independent of ℓ if $r_3 = a+$ satellite altitude (~ 6371 km $+$ 400 km). In any case, one would be ill advised to use these data to put the depth to the crustal magnetic sources at 10 km. However, the depth is clearly very shallow, so above $\ell \geq 15$ it is probably the crust that produces $\mathbf{B}_\ell$.

4.4.3 Crustal Signals

Implicit in the foregoing discussion is that the part of (4.4.30) given by r_2 (the crust) continues for $\ell \leq 12$, where it is so much smaller than the core field on $S(a)$, the earth's surface, as to be invisible. Similarly, the core field is presumed to be described by the r_1 part of (4.4.30) even for $\ell \geq 15$, where it is dominated on $S(a)$ by the crustal field. There is no observational evidence for either hypothesis beyond the data shown in Figure 4.4.1, the more complete data of Cain et al. (1989b), and a paper by Meyer et al. (1983), which models the magnetic susceptibility of the crust by assigning appropriate susceptibilities to known geological provinces and assuming all crustal magnetization to be induced. They find r_2 is the radius of earth and
$$\langle |\mathbf{B}_\ell^{(2)}|^2 \rangle_{S(r_2)}^{1/2} = 3\text{nT}.$$

Although the crustal terms $\langle |\mathbf{B}_\ell^{(2)}|^2 \rangle_{S(a)}$ at the earth's surface are quite small for each separate ℓ, they die away so slowly with ℓ that the sum (4.4.24) is quite large on $S(a)$. On $S(r)$ it is

$$\langle |\mathbf{B}^{(2)}|^2 \rangle_{S(r)} = \sum_{\ell=1}^{\infty} R_\ell^{(2)}(r) = \sum_{\ell=1}^{\infty} R_\ell^{(2)}(r_2) \left(\frac{r_2}{r}\right)^{2\ell+4}.$$

If we take seriously the suggestion that $R_\ell^{(2)}(r_2)$ is approximately independent of ℓ, then we can take it outside and sum the geometric series, obtaining

$$\langle|\mathbf{B}^{(2)}|^2\rangle_{S(r)} = R_\ell^2(r_2) \left(\frac{r_2}{r}\right)^6 \left[1 - \left(\frac{r_2}{r}\right)^2\right]^{-1}. \qquad (4.4.31)$$

The mean crustal field at the earth's surface, $\langle|\mathbf{B}^{(2)}|^2\rangle_{S(a)}^{1/2}$, is very dependent on the very small difference between r_2 and a, so the value of $\langle|\mathbf{B}^{(2)}|^2\rangle_{S(a)}^{1/2}$ obtained from Figure 4.4.1 or Cain et al. (1989b) should not be taken too seriously. For what they are worth, these values are

$$\langle|\mathbf{B}^{(2)}|^2\rangle_{S(a)}^{1/2} = 78 \text{ nT} \quad \text{(Cain et al., 1989b)}$$

$$\langle|\mathbf{B}^{(2)}|^2\rangle_{S(a)}^{1/2} = 37 \text{ nT} \quad \text{(Langel and Estes, 1982).}$$

$$(4.4.32)$$

At a satellite altitude of 400 km, $\langle|\mathbf{B}_\ell^{(2)}|^2\rangle_{(S(r)}$ dies away more rapidly with ℓ, the sums are smaller, and the agreement is better. For $r = a + 400$ km, we find

$$\langle|\mathbf{B}^{(2)}|^2\rangle_{S(r)}^{1/2} = 12 \text{ nT} \quad \text{(Cain et al., 1989b)}$$

$$\langle|\mathbf{B}^{(2)}|^2\rangle_{S(r)}^{1/2} = 15 \text{ nT} \quad \text{(Langel and Estes, 1982).}$$

$$(4.4.33)$$

The values (4.4.32) are only root-mean-squares. Individual local variations can be much larger. For example, there is the Bangui anomaly in Africa, a magnetic signal remaining after the field model up to $\ell^* \approx 15$ is subtracted (this defines an *anomaly*). It is about 500 nT in a region 700 km by 250 km (Regan and Marsh, 1982). Much larger anomalies on smaller spatial scales occur near iron ore bodies. For example, an anomaly over a pyrite-mineralized dike in Quebec was 10,000 nT in about 25 meters of horizontal extent (Telford et al., 1976, p. 193). Anomalies with such short length scales would be invisible from a satellite.

The foregoing discussion quantifies the problem of using observations to see the core's magnetic field. The contribution to **B** from high ℓ, due to local crustal anomalies, can be an appreciable fraction of **B**; if we try to model it with $\ell^* \leq 15$ (say), we will be attributing crustal signals to the core. These signals will appear to have ℓ not much smaller than ℓ^*, because they are uncorrelated from observatory to observatory, and when they are extrapolated down to the core–mantle boundary they will be amplified by the factor $(a/c)^{\ell+2}$ where $c =$ core radius. This factor is $(1.83)^{\ell+2}$. For $\ell = 10$, a signal of 10 nT at $r = a$ becomes a signal of 1.4×10^4 nT at the core's surface.

Langel et al. (1982) have suggested an ingenious solution to the observatory problem. A simplified description of their procedure is this. MAGSAT was up for about six months in 1979–1980. They used the MAGSAT data (altitude averaging ≈ 350 km) to find the Gauss coefficients up to $\ell^* = 13$. Then they calculated what **B** this truncated model would produce at each observatory. The difference between **B** observed by that observatory and **B** predicted by the satellite data is the sum of all the harmonics above $\ell = 13$ at the observatory, and represents the station correction at the observatory. The time changes in **B** are presumably due to motion in the fluid core, and would not be visible at $\ell \geq 15$, so the observatory station corrections are probably approximately independent of time. They can be used to correct the observatory data at earlier times before magnetic satellites or at later times when there are no magnetic satellites in orbit, as, for example, at the time of writing this book. MAGSAT was the last one until 1997, probably. Thus, better models of the core field can be constructed from the satellite data. The station corrections observed by Langel et al. (1982) were typically of the order of 100 nT, but some outliers were as large as 4000 nT (Scott Base and the South Pole).

The foregoing discussion suggests that we will not be able to observe the Gauss coefficients of the core field for $\ell \geq 15$ even with satellite data. Shure et al. (1985) have evidence that $\ell = 10$ may be the highest degree at which we can see the core's Gauss coefficients. On the basis of (4.4.33), they assume that the rms crustal field at MAGSAT altitudes is about 10 nT. They regard this as a random noise signal superposed on the core field, so they try to fit the satellite data with an rms error no more than 10 nT, with a core field that is as small as possible at the surface of the core. More precisely, they seek those Gauss coefficients g_ℓ^m (no ℓ^*, since ℓ goes to ∞) that will minimize the rms value of B_r on the core–mantle boundary and still fit the satellite data to 10 nT rms. They find serious deviations from the Gauss coefficients obtained by ordinary least squares fitting of a truncated series to satellite data, and these deviations are present for all $\ell > 10$.

4.4.4 Inferences about the Field on the Core: Averaging Kernels

The preceding discussion suggests that we will never know the Gauss coefficients of the core field for degree greater than, say, 12. If we know only the Gauss coefficients of the core up to $\ell = L$, what can we say about **B** at the surface of the core? To answer this question we study the way in which truncation of a spherical harmonic expansion distorts the representation of a function on a sphere. For simplicity, we concentrate on the radial component of the magnetic field, B_r.

Let c be the radius of the core. Since we are dealing with internal sources, (4.2.10) gives (in terms of an orthonormal basis $\{\beta_\ell^{-\ell}, \ldots, \beta_\ell^\ell\}$ for each $\mathcal{H}_\ell$)

$$B_r(c\hat{\mathbf{r}}) = \sum_{\ell=1}^{\infty} \sum_{m=-\ell}^{\ell} (\ell+1) \left(\frac{a}{c}\right)^{\ell+2} g_\ell^m \beta_\ell^m(\hat{\mathbf{r}}). \qquad (4.4.34)$$

If we write

$$f_\ell^m = (\ell+1) \left(\frac{a}{c}\right)^{\ell+2} g_\ell^m \qquad (4.4.35)$$

and $B_r(c\hat{\mathbf{r}}) = f(\hat{\mathbf{r}})$, then

$$f(\hat{\mathbf{r}}) = \sum_{\ell=0}^{\infty} \sum_{m=-\ell}^{\ell} f_\ell^m \beta_\ell^m(\hat{\mathbf{r}}). \qquad (4.4.36)$$

Thus the question becomes, What can we say about the function f on $S(1)$ if we know only its harmonic coefficients

$$f_\ell^m = \langle \overline{\beta_\ell^m} f \rangle_{S(1)} \qquad (4.4.37)$$

for $0 \le \ell \le L$? (The fact that in our magnetic problem f_0^0 is known to be 0 is irrelevant to the general question.)

If w_ℓ^m are any complex constants, we can calculate

$$\tilde{f}_L(\hat{\mathbf{r}}) = \sum_{\ell=0}^{L} \sum_{m=-\ell}^{\ell} w_\ell^m f_\ell^m \beta_\ell^m(\hat{\mathbf{r}}). \tag{4.4.38}$$

How can we best choose these arbitrary constants w_ℓ^m so that $\tilde{f}_L$ will give us a good approximate picture of f? One way is to choose them to minimize $\langle |f - \tilde{f}_L|^2 \rangle_{S(1)}$. That is, we can choose the w_ℓ^m in (4.4.38) so that $\tilde{f}_L$ is the best possible approximation to f in the rms sense. We have

$$f - \tilde{f}_L = \sum_{\ell=0}^{L} \sum_{m=-\ell}^{\ell} f_\ell^m (1 - w_\ell^m) \beta_\ell^m + \sum_{\ell=L+1}^{\infty} \sum_{m=-\ell}^{\ell} f_\ell^m \beta_\ell^m,$$

so, averaging the square over $S(1)$,

$$\langle |f - \tilde{f}_L|^2 \rangle_{S(1)} = \sum_{\ell=0}^{L} \sum_{m=-\ell}^{\ell} |f_\ell^m|^2 |1 - w_\ell^m|^2 + \sum_{\ell=L+1}^{\infty} \sum_{m=-\ell}^{\ell} |f_\ell^m|^2.$$

Clearly, the best-fitting $\tilde{f}_L$ has

$$w_\ell^m = 1. \tag{4.4.39}$$

That is, among all $\tilde{f}_L$ of the form (4.4.38), which we can calculate given only f_ℓ^m for $0 \le \ell \le L$, $|m| \le \ell$, the $\tilde{f}_L$ that best fits f in the least squares sense is simply the partial sum

$$\tilde{f}_L(\hat{\mathbf{r}}) = \sum_{\ell=0}^{L} \sum_{m=-\ell}^{\ell} f_\ell^m \beta_\ell^m(\hat{\mathbf{r}}). \tag{4.4.40}$$

Why would we choose any other weight w_ℓ^m? The manner in which $\tilde{f}_L$ approximates f can be understood in another way. Using (4.4.37), we can write (4.4.38) as

$$\tilde{f}_L(\hat{\mathbf{r}}) = \sum_{\ell=0}^{L} \sum_{m=-\ell}^{\ell} w_\ell^m \beta_\ell^m(\hat{\mathbf{r}}) \frac{1}{4\pi} \int_{S(1)} d^2\hat{\mathbf{s}} \, \overline{\beta_\ell^m(\hat{\mathbf{s}})} f(\hat{\mathbf{s}})$$

or

$$\tilde{f}_L(\hat{\mathbf{r}}) = \frac{1}{4\pi} \int_{S(1)} d^2\hat{\mathbf{s}} \, W_L(\hat{\mathbf{r}}, \hat{\mathbf{s}}) f(\hat{\mathbf{s}}) \qquad (4.4.41)$$

where

$$W_L(\hat{\mathbf{r}}, \hat{\mathbf{s}}) = \sum_{\ell=0}^{L} \sum_{m=-\ell}^{\ell} w_\ell^m \beta_\ell^m(\hat{\mathbf{r}}) \overline{\beta_\ell^m(\hat{\mathbf{s}})}. \qquad (4.4.42)$$

Suppose we always choose

$$w_0^0 = 1. \qquad (4.4.43)$$

Since $\beta_0^0(\hat{\mathbf{r}}) = \beta_0^0(\hat{\mathbf{s}})$,

$$\langle \beta_0^0(\hat{\mathbf{r}}) \overline{\beta_0^0(\hat{\mathbf{s}})} \rangle = \langle |\beta_0^0|^2 \rangle = 1$$

while

$$\langle \beta_0^0 \overline{\beta_\ell^m(\hat{\mathbf{s}})} \rangle_{\hat{\mathbf{s}}} = 0, \quad \text{so} \quad \langle \beta_\ell^m(\hat{\mathbf{r}}) \overline{\beta_\ell^m(\hat{\mathbf{s}})} \rangle_{\hat{\mathbf{s}}} = 0 \quad \text{for} \quad \ell \geq 1.$$

Therefore (4.4.43) implies

$$\langle W_L(\hat{\mathbf{r}}, \hat{\mathbf{s}}) \rangle_{\hat{\mathbf{s}}} = 1. \qquad (4.4.44)$$

Then (4.4.41) says that $\tilde{f}_L(\hat{\mathbf{r}})$ is a weighted average of the values of $f(\hat{\mathbf{s}})$. The weight function W_L in (4.4.41) is often called an *averaging window*. We can give the window different shapes by choosing different weights w_ℓ^m, always subject to (4.4.43).

Ideally, we would like to have

$$\frac{1}{4\pi}W_L(\hat{\mathbf{r}}, \hat{\mathbf{s}}) = \delta(\hat{\mathbf{s}} - \hat{\mathbf{r}}),$$

which is the Dirac delta function on $S(1)$. This goal is unattainable because, as we will show, $\delta(\hat{\mathbf{s}} - \hat{\mathbf{r}})$ has all wavelengths in its spherical harmonic expansion. Its harmonic coefficients are

$$4\pi\delta_\ell^m(\hat{\mathbf{r}}) = 4\pi\langle\overline{\beta_\ell^m(\hat{\mathbf{s}})}\delta(\hat{\mathbf{r}} - \hat{\mathbf{s}})\rangle_{\hat{\mathbf{s}}} = \overline{\beta_\ell^m(\hat{\mathbf{r}})}.$$

Formally (because the series on the right does not converge in any ordinary sense), we can write

$$4\pi\delta(\hat{\mathbf{s}} - \hat{\mathbf{r}}) = \sum_{\ell=0}^{\ell}\sum_{m=-\ell}^{\ell}\overline{\beta_\ell^m(\hat{\mathbf{r}})}\beta_\ell^m(\hat{\mathbf{s}}) = \sum_{\ell=0}^{\infty}(2\ell+1)P_\ell(\hat{\mathbf{r}}\cdot\hat{\mathbf{s}}). \quad (4.4.45)$$

The shortest wavelength available in (4.4.42) is $2\pi/(L+\frac{1}{2})$. The best we can probably do in our choice of w_ℓ^m will produce a $W_L(\hat{\mathbf{r}}, \hat{\mathbf{s}})$, which, for fixed $\hat{\mathbf{r}}$, has a tall, narrow peak when $\hat{\mathbf{s}}$ is near $\hat{\mathbf{r}}$ and which is small when $\hat{\mathbf{s}}$ is outside this peak. The peak width will probably be of the order of the shortest available wavelength, $\approx 2\pi/(L+\frac{1}{2})$. A slightly more quantitative estimate of this peak width is as follows: if the peak is circular with radius α radians, its area is $2\pi(1 - \cos\alpha)$ as a disk on $S(1)$. If we cover $S(1)$ with disks of this area, $2/(1 - \cos\alpha)$ disks are required (area of $S(1) = 4\pi$). If we know the average of f over each such disk, we have $2/(1 - \cos\alpha)$ independent data about f. But we know f_ℓ^m for $|m| \le \ell$, $0 \le \ell \le L$, and these are $(L+1)^2$ independent data about f. We can't expect a clever choice of w_ℓ^m to improve this situation, so we expect $2/(1 - \cos\alpha) \le (L+1)^2$, or

$$\sin\frac{\alpha}{2} \ge \frac{1}{L+1}.$$

For large L, this gives $\alpha \ge 2/(L+1)$ instead of $\alpha \ge \pi/(L+\frac{1}{2})$.

How do we choose w_ℓ^m to produce such windows? It will simplify the discussion to choose w_ℓ^m to depend only on ℓ, not m. Then (4.4.38) can be written

$$\tilde{f}_L(\hat{\mathbf{r}}) = \sum_{\ell=0}^{L} w_\ell \sum_{m=-\ell}^{\ell} f_\ell^m \beta_\ell^m(\hat{\mathbf{r}}) \qquad (4.4.46)$$

and (4.4.42) becomes

$$W_L(\hat{\mathbf{r}}, \hat{\mathbf{s}}) = \sum_{\ell=0}^{L} w_\ell \sum_{m=-\ell}^{\ell} \beta_\ell^m(\hat{\mathbf{r}})\overline{\beta_\ell^m(\hat{\mathbf{s}})}.$$

Invoking (3.5.13), the Addition Theorem for spherical harmonics, we conclude that

$$W_L(\hat{\mathbf{r}}, \hat{\mathbf{s}}) = \sum_{\ell=0}^{L}(2\ell+1)w_\ell P_\ell(\hat{\mathbf{r}} \cdot \hat{\mathbf{s}}) = \tilde{W}_L(\hat{\mathbf{r}} \cdot \hat{\mathbf{s}}). \qquad (4.4.47)$$

For any fixed $\hat{\mathbf{r}}$, the averaging window is axisymmetric about $\hat{\mathbf{r}}$, and its level lines as a function of $\hat{\mathbf{s}}$ are small circles centered on the point $\hat{\mathbf{r}}$. To make $\tilde{f}_L$ fit f best in the least squares sense, we have seen that we must take $w_\ell = 1$ for $0 \le \ell \le L$. In that case, (4.4.47) can be summed exactly:

$$\sum_{\ell=0}^{L}(2\ell+1)P_\ell(\mu) = (L+1)P_\ell^{(1,0)}(\mu) \qquad (4.4.48)$$

where $P_L^{(\alpha,\beta)}(\mu)$ is a Jacobi polynomial (Erdelyi et al., 1953). A plot of $\tilde{W}_L(\cos\theta)$ is given for $L = 10$ in Figure 4.4.2. In the special case of a window with $w_\ell = 1$, the averaging kernel (4.4.47) is called the *Dirichlet kernel*. Its behavior as $L \to \infty$ can be used to study the convergence of the partial sums in the spherical harmonic expansion for f. For $L = 10$, we see from the figure that the radius of the central peak is about $20°$. Our simple argument for α predicts $10°$.

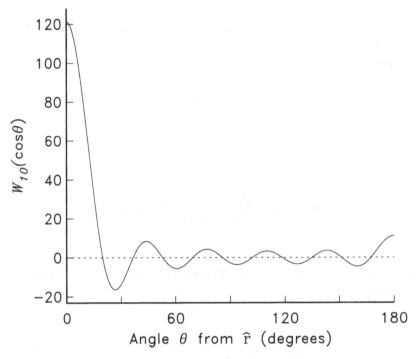

Figure 4.4.2: Dirichlet kernel (4.4.47) for $L = 10$.

Our overoptimistic result probably comes in part from assuming that one can cover $S(1)$ with *nonoverlapping* circular disks, and in part from the vagueness of the definition of peak width.

The sidebands (i.e., the subsidiary peaks) in Figure 4.4.2 are rather large. We might be willing to accept a wider peak in order to suppress the sidebands. One class of filters with well-controlled sidebands comes from Chebyshev polynomials (Erdelyi et al., 1953). Different choices of w_ℓ give a trade-off between sideband height and peak width. The techniques for choosing the w_ℓ to achieve various desirable properties in W_L have been developed in the study of linear inverse theory. This is the technology that was employed on this problem by John Booker in his Ph.D. thesis. The method used will not be pursued here, but see Booker (1968), Backus and Gilbert (1968), or Backus (1970a, b, c).

Evidently, our view of $B_r(c\hat{r})$ on the core–mantle boundary, if obtained from the Gauss coefficients with degrees $\leq L$, will be the blurred view obtained by a myopic observer who has forgotten his or her glasses. The "circle of confusion" for $L = 10$ will have a radius on the core of at least $20°$.

Exercises

1. Consider a massive body occupying a region V of finite size. Let $\rho(\mathbf{r})$ be the mass density of the body and let $\phi(\mathbf{r})$ be its gravitational potential. Choose an origin $\mathbf{0}$ (not necessarily in V) and a radius a large enough that V lies entirely inside $S(a)$. Let $\{\beta_l^{-l}, \ldots, \beta_l^l\}$ be an orthonormal basis for $\mathcal{H}_l$.
   (a) Show that there are coefficients g_l^m such that if $r > a$ then

$$\phi(r\hat{\mathbf{r}}) = a \sum_{l=0}^{\infty} \left(\frac{a}{r}\right)^{l+1} \sum_{m=-l}^{l} g_l^m \beta_l^m(\hat{\mathbf{r}}).$$

   (b) Calculate g_l^m explicitly in terms of $\rho(\hat{\mathbf{r}})$ as an integral over V.
   (c) Show that g_0^0 is independent of the choice of origin, but that g_1^{-1}, g_1^0, g_1^1 are not. Show that the origin can be chosen so that $g_1^{-1} = g_1^0 = g_1^1 = 0$.
   (d) How much of (c) remains true if ρ is a charge density instead of a mass density and ϕ is the electrostatic potential?

2. (a) For what ℓ do the members of $\mathcal{H}_\ell$ have wavelengths of about 400 km on $S(a)$, the surface of the earth?
   (b) Suppose the geomagnetic field $\mathbf{B}$ has only internal sources, and $\mathbf{B}_\ell$ is the part of $\mathbf{B}$ produced by spherical harmonics in $\mathcal{H}_\ell$. Suppose that on $S(a)$ the root-mean-square value of $|\mathbf{B}_\ell|$ is about 10 nT. Suppose that all the sources lie inside the ball of radius c, $0 < r < c$. Find the magnetic energy of $\mathbf{B}_\ell$ outside c as a function of c.
   (c) If the rest mass of the energy calculated in (b) is not to exceed the mass of the earth, show that c must be considerably larger than 3486 km; i.e., there must be sources outside the liquid core of the earth.

3. Suppose the geomagnetic field $\mathbf{B}$ has only internal sources. Suppose that the Y component of $\mathbf{B}$ is measured everywhere on $S(a)$, but the X and Z components are unknown (X, Y, Z are the conventional symbols used in geomagnetism; see Chapter 1). What can be

determined about the Gauss coefficients F_l^m if the geomagnetic potential is expanded as

$$\Psi(r\hat{\mathbf{r}}) = a \sum_{l=1}^{\infty} \sum_{m=-l}^{l} F_l^m \left(\frac{a}{r}\right)^{l+1} Y_l^m(\hat{\mathbf{r}})$$

where the Y_l^m are fully normalized complex spherical harmonics?

5

THE MIE REPRESENTATION

5.1 The Helmholtz Representation Theorem

The more familiar form of the Helmholtz Representation Theorem (see section 2.1) decomposes a vector field into the sum of two parts, one the gradient of a scalar, the other the curl of a vector field. In this section we develop an analog to this theorem for tangent vector fields on the surface of a sphere, in which the gradient is replaced by the surface operator ∇_1 and the curl by the surface curl Λ. It is then possible to extend this representation to vector fields that are not tangential, and we do this. The result is a formula for the vector field in terms of three scalar fields, which are independent of the coordinate reference frame. An obvious alternative way of writing the vector field in terms of three scalar fields, namely the components, is clearly not coordinate-independent. The new representation lays the groundwork for the main subject of this chapter, the *Mie representation*, which applies to special kinds of vector fields, to be introduced in the next section.

5.1.1 Solving the Surface Form of Poisson's Equation

Poisson's equation in electrostatics or elementary gravity takes the form

$$\nabla^2 V = f.$$

Its solutions are the potentials of charge or matter distributions, fields of fundamental importance to classical physics. We now study the form of this equation on the surface of a sphere, replacing the Laplacian by the Beltrami operator. The physical significance of this equation is unimportant for us, but its solution will be central to the representations that are the main subject of this chapter.

Given a scalar field f on $S(r)$, we want to find the general solution to

$$\nabla_1^2 g = f. \tag{5.1.1}$$

That is, we want to find all scalar fields g on $S(r)$ that satisfy (5.1.1). The following theorem supplies the condition for existence and the statement of the uniqueness of solutions.

Theorem 1: Suppose f is a scalar field on $S(r)$. Equation (5.1.1) has no solution g unless

$$\langle f \rangle_{S(r)} = 0. \tag{5.1.2}$$

If f satisfies (5.1.2), then there is exactly one solution g of (5.1.1) with the property

$$\langle g \rangle_{S(r)} = 0. \tag{5.1.3}$$

We call this solution

$$g = \nabla_1^{-2} f. \tag{5.1.4}$$

Every other solution g of (5.1.1) is of the form $g = \nabla_1^{-2} f + C$ where C is a constant, and any C gives a solution of (5.1.1).

Proof: Choose an orthonormal basis $\{\beta_\ell^{-\ell}, \ldots, \beta_\ell^\ell\}$ for each $\mathcal{H}_\ell$. Expand f and g in spherical harmonics on $S(r)$ as

$$f(r\hat{\mathbf{r}}) = \sum_{\ell=0}^{\infty} \sum_{m=-\ell}^{\ell} f_\ell^m(r) \beta_\ell^m(\hat{\mathbf{r}})$$

$$g(r\hat{\mathbf{r}}) = \sum_{\ell=0}^{\infty} \sum_{m=-\ell}^{\ell} g_\ell^m(r) \beta_\ell^m(\hat{\mathbf{r}}).$$

Here, as usual, the expansion coefficients are given by

$$f_\ell^m(r) = \langle \overline{\beta_\ell^m(\hat{\mathbf{r}})} f(r\hat{\mathbf{r}}) \rangle_{S(1)} = \frac{1}{4\pi} \int_{S(1)} d^2\hat{\mathbf{r}} \ \overline{\beta_\ell^m(\hat{\mathbf{r}})} f(r\hat{\mathbf{r}}) \tag{5.1.5}$$

and similarly for g. Now $\beta_0^0(\hat{\mathbf{r}})$ is a constant, so

$$f_0^0(r) = \frac{\overline{\beta_0^0}}{4\pi} \int_{S(1)} d^2\hat{\mathbf{r}} \ f(r\hat{\mathbf{r}})$$

$$= \frac{\overline{\beta_0^0}}{4\pi r^2} \int_{S(r)} d^2\mathbf{r} \ f(\mathbf{r}) = \overline{\beta_0^0} \langle f \rangle_{S(r)}.$$

Thus f satisfies (5.1.2) if and only if $f_0^0(r) = 0$. Suppose that f and g satisfy (5.1.1). Then

$$f_\ell^m(r) = \langle \overline{\beta_\ell^m(\hat{\mathbf{r}})} \nabla_1^2 g(r\hat{\mathbf{r}}) \rangle_{S(1)} = \langle \nabla_1^2 \overline{\beta_\ell^m(\hat{\mathbf{r}})} g(r\hat{\mathbf{r}}) \rangle_{S(1)}$$

$$= \ell(\ell+1) \langle \overline{\beta_\ell^m(\hat{\mathbf{r}})} g(r\hat{\mathbf{r}}) \rangle_{S(1)} = -\ell(\ell+1) g_\ell^m(r).$$

Conversely, if $f_\ell^m(r) = -\ell(\ell+1) g_\ell^m(r)$, then one can run this argument backward and conclude that

$$f_\ell^m(r) = \langle \overline{\beta_\ell^m(\hat{\mathbf{r}})} \nabla_1^2 g(r\hat{\mathbf{r}}) \rangle_{S(1)},$$

so from (5.1.5),

$$\langle \overline{\beta_\ell^m(\hat{\mathbf{r}})} [f - \nabla_1^2 g] \rangle_{S(1)} = 0$$

for all ℓ, m. But then $f - \nabla_1^2 g = 0$. It follows that f and g satisfy (5.1.1) if and only if

$$f_\ell^m(r) = -\ell(\ell+1) g_\ell^m(r) \quad \text{for } \ell = 0, 1, 2, \ldots, m = -\ell, \ldots, \ell. \quad (5.1.6)$$

If we are given an f satisfying (5.1.2), then we can define $g_\ell^m(r)$ for $\ell \geq 1$ by (5.1.6). If we take $g_0^0(r) = 0$, then g solves both (5.1.1) and (5.1.3). Moreover, if g solves both (5.1.1) and (5.1.3), its g_ℓ^m for $\ell \geq 1$ must be given by (5.1.6) and its $g_0^0(r)$ must vanish. Therefore, when f satisfies (5.1.2), then (5.1.1) and (5.1.3) have exactly one solution g, which we label $\nabla_1^{-2} f$. As we see from (5.1.6), all other solutions g of (5.1.1) can differ from $\nabla_1^{-2} f$ only in having $g_0^0(r) \neq 0$. But since β_0^0 is constant, these solutions differ from $\nabla_1^{-2} f$ by a constant. $\triangleleft$

The proof has furnished a method of actually calculating the solutions, through a spherical harmonic expansion of g. But there is a more compact and revealing solution, which we give next.

5.1.2 Integral Form of the Solution

For the sake of definiteness we set $r = 1$ and work on the surface $S(1)$. We write the solution as a spherical harmonic expansion in an orthonormal basis as before:

$$g(\hat{\mathbf{r}}) = \sum_{\ell=0}^{\infty} \sum_{m=-\ell}^{\ell} g_\ell^m \beta_\ell^m(\hat{\mathbf{r}}).$$

We have seen that $g_0^0 = 0$, and for $\ell > 0$ we have from (5.1.6) and (5.1.5):

$$g_\ell^m = -\frac{f_\ell^m}{\ell(\ell+1)}$$

$$= -\frac{1}{4\pi\ell(\ell+1)} \int_{S(1)} d^2\hat{\mathbf{s}} \ \overline{\beta_\ell^m(\hat{\mathbf{s}})} f(\hat{\mathbf{s}}).$$

Next we substitute this integral for the coefficients into the expansion of $g(\hat{\mathbf{r}})$:

$$g(\hat{\mathbf{r}}) = -\frac{1}{4\pi} \sum_{\ell=1}^{\infty} \frac{1}{\ell(\ell+1)} \sum_{m=-\ell}^{\ell} \beta_\ell^m(\hat{\mathbf{r}}) \int_{S(1)} d^2\hat{\mathbf{s}} \ \overline{\beta_\ell^m(\hat{\mathbf{s}})} f(\hat{\mathbf{s}})$$

$$= -\frac{1}{4\pi} \sum_{\ell=1}^{\infty} \frac{1}{\ell(\ell+1)} \left[\int_{S(1)} d^2\hat{\mathbf{s}} \ \sum_{m=-\ell}^{\ell} \beta_\ell^m(\hat{\mathbf{r}})\overline{\beta_\ell^m(\hat{\mathbf{s}})} f(\hat{\mathbf{s}}) \right]$$

$$= -\frac{1}{4\pi} \sum_{\ell=1}^{\infty} \frac{1}{\ell(\ell+1)} \left[\int_{S(1)} d^2\hat{\mathbf{s}} \ (2\ell+1)P_\ell(\hat{\mathbf{r}} \cdot \hat{\mathbf{s}})f(\hat{\mathbf{s}}) \right].$$

$$(5.1.7)$$

Notice how we used the Spherical Harmonic Addition Theorem (3.5.13) to perform the sum over m under the integral. We wish to interchange the sum over ℓ and the integral, which requires justification: all is well by the Lebesgue Dominated Convergence Theorem (Korevaar, 1968) if we can show the following series to be convergent:

$$S = \sum_{\ell=1}^{\infty} \frac{2\ell+1}{\ell(\ell+1)} \left[\int_{S(1)} d^2\hat{\mathbf{s}} \ |P_\ell(\hat{\mathbf{r}} \cdot \hat{\mathbf{s}})f(\hat{\mathbf{s}})| \right].$$

We apply Schwarz's inequality:

$$\frac{1}{4\pi} \int_{S(1)} d^2\hat{\mathbf{s}} \ |P_\ell(\hat{\mathbf{r}} \cdot \hat{\mathbf{s}})| \ |f(\hat{\mathbf{s}})| \leq \|P_\ell\| \cdot \|f\|$$

where the norm is the familiar one of Chapter 3: $\|f\| = \langle |f|^2 \rangle_{S(1)}^{1/2}$. Recall that $\sqrt{2\ell+1}P_\ell = Y_\ell^0$ and by its construction $\|Y_\ell^0\| = 1$; thus $\|P_\ell\| = (2\ell+1)^{-1/2}$. We have assumed f is at least bounded, so its norm exists. Thus the convergence of the series for S comes down to the convergence of

$$\sum_{\ell=1}^{\infty} \frac{\sqrt{2\ell+1}}{\ell(\ell+1)} < \sum_{\ell=1}^{\infty} \frac{\sqrt{2\ell+2}}{\ell(\ell+1)} < \sum_{\ell=1}^{\infty} \frac{\sqrt{2}}{\ell(\ell+1)^{1/2}} < \sum_{\ell=1}^{\infty} \frac{\sqrt{2}}{\ell^{3/2}}.$$

The series on the right converges by the integral test, so we may interchange the integral and the infinite sum in (5.1.7). Thus we arrive at

$$g(\hat{\mathbf{r}}) = -\frac{1}{4\pi} \int_{S(1)} d^2\hat{\mathbf{s}} \; f(\hat{\mathbf{s}}) \left[\sum_{\ell=1}^{\infty} \frac{2\ell+1}{\ell(\ell+1)} P_\ell(\hat{\mathbf{r}} \cdot \hat{\mathbf{s}}) \right]. \qquad (5.1.8)$$

The next task is to sum the series for the function inside the brackets, which we name

$$K(\mu) = \sum_{\ell=1}^{\infty} \frac{2\ell+1}{\ell(\ell+1)} P_\ell(\mu). \qquad (5.1.9)$$

The standard approach for finding an elementary expression for (5.1.9) is to appeal to (3.5.23), the generating function for Legendre polynomials; this is the first exercise at the end of the chapter. Instead we exploit the recurrence relations studied in section 7 of Chapter 3. If we combine (3.7.14) and (3.7.38) we find

$$(2\ell+1)s^2 \partial_\mu P_\ell^m = (\ell+1)(\ell+m)P_{\ell-1}^m - \ell(\ell-m+1)P_{\ell+1}^m$$

where we recall $s^2 = 1 - \mu^2$. Specializing to $m = 0$ and rearranging slightly we see on the left a term with an obvious connection to (5.1.9):

$$\frac{2\ell+1}{\ell(\ell+1)} \partial_\mu P_\ell = -\frac{P_{\ell+1}(\mu) - P_{\ell-1}(\mu)}{1 - \mu^2}.$$

We sum from $\ell = 1$ to L:

$$\sum_{\ell=1}^{L} \frac{2\ell+1}{\ell(\ell+1)} \partial_\mu P_\ell = -\frac{1}{1-\mu^2} \sum_{\ell=1}^{L} [P_{\ell+1}(\mu) - P_{\ell-1}(\mu)]$$

$$= -\frac{1}{1-\mu^2} \left[\left\{ \sum_{\ell=1}^{L-2} P_{\ell+1} + P_L + P_{L+1} \right\} - \left\{ P_0 + P_1 + \sum_{\ell=3}^{L} P_{\ell-1} \right\} \right]$$

$$= \frac{1}{1-\mu^2} \left[P_0 + P_1 - P_L - P_{L+1} + \sum_{\ell=2}^{L-1} P_\ell - \sum_{\ell=2}^{L-1} P_\ell \right]$$

$$= \frac{1}{1-\mu} - \frac{P_L + P_{L+1}}{1-\mu^2},$$

since $P_0(\mu) = 1$ and $P_1(\mu) = \mu$. Now we integrate with respect to μ and then we have from (5.1.9) that

$$K(\mu) = C - \ln(1-\mu) + \lim_{L\to\infty} \int_0^\mu dt \, \frac{P_L(t) + P_{L+1}(t)}{1-t^2} \qquad (5.1.10)$$

where C is a constant of integration. From the asymptotic theory of section 3.6 we find when we set $m = 0$ that for t in the interval $(-1+\epsilon, \, 1-\epsilon)$ with $\epsilon > 0$, $P_L(t)$ tends to zero uniformly like $L^{-1/2}$ as L tends to infinity. Thus the final term in (5.1.10) vanishes if $\mu \neq \pm 1$. We may write (5.1.8) thus:

$$g(\hat{\mathbf{r}}) = \langle K(\hat{\mathbf{r}} \cdot \hat{\mathbf{s}}) f(\hat{\mathbf{s}}) \rangle_{\hat{\mathbf{s}}},$$

and since we have insisted in (5.1.2) that $\langle f \rangle_{S(1)} = 0$, this equation shows that any constant C in (5.1.10) would serve. It might be convenient to choose $C = 0$, but it is more satisfying if $\langle K \rangle_{S(1)} = 0$, a property evident from the spherical harmonic expansion. By (5.1.10) this condition leads to

$$0 = \int_{-1}^{1} K(\mu) \, d\mu = \int_{-1}^{1} C \, d\mu - \int_{-1}^{1} \ln(1-\mu) \, d\mu.$$

Upon performing the integrals, we find $C = \ln 2 - 1$, so that choice of C yields

$$K(\mu) = -\ln\left(\frac{1-\mu}{2}\right) - 1.$$

With this definition of K we arrive at the final result:

$$\nabla_1^{-2} f(\hat{\mathbf{r}}) = \langle K(\hat{\mathbf{r}} \cdot \hat{\mathbf{s}}) f(\hat{\mathbf{s}}) \rangle_{\hat{\mathbf{s}}}. \qquad (5.1.11)$$

5.1.3 The Helmholtz Representation Theorem on $S(r)$ and $S(a,b)$

Now we use Theorem 1 to find a representation of a tangent vector field on a sphere in terms of gradients of scalars. Before continuing, it might be a good idea for the reader to refer to sections 7.2.12 and 7.4 of the Mathematical Appendix.

Theorem 2: (Helmholtz Theorem for tangent vector fields) Suppose $\mathbf{v}_S$ is a tangent vector field on $S(r)$. Then there are unique scalar fields g and h on $S(r)$ with these properties:

$$\langle g \rangle_{S(r)} = \langle h \rangle_{S(r)} = 0 \qquad (5.1.12)$$

$$\mathbf{v}_S = \boldsymbol{\nabla}_1 g + \boldsymbol{\Lambda} h. \qquad (5.1.13)$$

Proof: First we consider the uniqueness question. If g and h satisfy (5.1.13), then because $\boldsymbol{\nabla}_1 \cdot \boldsymbol{\Lambda} = \boldsymbol{\Lambda} \cdot \boldsymbol{\nabla}_1 = 0$ from (7.2.119, 7.2.120) and $\boldsymbol{\Lambda} \cdot \boldsymbol{\Lambda} = \boldsymbol{\nabla}_1 \cdot \boldsymbol{\nabla}_1$ from (7.2.118), we have

$$\nabla_1^2 g = \boldsymbol{\nabla}_1 \cdot \mathbf{v}_S \qquad (5.1.14)$$

$$\nabla_1^2 h = \boldsymbol{\Lambda} \cdot \mathbf{v}_S. \qquad (5.1.15)$$

But these equations together with (5.1.12) uniquely determine g and h by Theorem 1.

Next we treat the existence of the scalars. We appeal to theorems for general surface operators $\boldsymbol{\nabla}_S$ and $\boldsymbol{\Lambda}_S$ treated in section 7.4.5: recall that on the spherical surface $S(r)$, we have $r\boldsymbol{\nabla}_S = \boldsymbol{\nabla}_1$ and $r\boldsymbol{\Lambda}_S = \boldsymbol{\Lambda}$. According to (7.4.21) applied to the surface $S = S(r)$, $\langle \boldsymbol{\Lambda} \cdot \mathbf{v}_S \rangle_{S(r)} = 0$. Therefore, by Theorem 1 we can find an h that solves (5.1.15) and (5.1.12). This h also satisfies $\boldsymbol{\Lambda} \cdot \boldsymbol{\Lambda} h = \boldsymbol{\Lambda} \cdot \mathbf{v}_S$, because, as we noted earlier, $\boldsymbol{\Lambda} \cdot \boldsymbol{\Lambda} = \boldsymbol{\nabla}_1 \cdot \boldsymbol{\nabla}_1$. Therefore,

$$\boldsymbol{\Lambda} \cdot (\mathbf{v}_S - \boldsymbol{\Lambda} h) = 0.$$

By Theorem 4 in subsection 7.4.6 specialized to the sphere, there is a scalar field g on $S(r)$ such that $\mathbf{v}_S - \boldsymbol{\Lambda} h = \boldsymbol{\nabla}_1 g$. But this is (5.1.13). If g does not satisfy equation (5.1.12), replace it by $g - \langle g \rangle_{S(r)} = g'$. Since $\langle g \rangle_{S(r)}$ is a constant, $\nabla_1 g = \nabla_1 g'$, and obviously $\langle g' \rangle_{S(r)} = 0$. ◁

The representation can easily be extended to vector fields on $S(r)$ that are not tangent to $S(r)$.

Theorem 3: Suppose **v** is a vector field on $S(r)$. Then there are unique scalar fields f, g, h on $S(r)$ such that

$$\langle g \rangle_{S(r)} = \langle h \rangle_{S(r)} = 0 \tag{5.1.16}$$

and

$$\mathbf{v} = \hat{\mathbf{r}} f + \nabla_1 g + \mathbf{\Lambda} h. \tag{5.1.17}$$

Note that (5.1.16) and (5.1.17) constitute the Helmholtz representation for **v**; then f, g, h are the Helmholtz scalars for **v**. Therefore, explicitly:

$$\left. \begin{aligned} f &= \hat{\mathbf{r}} \cdot \mathbf{v} \\ \nabla_1^2 g &= \nabla_1 \cdot \mathbf{v}_S \\ \nabla_1^2 h &= \mathbf{\Lambda} \cdot \mathbf{v}_S \end{aligned} \right\}. \tag{5.1.18}$$

Proof: If (5.1.17) is to work, we must take

$$f = v_r = \hat{\mathbf{r}} \cdot \mathbf{v}$$

because $\hat{\mathbf{r}} \cdot \nabla_1 g = \hat{\mathbf{r}} \cdot \mathbf{\Lambda} h = 0$. Then we must have $\mathbf{v}_S = \mathbf{v} - \hat{\mathbf{r}} v_r = \nabla_1 g + \mathbf{\Lambda} h$, so we can use Theorem 2. ◁

The greatest advantage of (5.1.17) is that it expresses a vector field **v** on $S(r)$ in terms of scalar fields on $S(r)$ without introducing a coordinate system on $S(r)$. The fields f, g, h are just as smooth as **v** (in fact g and h are a little smoother). There are no singularities at the north and south poles, such as occur when **v** is expressed in terms of its components v_r, v_θ, v_λ.

We noted at the end of section 7.2 the strong formal (algebraic) resemblance of $\hat{\mathbf{r}}, \nabla_1, \mathbf{\Lambda}$ to an orthogonal triple of vectors. The analogy continues in (5.1.17). The vector **v** looks formally like a "linear combination" of $\hat{\mathbf{r}}, \nabla_1, \mathbf{\Lambda}$. The term $\hat{\mathbf{r}} f$ is orthogonal to $\nabla_1 g$ and $\mathbf{\Lambda} h$ in the ordinary sense. But the two terms $\nabla_1 g$ and $\mathbf{\Lambda} h$ are orthogonal to each other only on average. That is,

$$\langle \nabla_1 g \cdot \mathbf{\Lambda} h \rangle_{S(r)} = 0. \tag{5.1.19}$$

To prove this, we first note the identity for any tangent vector field $\mathbf{v}_S$:

$$\boldsymbol{\nabla}_1 \cdot (\mathbf{v}_S \boldsymbol{\nabla}_1 g) = g \boldsymbol{\nabla}_1 \cdot \mathbf{v}_S + \mathbf{v}_S \cdot \boldsymbol{\nabla}_1 g.$$

If we average over $S(r)$ and apply (7.4.21) on the sphere we see

$$0 = \langle \boldsymbol{\nabla}_1 g \cdot \mathbf{v}_S \rangle_{S(r)} + \langle g(\boldsymbol{\nabla}_1 \cdot \mathbf{v}_S) \rangle_{S(r)}$$

and, substituting $\mathbf{v}_S = \Lambda h$, we observe

$$\langle \boldsymbol{\nabla}_1 g \cdot \Lambda h \rangle_{S(r)} = -\langle g(\boldsymbol{\nabla}_1 \cdot \Lambda h) \rangle_{S(r)}.$$

Because $\boldsymbol{\nabla}_1 \cdot \Lambda = 0$ this proves (5.1.19).

Equation (5.1.19) means that $\langle |\mathbf{v}|^2 \rangle_{S(r)}$ is easy to compute. We have

$$\langle |\mathbf{v}|^2 \rangle_{S(r)} = \langle |f|^2 \rangle_{S(r)} + \langle \overline{\boldsymbol{\nabla}_1 g} \cdot \boldsymbol{\nabla}_1 g \rangle_{S(r)} + \langle \overline{\Lambda h} \cdot \Lambda h \rangle_{S(r)} \quad (5.1.20)$$

or, using the now familiar identities,

$$\langle |\mathbf{v}|^2 \rangle_{S(r)} = \langle |f|^2 \rangle_{S(r)} - \langle \bar{g} \nabla_1^2 g \rangle_{S(r)} - \langle \bar{h} \nabla_1^2 h \rangle_{S(r)}. \quad (5.1.21)$$

Finally, we can extend this type of representation to fields not confined to a spherical surface. If $\mathbf{v}$ is a vector field in a shell $S(a, b)$, then on every $S(r)$ in the shell, $\mathbf{v}$ has a representation (5.1.17). Of course the scalar fields f, g, h will be different on each $S(r)$. That is, f, g, h will depend on r as well as $\hat{\mathbf{r}}$. With this understanding, we can use (5.1.17) to represent an arbitrary vector field in a shell in terms of three scalar fields in the shell. These fields are uniquely determined in $S(a, b)$ by $\mathbf{v}$ because (5.1.16) are required to hold on $S(r)$ for each r in $a < r < b$.

5.1.4 Divergence and Curl in the Helmholtz Representation

In order to work effectively with (5.1.17), we need to be able to perform the operations of vector calculus on it. We suppose that (5.1.17) holds in the shell $S(a, b)$, and we calculate $\boldsymbol{\nabla} \cdot \mathbf{v}$ and $\boldsymbol{\nabla} \times \mathbf{v}$. The divergence is straightforward.

$$\boldsymbol{\nabla} \cdot \mathbf{v} = \left(\partial_r \hat{\mathbf{r}} + \frac{1}{r} \boldsymbol{\nabla}_1 \right) \cdot \mathbf{v}$$

$$= \partial_r (\hat{\mathbf{r}} \cdot \mathbf{v}) + \frac{1}{r} (\boldsymbol{\nabla}_1 \cdot \mathbf{v})$$

$$= \partial_r f + \frac{1}{r} \boldsymbol{\nabla}_1 \cdot (\hat{\mathbf{r}} f + \boldsymbol{\nabla}_1 g + \Lambda h).$$

But

$$\mathbf{\nabla}_1 \cdot \mathbf{\Lambda} h = 0$$

$$\mathbf{\nabla}_1 \cdot \mathbf{\nabla}_1 g = \nabla_1^2 g$$

$$\mathbf{\nabla}_1 \cdot (\hat{\mathbf{r}} f) = (\mathbf{\nabla}_1 \cdot \hat{\mathbf{r}}) f + \hat{\mathbf{r}} \cdot \mathbf{\nabla}_1 f = 2f.$$

Thus substituting these

$$\begin{aligned}
\mathbf{\nabla} \cdot \mathbf{v} &= \partial_r f + \frac{2}{r} f + \frac{1}{r} \nabla_1^2 g \\
&= \frac{1}{r^2} \partial_r r^2 f + \frac{1}{r} \nabla_1^2 g.
\end{aligned} \tag{5.1.22}$$

The curl is more complicated. Obviously we would like to express the result, which is a vector field, as a Helmholtz expansion. We note the separate pieces:

$$\mathbf{\nabla} \times (\hat{\mathbf{r}} f) = \mathbf{\nabla} \times \mathbf{r}^M \left(\frac{f}{r} \right) = -\mathbf{\Lambda} \frac{f}{r}. \tag{5.1.23}$$

Also,

$$\mathbf{\nabla} g = \hat{\mathbf{r}} \partial_r g + \frac{1}{r} \mathbf{\nabla}_1 g = \hat{\mathbf{r}} \partial_r g + \mathbf{\nabla}_1 \frac{g}{r}.$$

If we replace g by rg in this formula we find

$$\mathbf{\nabla}_1 g = \mathbf{\nabla}(rg) - \hat{\mathbf{r}} \partial_r (rg).$$

Thus, from (5.1.23),

$$\mathbf{\nabla} \times \mathbf{\nabla}_1 g = \mathbf{\Lambda} \left[\frac{1}{r} \partial_r (rg) \right]. \tag{5.1.24}$$

Finally,

$$\begin{aligned}
\mathbf{\nabla} \times \mathbf{\Lambda} h &= \left[\hat{\mathbf{r}} \partial_r + \mathbf{\nabla}_1 \left(\frac{1}{r} \right)^M \right] \times \mathbf{\Lambda} h \\
&= \hat{\mathbf{r}} \times \partial_r \mathbf{\Lambda} h + \mathbf{\nabla}_1 \times \left(\frac{1}{r} \right)^M \mathbf{\Lambda} h \\
&= \hat{\mathbf{r}} \times \mathbf{\Lambda} \partial_r h + \mathbf{\nabla}_1 \times \mathbf{\Lambda} \frac{h}{r} \\
&= -\mathbf{\nabla}_1 (\partial_r h) + \mathbf{\nabla}_1 \times \left[\hat{\mathbf{r}} \times \mathbf{\nabla}_1 \frac{h}{r} \right].
\end{aligned}$$

We expand the second term separately: if $\mathbf{u}_S$ is any tangent vector field on $S(r)$,

$$\boldsymbol{\nabla}_1 \times (\hat{\mathbf{r}} \times \mathbf{u}_S) = [\boldsymbol{\nabla}_1 \cdot \mathbf{u}_S + \mathbf{u}_S \cdot \boldsymbol{\nabla}_1]\hat{\mathbf{r}} - [(\boldsymbol{\nabla}_1 \cdot \hat{\mathbf{r}}) + \hat{\mathbf{r}} \cdot \boldsymbol{\nabla}_1]\mathbf{u}_S$$

$$= \hat{\mathbf{r}}\boldsymbol{\nabla}_1 \cdot \mathbf{u}_S + \mathbf{u}_S - 2\mathbf{u}_S$$

$$= \hat{\mathbf{r}}\boldsymbol{\nabla}_1 \cdot \mathbf{u}_S - \mathbf{u}_S.$$

With $\mathbf{u}_S = \boldsymbol{\nabla}_1(h/r)$ we get

$$\boldsymbol{\nabla}_1 \times \left[\hat{\mathbf{r}} \times \boldsymbol{\nabla}_1 \frac{h}{r}\right] = \hat{\mathbf{r}}\boldsymbol{\nabla}_1^2\frac{h}{r} - \boldsymbol{\nabla}_1\frac{h}{r} = \frac{\hat{\mathbf{r}}}{r}\boldsymbol{\nabla}_1^2 h - \boldsymbol{\nabla}_1\frac{h}{r}.$$

Then

$$\boldsymbol{\nabla} \times \boldsymbol{\Lambda}h = \hat{\mathbf{r}}\frac{1}{r}\boldsymbol{\nabla}_1^2 h - \boldsymbol{\nabla}_1\left(\frac{h}{r} + \partial_r h\right),$$

so

$$\boldsymbol{\nabla} \times \boldsymbol{\Lambda}h = \hat{\mathbf{r}}\left(\frac{1}{r}\boldsymbol{\nabla}_1^2 h\right) - \boldsymbol{\nabla}_1\left(\frac{1}{r}\partial_r rh\right). \tag{5.1.25}$$

Combining the three formulas, if

$$\mathbf{v} = \hat{\mathbf{r}}f + \boldsymbol{\nabla}_1 g + \boldsymbol{\Lambda}h \quad \text{in} \quad S(a,b)$$

then

$$\boldsymbol{\nabla} \times \mathbf{v} = \hat{\mathbf{r}}\left[\frac{1}{r}\boldsymbol{\nabla}_1^2 h\right] + \boldsymbol{\nabla}_1\left[-\frac{1}{r}\partial_r rh\right] + \boldsymbol{\Lambda}\left[\frac{1}{r}\partial_r rg - \frac{f}{r}\right]. \tag{5.1.26}$$

This is almost but not quite the Helmholtz representation for $\boldsymbol{\nabla} \times \mathbf{v}$. The problem is whether the scalars

$$\tilde{g} = -\frac{1}{r}\partial_r rh \quad \text{and} \quad \tilde{h} = \frac{1}{r}(\partial_r rg - f)$$

satisfy $\langle \tilde{g} \rangle_{S(r)} = 0$, $\langle \tilde{h} \rangle_{S(r)} = 0$ on every $S(r)$ in $a < r < b$. First we show that $\tilde{g}$ qualifies. Since $\langle h \rangle_{S(r)} = 0$ for all r in the interval it follows that

$$\int_{S(1)} d^2\hat{\mathbf{r}}\, h(r\hat{\mathbf{r}}) = 0.$$

As this result remains true for every r, we may multiply by r or differentiate with respect to r and so on: therefore

$$\partial_r \left[r \int_{S(1)} d^2\hat{\mathbf{r}} \; h(r\hat{\mathbf{r}}) \right] = 0.$$

The integration is over $S(1)$, so we may move the functions of r under the integral:

$$\int_{S(1)} d^2\hat{\mathbf{r}} \; \frac{1}{r} \partial_r r h(r\hat{\mathbf{r}}) = 0,$$

so finally

$$\left\langle \frac{1}{r} \partial_r r h \right\rangle_{S(r)} = 0. \qquad (5.1.27)$$

Thus $-\frac{1}{r}\partial_r r h$ is the Helmholtz scalar $\tilde{g}$ in (5.1.26). The same argument shows that

$$\left\langle \frac{1}{r} \partial_r r g \right\rangle_{S(r)} = 0,$$

so

$$\langle \tilde{h} \rangle_{S(r)} = -\frac{1}{r} \langle f \rangle_{S(r)}.$$

But there is no reason to expect $\langle f \rangle_{S(r)} = 0$. Therefore, we adjust the scalar to fix the problem: the Helmholtz representation for $\nabla \times \mathbf{v}$,

$$\nabla \times \mathbf{v} = \hat{\mathbf{r}} \tilde{f} + \nabla_1 \tilde{g} + \Lambda \tilde{h} \qquad (5.1.28)$$

with $\langle \tilde{g} \rangle_{S(r)} = \langle \tilde{h} \rangle_{S(r)} = 0$ when $a < r < b$ is given in terms of the Helmholtz scalars for $\mathbf{v}$ by

$$\tilde{f} = \frac{1}{r} \nabla_1^2 h \qquad (5.1.29)$$

$$\tilde{g} = -\frac{1}{r} \partial_r r h \qquad (5.1.30)$$

$$\tilde{h} = \frac{1}{r} (\partial_r r g - f + \langle f \rangle_{S(r)}). \qquad (5.1.31)$$

5.2 The Mie Representation of Vector Fields

In this section we introduce three kinds of divergence-free vector fields: solenoidal, toroidal, and poloidal fields. We will see that every solenoidal field can be written uniquely as the sum of a toroidal and a poloidal part. The toroidal and poloidal fields are each associated with a scalar field from which the fields can be derived via appropriate curl operations. These scalars are unique and are related to the Helmholtz scalars of the previous section. Solenoidal fields deserve all this attention because the geomagnetic field is an example of one, not only in the atmospheric cavity (where we represented it as the gradient of a potential) but in regions such as the core, where electric current flows. Applications appear in the next three sections of the chapter.

5.2.1 Solenoidal Vector Fields

A vector field $\mathbf{v}$ on an open subset U of R^3 is said to be *solenoidal* in U if $\oint_S dA\hat{\mathbf{n}} \cdot \mathbf{v} = 0$ for every closed surface S lying entirely in U. Here $\hat{\mathbf{n}}$ is the unit outward normal to S.

The following important consequence is a point-wise property of solenoidal fields, not very obvious from the definition: if $\mathbf{v}$ is solenoidal in U, then $\nabla \cdot \mathbf{v} = 0$ in U.

Proof: Let B be any small open ball in U. Let ∂B be its boundary. By hypothesis, $\int_{\partial B} d^2\mathbf{r}\, \hat{\mathbf{n}} \cdot \mathbf{v} = 0$. By Gauss' Theorem, $\int_B d^3\mathbf{r}\, (\nabla \cdot \mathbf{v}) = 0$. Therefore $\nabla \cdot \mathbf{v}$ is a scalar field whose integral over every ball B in U vanishes. By taking very small B, and assuming $\nabla \cdot \mathbf{v}$ continuous, we conclude $\nabla \cdot \mathbf{v} = 0$.   ◁

The converse is false, so that a divergence-free field is not necessarily solenoidal. Here is an example: suppose $0 < a < b$. Then $\mathbf{v} = \hat{\mathbf{r}}/r^2$ satisfies $\nabla \cdot \mathbf{v} = 0$ in $S(a,b)$, but $\int_{S(r)} d^2\mathbf{r}\, \hat{\mathbf{r}} \cdot \mathbf{v} \neq 0$, so $\mathbf{v}$ is not solenoidal in $S(a,b)$.

However, we do have the following, which by adding one integral condition to the vanishing divergence, guarantees the solenoidal property: suppose $0 < a < b$ and $\nabla \cdot \mathbf{v} = 0$ in $S(a,b)$ and $\langle v_r \rangle_{S(r_1)} = 0$ for one r_1 in $a < r_1 < b$. Then $\mathbf{v}$ is solenoidal in $S(a,b)$.

Proof: If S is a closed surface in $S(a,b)$, there are only two possibilities: (1) there is a subset U of $S(a,b)$ whose boundary ∂U is just S, or (2) there is a subset U of $S(a,b)$ whose boundary is S together with $S(a)$; see Figure 5.2.1. In Case 1 $\nabla \cdot \mathbf{v} = 0$

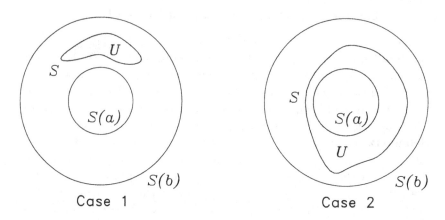

Figure 5.2.1: Subset U of $S(a,b)$ for Cases 1 and 2. The subset U is shaded in both cases.

and Gauss' Theorem give $\int_S d^2\mathbf{r}\ \hat{\mathbf{n}}\cdot\mathbf{v} = 0$. In Case 2, by applying Gauss' Theorem to the shell $S(r,r_1)$ or $S(r_1,r)$, we conclude that $\langle v_r \rangle_{S(r)} = 0$ for all r in $a < r < b$. Then we make r so close to a that $S(r)$ lies inside S, and we apply Gauss' Theorem to the volume bounded by $S(r)$ and S.   ◁

Solenoidal vector fields are of interest to us because the geomagnetic field is solenoidal. There are no isolated monopoles. This follows from Maxwell's equations because $\nabla \cdot \mathbf{B} = 0$ *everywhere* in R^3, not just in a shell. Hence $\int_S d^2\mathbf{r}\ \hat{\mathbf{n}}\cdot\mathbf{B} = 0$ for every closed surface S in R^3. Notice in contrast that the gravitational field in the atmosphere is divergence-free (neglecting the density of the air), but not solenoidal.

When there are currents present, $\nabla \times \mathbf{B} \neq \mathbf{0}$, so we can no longer write $\mathbf{B} = -\nabla\psi$. The utility of this formula arose from its ability to represent the vector field $\mathbf{B}$ in terms of a single scalar field. When $\nabla \times \mathbf{B} \neq 0$, we need a replacement. We have to make do with the weaker condition $\nabla \cdot \mathbf{B} = 0$, which we will see makes it possible to represent $\mathbf{B}$ in terms of two scalar fields. This representation of $\mathbf{B}$ is called the *Mie representation*, after its originator. See Backus (1986) for a survey and bibliography. A statement of the theorem is as follows:

Theorem 1: (Mie Representation Theorem) Suppose $0 \leq a < b \leq \infty$, and $\mathbf{v}$ is a solenoidal vector field in $S(a,b)$. Then there are unique scalar fields p and q in $S(a, b)$ such that

$$\langle p \rangle_{S(r)} = \langle q \rangle_{S(r)} = 0 \quad \text{if} \quad a < r < b \qquad (5.2.1)$$

and

$$\mathbf{v} = \boldsymbol{\nabla} \times \boldsymbol{\Lambda} p + \boldsymbol{\Lambda} q. \qquad (5.2.2)$$

The field $\boldsymbol{\nabla} \times \boldsymbol{\Lambda} p$ is called *poloidal*, and is the poloidal part of $\mathbf{v}$. The field $\boldsymbol{\Lambda} q$ is called *toroidal*, and is the toroidal part of $\mathbf{v}$.

Proof: The Helmholtz representation of $\mathbf{v}$ in $S(a, b)$ is

$$\mathbf{v} = \hat{\mathbf{r}} f + \boldsymbol{\nabla}_1 g + \boldsymbol{\Lambda} h.$$

Then by (5.1.22)

$$\boldsymbol{\nabla} \cdot \mathbf{v} = \frac{1}{r} \left[\frac{1}{r} \partial_r r^2 f + \nabla_1^2 g \right].$$

Thus the fact that $\mathbf{v}$ is solenoidal in $S(a, b)$ means that

$$\langle f \rangle_{S(r)} = 0 \quad \text{for all } r \text{ in } a < r < b \qquad (5.2.3)$$

and also

$$\partial_r (r^2 f) + \nabla_1^2 (rg) = 0 \quad \text{in} \quad S(a, b). \qquad (5.2.4)$$

If $\mathbf{v}$ were to be represented in the form (5.2.2), then (5.1.25) shows that the Helmholtz representation of $\mathbf{v}$ is

$$\mathbf{v} = \hat{\mathbf{r}} \frac{1}{r} (\nabla_1^2 p) - \boldsymbol{\nabla}_1 \left(\frac{1}{r} \partial_r rp \right) + \boldsymbol{\Lambda} q. \qquad (5.2.5)$$

In that case, we must have (because of the uniqueness of the Helmholtz scalars for $\mathbf{v}$)

$$\nabla_1^2 p = rf \qquad (5.2.6)$$

$$\partial_r (rp) = -rg \qquad (5.2.7)$$

$$q = h. \qquad (5.2.8)$$

Equations (5.2.6) and (5.2.1) determine p in $S(a, b)$, and (5.2.8) determines q. This shows that if p and q exist, they are unique.

Do they exist? Because of (5.2.3), we can solve (5.2.6) for p. The solution is

$$p = \nabla_1^{-2}(rf). \tag{5.2.9}$$

Let us choose p and q according to (5.2.9) and (5.2.8) and define the field $\mathbf{u}$ in $S(a, b)$ by

$$\mathbf{u} = \nabla \times \Lambda p + \Lambda q.$$

Then, as noted in (5.2.5),

$$\mathbf{u} = \hat{\mathbf{r}} \frac{1}{r} \nabla_1^2 p - \nabla_1 \frac{1}{r} \partial_r rp + \Lambda q.$$

This field is

$$\mathbf{u} = \hat{\mathbf{r}} f - \nabla_1 \left(\frac{1}{r} \partial_r rp \right) + \Lambda h.$$

We will show that $\mathbf{v} = \mathbf{u}$ if we can show that

$$g = -\frac{1}{r} \partial_r rp,$$

that is,

$$rg + \partial_r rp = 0 \qquad \text{for all } r \text{ in } a < r < b. \tag{5.2.10}$$

Therefore, since $\langle g \rangle_{S(r)} = 0$ and $\langle p \rangle_{S(r)} = 0$, $\langle rg + \partial_r rp \rangle_{S(r)} = 0$. Hence, to prove (5.2.10) it suffices to prove

$$\nabla_1^2 [rg + \partial_r rp] = 0$$

or

$$\nabla_1^2 (rg) + \partial_r r \nabla_1^2 p = 0.$$

But p solves (5.2.6). Therefore what we must prove is

$$\nabla_1^2 (rg) + \partial_r r^2 f = 0.$$

This is precisely the condition (5.2.4), resulting from $\nabla \cdot \mathbf{v} = 0$. ◁

Corollary 1.1: If $\mathbf{v}$ is solenoidal in $S(a,b)$ and we know $\mathbf{v}$ on one surface $S(r)$, then on that $S(r)$ we can find $p, \partial_r p$, and q for $\mathbf{v}$.

Proof: From (5.2.5) we know $\nabla_1^2 p$, $\frac{1}{r}\partial_r rp$, and q on $S(r)$. But knowing ∇_1^2 gives us p because of (5.2.1), and then we get $\partial_r p$ from $\frac{1}{r}\partial_r rp = \partial_r p + \frac{1}{r}p$. ◁

5.2.2 Poloidal and Toroidal Fields

In subsections (5.2.1) and (5.2.2), the scalar fields p and q are called the *poloidal* and the *toroidal scalars* for $\mathbf{v}$. They are the *Mie scalars* of $\mathbf{v}$. The vector fields $\nabla \times \Lambda p$ and Λq are called the *poloidal* and *toroidal parts* of $\mathbf{v}$. The toroidal and poloidal parts of $\mathbf{v}$ deserve study in their own right. So now we study poloidal and toroidal fields in general.

We define a vector field $\mathbf{T}$ on $S(a,b)$ to be *toroidal* if there is a scalar field q' on $S(a,b)$, such that

$$\mathbf{T} = \Lambda q'. \tag{5.2.11}$$

Furthermore, we define a vector field $\mathbf{P}$ on $S(a,b)$ as *poloidal* if it is the curl of a toroidal field, i.e., if there is a scalar field p' on $S(a,b)$ such that

$$\mathbf{P} = \nabla \times \Lambda p'. \tag{5.2.12}$$

Remark 1: If $\mathbf{T}$ is toroidal in $S(a,b)$, then there is a unique scalar field q on $S(a,b)$ such that

$$\mathbf{T} = \Lambda q \quad \text{and} \quad \langle q \rangle_{S(r)} = 0 \quad \text{for } a < r < b. \tag{5.2.13}$$

Proof: Uniqueness is obvious, because (5.2.13) is the Mie representation of $\mathbf{T}$. To prove existence, suppose $\mathbf{T}$ is given by (5.2.11). Let $q = q' - \langle q' \rangle_{S(r)}$. Since $\langle q' \rangle_{S(r)}$ depends only on r, not θ or λ, $\Lambda \langle q' \rangle_{S(r)} = 0$. Hence $\Lambda q' = \Lambda q$. ◁

Remark 2: If $\mathbf{P}$ is poloidal in $S(a,b)$, then there is a unique scalar field p on $S(a,b)$ such that

$$\mathbf{P} = \nabla \times \Lambda p \quad \text{and} \quad \langle p \rangle_{S(r)} = 0 \quad \text{for} \quad a < r < b. \tag{5.2.14}$$

Proof: Again, uniqueness follows from the Mie representation theorem. Existence follows from applying Remark 1 to the toroidal field $\Lambda p'$. ◁

Now we can assert a slightly stronger version of Theorem 1:
Theorem 2: If $\mathbf{v}$ is a solenoidal vector field on $S(a,b)$, then there is
a unique poloidal field $\mathbf{P}$ and a unique toroidal field $\mathbf{T}$ such that

$$\mathbf{v} = \mathbf{P} + \mathbf{T}. \tag{5.2.15}$$

Proof: Existence follows from the Mie representation theorem for
$\mathbf{v}$. To prove uniqueness, suppose $\mathbf{v}$ can be written as in (5.2.15),
with $\mathbf{P}$ and $\mathbf{T}$ given by (5.2.12) and (5.2.11). Then $\mathbf{P}$ and $\mathbf{T}$ can
also be written as in (5.2.14) and (5.2.13). But then (5.2.15) is
the Mie representation of $\mathbf{v}$, so p and q in (5.2.14) and (5.2.13)
are determined uniquely by $\mathbf{v}$. Thus $\mathbf{P}$ and $\mathbf{T}$ are also determined
uniquely by $\mathbf{v}$.   ◁

Corollary 2.1: A vector field that is both poloidal and toroidal
in $S(a,b)$ vanishes.
Proof: If $\mathbf{v} = \mathbf{P} + 0 = 0 + \mathbf{T}$ then applying Theorem 2, $\mathbf{P} = 0$ and
$0 = \mathbf{T}$.   ◁

Obviously, the curl of a toroidal field is poloidal. The following
formula shows that the curl of a poloidal field is toroidal: if p is a
scalar field on $S(a,b)$ then

$$\boldsymbol{\nabla} \times (\boldsymbol{\nabla} \times \boldsymbol{\Lambda}p) = -\boldsymbol{\Lambda}(\nabla^2 p). \tag{5.2.16}$$

Proof: By (7.2.17) for any vector field $\mathbf{v}$, $\boldsymbol{\nabla} \times (\boldsymbol{\nabla} \times \mathbf{v}) = \boldsymbol{\nabla}(\boldsymbol{\nabla} \cdot \mathbf{v}) - \nabla^2\mathbf{v}$. If $\mathbf{v} = \boldsymbol{\Lambda}p$, then $\boldsymbol{\nabla} \cdot \mathbf{v} = 0$ and $\nabla^2\mathbf{v} = \nabla^2\boldsymbol{\Lambda}p = \boldsymbol{\Lambda}\nabla^2 p$
because of (3.4.6).   ◁

Now we have a test for poloidalness or toroidalness.
Theorem 3: A vector field $\mathbf{T}$ in $S(a,b)$ is toroidal if and only if
it is solenoidal and tangential, that is, $\hat{\mathbf{r}} \cdot \mathbf{T} = 0$.
Proof: If $\mathbf{T}$ is toroidal, $\boldsymbol{\nabla} \cdot \mathbf{T} = \boldsymbol{\nabla} \cdot \boldsymbol{\Lambda}q = 0$ and $\hat{\mathbf{r}} \cdot \mathbf{T} = \hat{\mathbf{r}} \cdot \boldsymbol{\Lambda}q = 0$.
But then $\langle T_r \rangle_{S(r)} = 0$ for $a < r < b$, so $\mathbf{T}$ is solenoidal. Conversely,
if $\mathbf{T}$ is solenoidal it has a Mie representation,

$$\mathbf{T} = \boldsymbol{\nabla} \times \boldsymbol{\Lambda}p + \boldsymbol{\Lambda}q \quad \text{and} \quad \langle p \rangle_{S(r)} = \langle q \rangle_{S(r)} = 0$$
$$\mathbf{T} = \hat{\mathbf{r}}\left(\frac{1}{r}\nabla_1^2 p\right) - \boldsymbol{\nabla}_1\left(\frac{1}{r}\partial_r rp\right) + \boldsymbol{\Lambda}q.$$

If $\hat{\mathbf{r}} \cdot \mathbf{T} = 0$, then $\nabla_1^2 p = 0$, so $p = 0$, so $\mathbf{T} = \boldsymbol{\Lambda}q$.   ◁

Theorem 4: A vector field $\mathbf{P}$ in $S(a, b)$ is poloidal if and only if it is solenoidal and its curl is tangential (i.e., $\hat{\mathbf{r}} \cdot \boldsymbol{\nabla} \times \mathbf{P} = 0$).

Proof: If $\mathbf{P}$ is poloidal, clearly $\boldsymbol{\nabla} \cdot \mathbf{P} = 0$. Also, $\hat{\mathbf{r}} \cdot \mathbf{P} = \frac{1}{r}\nabla_1^2 p$, so $\langle \hat{\mathbf{r}} \cdot p \rangle_{S(r)} = \frac{1}{r}\langle \nabla_1^2 p \rangle_{S(r)} = 0$, so $\mathbf{P}$ is solenoidal. By equation (5.2.16) $\hat{\mathbf{r}} \cdot \boldsymbol{\nabla} \times \mathbf{P} = -\hat{\mathbf{r}} \cdot \boldsymbol{\Lambda}(\nabla^2 p) = 0$. Conversely, if $\mathbf{P}$ is solenoidal, it has a Mie representation

$$\mathbf{P} = \boldsymbol{\nabla} \times \boldsymbol{\Lambda} p + \boldsymbol{\Lambda} q \quad \text{and} \quad \langle p \rangle_{S(r)} = \langle q \rangle_{S(r)} = 0.$$

Then
$$\boldsymbol{\nabla} \times \mathbf{P} = \boldsymbol{\nabla} \times \boldsymbol{\Lambda} q - \boldsymbol{\Lambda}\nabla^2 p$$
$$= \hat{\mathbf{r}}\frac{1}{r}\nabla_1^2 q - \boldsymbol{\nabla}_1\frac{1}{r}\partial_r rq - \boldsymbol{\Lambda}(\nabla^2 p)\overset{\cdot}{}$$

If also $\hat{\mathbf{r}} \cdot \boldsymbol{\nabla} \times \mathbf{P} = 0$, then $\nabla_1^2 q = 0$, so $q = 0$, so $\mathbf{P} = \boldsymbol{\nabla} \times \boldsymbol{\Lambda} p$. ◁

Corollary 4.1: A vector field $\mathbf{P}$ in $S(a, b)$ is poloidal if and only if it is solenoidal and its curl is toroidal.

Proof: Sufficiency follows from (5.2.16). Necessity follows from Theorems 4 and 2. ◁

Finally we note properties of the radial components of toroidal and poloidal fields. First, a toroidal field has no radial component. This is Theorem 3. Second, on each $S(r)$ in $S(a, b)$ the poloidal scalar of a poloidal field $\mathbf{P}$ is determined by P_r, and if P_r is known throughout $S(a, b)$, $\mathbf{P}$ is determined there.

Proof: The poloidal part follows from the unique solution of

$$\nabla_1^2 p = rP_r \quad \text{and} \quad \langle p \rangle_{S(r)} = 0.$$

Note that P_r can be chosen arbitrarily subject to $\langle P_r \rangle_{S(r)} = 0$. ◁

5.2.3 Continuity of the Mie Scalars

Suppose $\mathbf{v}$ is solenoidal in $S(a, c)$. Then we write the usual Mie representation:

$$\mathbf{v} = \boldsymbol{\nabla} \times \boldsymbol{\Lambda} p + \boldsymbol{\Lambda} q$$
$$= \hat{\mathbf{r}}\frac{1}{r}\nabla_1^2 p - \boldsymbol{\nabla}_1\frac{1}{r}\partial_r rp + \boldsymbol{\Lambda} q. \tag{5.2.17}$$

This is a Helmholtz representation for $\mathbf{v}$. Therefore from the explicit solution for the Helmholtz scalars, (5.1.18):

$$\nabla_1^2 p = r v_r$$

$$\nabla_1^2 \left(\frac{1}{r} \partial_r r p \right) = -\nabla_1 \cdot \mathbf{v}_S$$

$$\nabla_1^2 q = \mathbf{\Lambda} \cdot \mathbf{v}_S.$$

Hence

$$p = \nabla_1^{-2}[r v_r] \qquad\qquad (5.2.18)$$

$$\frac{1}{r} \partial_r r p = -\nabla_1^{-2}[\nabla_1 \cdot \mathbf{v}_S] \qquad\qquad (5.2.19)$$

$$q = \nabla_1^{-2}[\mathbf{\Lambda} \cdot \mathbf{v}_S]. \qquad\qquad (5.2.20)$$

Now suppose $a < b < c$ and that $\mathbf{v}$ has a jump discontinuity across $S(b)$. Equation (5.2.18) holds on $S(b + \epsilon)$ and $S(b - \epsilon)$ for any $\epsilon > 0$, no matter how small. Hence

$$p^+ = \nabla_1^{-2}[r v_r]^+ \quad \text{and} \quad p^- = \nabla_1^{-2}[r v_r]^-.$$

Thus subtracting we find an equation for the jump in the poloidal scalar:

$$[p]_-^+ = \nabla_1^{-2}[r v_r]_-^+. \qquad\qquad (5.2.21)$$

Similarly

$$\left[\frac{1}{r} \partial_r r p \right]_-^+ = -\nabla_1^{-2}(\nabla_1 \cdot [\mathbf{v}_S]_-^+) \qquad\qquad (5.2.22)$$

$$[q]_-^+ = \nabla_1^{-2}(\mathbf{\Lambda} \cdot [\mathbf{v}_S]_-^+). \qquad\qquad (5.2.23)$$

These are special cases of the following theorem, whose proof we omit:

Theorem: If $\mathbf{v}$ is solenoidal in $S(a, c)$ and v_r is continuous there, its poloidal scalar is continuous there. If $\mathbf{v}$ is continuous there, then its poloidal scalar is continuously differentiable and its toroidal scalar is continuous there.

In all of the foregoing, we have discussed the Mie representation of $\mathbf{v}$ in a shell $S(a, b)$. This shell never includes $\mathbf{0}$ even if $a = 0$ because it has $a < r < b$. What if $\mathbf{v}$ is solenoidal in a ball $B(b)$? This will happen whenever $\nabla \cdot \mathbf{v} = 0$ in $B(b)$; no other condition is required now. In $S(0, b)$ we have

$$\mathbf{v} = \mathbf{P} + \mathbf{T} = \nabla \times \Lambda p + \Lambda q.$$

The integral solution (5.1.11) can be used to show that if $\mathbf{v}$ is N-times continuously differentiable in $B(b)$ then in that ball: (a) p is N-times continuously differentiable so $\mathbf{P}$ is $N - 2$ times continuously differentiable; (b) q is $N - 1$ times continuously differentiable, so $\mathbf{T}$ is $N - 2$ times continuously differentiable. For details see Backus (1986).

5.2.4 Summary

We list here a summary of the most important of the foregoing results for easy reference:
A solenoidal field $\mathbf{v}$ is a divergence-free vector field, with the additional property that there is no net flux out of any closed surface formed inside the region where it is defined. The magnetic field $\mathbf{B}$ is an example of a solenoidal field. In the following we consider the solenoidal field to be defined in a spherical shell $S(a, b)$ with its center at the origin.
Every solenoidal field can be uniquely decomposed into the sum of a toroidal field $\mathbf{T}$ and poloidal field $\mathbf{P}$:

$$\mathbf{v} = \mathbf{T} + \mathbf{P}$$

where the two parts are themselves solenoidal.
Toroidal fields can be uniquely associated with a scalar field q and are derived from q by the relation:

$$\mathbf{T} = \Lambda q$$

and $\langle q \rangle_{S(r)} = 0$ for $a < r < b$. On each spherical surface $S(r)$ within $S(a, b)$ the toroidal scalar for the toroidal part of $\mathbf{v}$ obeys the differential equation:

$$\nabla_1^2 q = \Lambda \cdot \mathbf{v}_S \tag{5.2.24}$$

where $\mathbf{v}_S$ is the tangent vector field associated with $\mathbf{v}$ on $S(r)$: $\mathbf{v}_S = \mathbf{v} - \hat{\mathbf{r}}\hat{\mathbf{r}} \cdot \mathbf{v}$. The differential equation and condition of averaging to zero are sufficient to determine the scalar field q uniquely from $\mathbf{v}$; see (5.1.11).

A toroidal field has no radial component: this means $\hat{\mathbf{r}} \cdot \mathbf{T} = 0$ at every point. The curl of a toroidal field is always a poloidal field.

A poloidal field is defined to be the curl of a toroidal field: the associated scalar field is p:

$$\left. \begin{aligned} \mathbf{P} &= \boldsymbol{\nabla} \times \boldsymbol{\Lambda} p \\ &= \frac{\hat{\mathbf{r}}}{r} \nabla_1^2 p - \boldsymbol{\nabla}_1 r^{-1} \partial_r r p \end{aligned} \right\}, \qquad (5.2.25)$$

and of course $\langle p \rangle_{S(r)} = 0$ for $a < r < b$. The second form of the equation for $\mathbf{P}$ is simply an alternative expansion of $\boldsymbol{\nabla} \times \boldsymbol{\Lambda} p$. On each spherical surface the poloidal scalar p for the poloidal part of $\mathbf{v}$ obeys the differential equation:

$$\nabla_1^2 p = r \hat{\mathbf{r}} \cdot \mathbf{v}, \qquad (5.2.26)$$

and this equation and the averaging condition serve to define uniquely the scalar field p from $\mathbf{v}$. In addition, p obeys on each $S(r)$ in the shell another differential equation:

$$\nabla_1^2 \left(\frac{1}{r} \partial_r r p \right) = -\boldsymbol{\nabla} \cdot \mathbf{v}_S. \qquad (5.2.27)$$

Complete reciprocity obtains: the curl of a poloidal field is always a toroidal field. The corresponding toroidal scalar is $-\nabla^2 p$.

5.3 Application to Sources

The first application of our new representation is to the problem considered in Chapter 4 of the sources of a magnetic field observed in a shell. We showed there that it was possible to divide the observed field into parts of internal and external origin; that is, the sources might lie inside the inner surface or beyond the outer one. This resolution depended on there being no electric currents or other sources within the shell. First we provide the same resolution in the Mie representation. Then we generalize the idea as far as possible, allowing currents to flow inside the shell; this is possible with the Mie representation because it holds for all solenoidal fields.

5.3.1 Mie Sources of a Magnetic Field

Suppose that $\mathbf{B}$ is the magnetic field and $\mathbf{J} = \mathbf{J}^{(F)} + \boldsymbol{\nabla} \times \mathbf{M}$ is the total current (including averaging electron current $\boldsymbol{\nabla} \times \mathbf{M}$ in neutral molecules) in the shell $S(a, c)$. We accept the pre-Maxwell equations, so

$$\boldsymbol{\nabla} \times \mathbf{B} = \mu_0 \mathbf{J}. \tag{5.3.1}$$

Taking the curl of the Mie representation for $\mathbf{B}$,

$$\mathbf{B} = \boldsymbol{\nabla} \times \boldsymbol{\Lambda} p + \boldsymbol{\Lambda} q = \mathbf{P} + \mathbf{T} \tag{5.3.2}$$

we deduce from (5.2.16) that

$$\mu_0 \mathbf{J} = \boldsymbol{\nabla} \times \boldsymbol{\Lambda} q - \boldsymbol{\Lambda} \nabla^2 p. \tag{5.3.3}$$

Thus $\mathbf{J}$ is also solenoidal in $S(a, c)$. If we write its Mie representation (5.3.2) as

$$\mathbf{J} = \boldsymbol{\nabla} \times \boldsymbol{\Lambda} \tilde{p} + \boldsymbol{\Lambda} \tilde{q} \tag{5.3.4}$$

then obviously

$$q = \mu_0 \tilde{p} \tag{5.3.5}$$

$$\nabla^2 p = -\mu_0 \tilde{q} = -\frac{1}{\epsilon_0} \left(\frac{\tilde{q}}{c^2} \right). \tag{5.3.6}$$

These equations tell us that poloidal currents produce toroidal magnetic fields, and toroidal currents produce poloidal magnetic fields. Furthermore, the toroidal magnetic scalar is just μ_0 times the poloidal current scalar, and the poloidal magnetic scalar is numerically equal to the electrostatic potential produced by charge density $\tilde{q}/c^2$ where $\tilde{q}$ is the toroidal current scalar and c is the speed of light. The poloidal magnetic field *is not* $-\boldsymbol{\nabla} p$; it is $\boldsymbol{\nabla} \times \boldsymbol{\Lambda} p$.

We can express the toroidal magnetic field directly in terms of the current as follows: applying (5.2.26) to the poloidal part of $\mathbf{J}$ we see that the poloidal scalar $\tilde{p}$ in (5.3.4) obeys

$$\nabla_1^2 \tilde{p} = r J_r$$

$$\tilde{p} = \nabla_1^{-2} r J_r.$$

Then through (5.3.5), the toroidal part of the magnetic field is given by

$$\mathbf{T} = \mathbf{\Lambda}q = \mu_0 \mathbf{\Lambda}[\nabla_1^{-2}(r J_r)]. \tag{5.3.7}$$

Equation (5.3.7) shows that if $\mathbf{J} = \mathbf{0}$ in $S(a,c)$, then there can be no toroidal magnetic field there. But (5.3.7) shows something more remarkable. If $a < b < c$ and $J_r = 0$ on $S(b)$, then $\mathbf{T} = \mathbf{0}$ on $S(b)$. That is,

Theorem 1: If no current crosses a spherical surface $S(b)$, the magnetic field on that surface is purely poloidal.

The physical reason for this result is easy to see. If $J_r = 0$ on $S(b)$, then by Stokes' Theorem (7.3.1) $\oint d\ell \hat{\tau} \cdot \mathbf{B}_S = 0$ around any closed curve on $S(b)$. Then $\mathbf{B}_S = \nabla_1 \phi$ for some scalar ϕ; $\mathbf{B}_S$ contains no toroidal part, $\mathbf{\Lambda}q$, on $S(b)$.

If there are no currents in $S(a,b)$, we now have two ways of representing $\mathbf{B}$ there in terms of a single scalar:

$$\mathbf{B} = -\nabla\psi \qquad \nabla^2\psi = 0 \qquad \text{(Gauss representation)} \tag{5.3.8}$$

$$\mathbf{B} = \nabla \times \mathbf{\Lambda}p \qquad \nabla^2 p = 0 \qquad \text{(Mie representation)}. \tag{5.3.9}$$

The obvious question is, How are ψ and p related? We can find out by writing out (5.3.8) and (5.3.9) in their Helmholtz forms. We have

$$\mathbf{B} = -\hat{\mathbf{r}}\partial_r\psi - \nabla_1 \frac{1}{r}\psi \tag{5.3.10}$$

$$\mathbf{B} = \hat{\mathbf{r}}\frac{1}{r}\nabla_1^2 p - \nabla_1 \frac{1}{r}\partial_r r p. \tag{5.3.11}$$

Since we have agreed to choose the constant of integration in ψ so that

$$\langle\psi\rangle_{S(r)} = 0$$

for $a < r < b$, therefore both (5.3.10) and (5.3.11) are the Helmholtz representation for $\mathbf{B}$. Since the Helmholtz scalars are unique,

$$-\partial_r\psi = \frac{1}{r}\nabla_1^2 p \quad \text{and} \quad \frac{1}{r}\partial_r r p = \frac{1}{r}\psi.$$

Thus

$$\psi = \partial_r(rp) \tag{5.3.12}$$

$$p = -\nabla_1^{-2}(r\partial_r\psi). \tag{5.3.13}$$

In particular, we can express p in terms of the Gauss coefficients. If

$$\psi = a \sum_{\ell=1}^{\infty} \sum_{m=-\ell}^{\ell} \left[g_\ell^m \left(\frac{a}{r}\right)^{\ell+1} + k_\ell^m \left(\frac{r}{a}\right)^{\ell} \right] \beta_\ell^m(\hat{\mathbf{r}}) \tag{5.3.14}$$

then

$$r\partial_r\psi = a \sum_{\ell=1}^{\infty} \sum_{m=-\ell}^{\ell} \left[-(\ell+1)g_\ell^m \left(\frac{a}{r}\right)^{\ell+1} + \ell k_\ell^m \left(\frac{r}{a}\right)^{\ell} \right] \beta_\ell^m(\hat{\mathbf{r}}).$$

Since

$$\nabla_1^{-2}\beta_\ell^m = -\frac{1}{\ell(\ell+1)}\beta_\ell^m$$

we have, finally,

$$p(r\hat{\mathbf{r}}) = a \sum_{\ell=1}^{\infty} \sum_{m=-\ell}^{\ell} \left[\frac{1}{\ell}g_\ell^m \left(\frac{a}{r}\right)^{\ell+1} - \frac{1}{(\ell+1)}k_\ell^m \left(\frac{r}{a}\right)^{\ell} \right] \beta_\ell^m(\hat{\mathbf{r}}). \tag{5.3.15}$$

The advantage of the Mie over the Gauss representation is that Mie continues into regions where $\mathbf{J} \neq \mathbf{0}$, while Gauss does not. There is simply no way of adding an extra term to (5.3.8) to make it work where $\mathbf{J} \neq \mathbf{0}$, while (5.3.9) is merely a special case of (5.3.2), with $q = 0$ in $S(a,c)$ and with p continuing smoothly out of $S(a,c)$ and satisfying $\nabla^2 p = 0$ in $S(a,c)$. This generality will prove very useful when we study systems in which the fields above the earth must be connected to the currents inside it, for example, in section 5.4.

5.3.2 Internal and External Fields: A Complication

It will be recalled from Chapter 4 that if $S(a,c)$ had no currents in it, then the Gauss representation of $\mathbf{B}$ made it possible to use measurements of $\mathbf{B}$ on $S(a)$ alone to separate $\mathbf{B}$ into two parts on $S(a)$,

$$\mathbf{B} = \mathbf{B}^{I(a)} + \mathbf{B}^{E(a)} \tag{5.3.16}$$

where $\mathbf{B}^{I(a)}$ is produced by sources inside $S(a)$ and $\mathbf{B}^{E(a)}$ is produced by sources outside it. The Mie representation makes such a breakdown possible even when $S(a, c)$ contains currents. Such a separation has applications for magnetic field measurements in satellites that might be orbiting in regions where significant currents are flowing.

There is, however, a complication. What do we mean by the magnetic field produced by the currents inside $S(a)$? To illustrate the difficulty, consider an arbitrary region V. Suppose space is filled with currents of density $\mathbf{J}$. What part of the resulting magnetic field B is produced by the currents inside V? The obvious definition is

$$\mathbf{B}^{(I)}(\mathbf{r}) = \frac{\mu_0}{4\pi} \int_V d^3s\ \boldsymbol{\nabla}_{\mathbf{r}} \frac{1}{|\mathbf{r} - \mathbf{s}|} \times \mathbf{J}(\mathbf{s}), \qquad (5.3.17)$$

the result of adding up the Biot–Savart contributions of the individual elements of current in V. Equation (5.3.17) is equivalent to

$$\mathbf{B}^{(I)}(\mathbf{r}) = \boldsymbol{\nabla} \times \mathbf{A}^{(I)}(\mathbf{r}) \qquad (5.3.18)$$

where

$$\mathbf{A}^{(I)}(\mathbf{r}) = \frac{\mu_0}{4\pi} \int_V d^3s\ \frac{1}{|\mathbf{r} - \mathbf{s}|} \mathbf{J}(\mathbf{s}). \qquad (5.3.19)$$

What are the sources of $\mathbf{B}^{(I)}$? Since $\boldsymbol{\nabla} \cdot \mathbf{B}^{(I)} = 0$, we want to compute $\boldsymbol{\nabla} \times \mathbf{B}^{(I)}$; that is $\mathbf{B}^{(I)}$'s only source. We have

$$\boldsymbol{\nabla} \times \mathbf{B}^{(I)} = \boldsymbol{\nabla} \times (\boldsymbol{\nabla} \times \mathbf{A}^{(I)}) = \boldsymbol{\nabla}(\boldsymbol{\nabla} \cdot \mathbf{A}^{(I)}) - \nabla^2 \mathbf{A}^{(I)}. \qquad (5.3.20)$$

The kth component of $\mathbf{A}^{(I)}$ is given by (5.3.19) to be the electrostatic potential due to charge density J_k/c^2. Thus, because of the well-known relation between the Coulomb integral and the Poisson equation, (5.3.19) implies

$$\begin{aligned}
\mathbf{A}^{(I)} &= -\mu_0 \mathbf{J} && \text{in } V \\
&= 0 && \text{outside } V.
\end{aligned}$$

Thus (5.3.20) becomes

$$\begin{aligned}
\boldsymbol{\nabla} \times \mathbf{B}^{(I)} &= \mu_0\ \mathbf{J} + \boldsymbol{\nabla}(\boldsymbol{\nabla} \cdot \mathbf{A}^{(I)}) && \text{in } V \\
&= \phantom{\mu_0\ \mathbf{J} +} \boldsymbol{\nabla}(\boldsymbol{\nabla} \cdot \mathbf{A}^{(I)}) && \text{outside } V.
\end{aligned} \qquad (5.3.21)$$

If we can show that $\nabla(\nabla \cdot \mathbf{A}^{(I)}) = 0$ in R^3, then $\nabla \times \mathbf{B}^{(I)}$ will have all its sources in V, as we had hoped.

To calculate $\nabla(\nabla \cdot \mathbf{A}^{(I)})$, we write (5.3.19) as

$$A_k^{(I)}(\mathbf{r}) = \frac{\mu_0}{4\pi} \int_V d^3s \, \frac{1}{|\mathbf{r} - \mathbf{s}|} J_k(\mathbf{s}).$$

Then

$$\partial_j A_k^{(I)}(\mathbf{r}) = \frac{\mu_0}{4\pi} \int_V d^3s \left[\frac{\partial}{\partial r_j} \frac{1}{|\mathbf{r} - \mathbf{s}|} \right] J_k(\mathbf{s})$$

$$= -\frac{\mu_0}{4\pi} \int_V d^3s \left[\frac{\partial}{\partial s_j} \frac{1}{|\mathbf{r} - \mathbf{s}|} \right] J_k(\mathbf{s})$$

$$= \frac{\mu_0}{4\pi} \int_V d^3s \left\{ \frac{\partial}{\partial s_j} \left[\frac{J_k(\mathbf{s})}{|\mathbf{r} - \mathbf{s}|} \right] - \frac{1}{|\mathbf{r} - \mathbf{s}|} \frac{\partial}{\partial s_j} J_k(\mathbf{s}) \right\}.$$

Now set $j = k$ and sum to obtain

$$\nabla \cdot \mathbf{A}^{(I)}(\mathbf{r}) = -\frac{\mu_0}{4\pi} \int_V d^3s \, \nabla_\mathbf{s} \cdot \left[\frac{\mathbf{J}(\mathbf{s})}{|\mathbf{r} - \mathbf{s}|} \right] + \frac{\mu_0}{4\pi} \int_V d^3s \, \frac{\nabla \cdot \mathbf{J}(\mathbf{s})}{|\mathbf{r} - \mathbf{s}|}.$$

Since $\nabla \cdot \mathbf{J} = 0$, the second integrand vanishes. The first integral is evaluated by Gauss' Theorem to give

$$\nabla \cdot \mathbf{A}^{(I)}(\mathbf{r}) = -\frac{\mu_0}{4\pi} \int_V d^2s \, \frac{\hat{\mathbf{n}} \cdot \mathbf{J}(\mathbf{s})}{|\mathbf{r} - \mathbf{s}|}. \tag{5.3.22}$$

Thus $-\nabla \cdot \mathbf{A}^{(I)}$ is the electrostatic potential produced by the surface charge density $\hat{\mathbf{n}} \cdot \mathbf{J}/c^2$ on ∂V, where c is the speed of light. Then $\nabla(\nabla \cdot \mathbf{A}^{(I)})$ is the electric field produced by this charge density. In general it will be different from $\mathbf{0}$ everywhere, so (5.3.21) shows that the sources of the $\mathbf{B}^{(I)}$ defined by (5.3.17) are everywhere, not just in V. The only exception occurs when

$$\hat{\mathbf{n}} \cdot \mathbf{J} = 0 \quad \text{on} \quad \partial V. \tag{5.3.23}$$

In that case, $\nabla(\nabla \cdot \mathbf{A}^{(I)}) = \mathbf{0}$, and (5.3.21) shows that $\mathbf{B}^{(I)}$ defined by (5.3.17) does have its sources inside V.

5.3.3 Separation of Poloidal Fields

To resolve this complication, let the total current be given by (5.3.4). Then the toroidal current is $\Lambda\tilde{q}$. Define

$$\mathbf{P}^{I(a)} = \frac{\mu_0}{4\pi} \int_{B(a)} d^3\mathbf{s}\, \boldsymbol{\nabla}_{\mathbf{r}} \frac{1}{|\mathbf{r}-\mathbf{s}|} \times \Lambda\tilde{q} \qquad (5.3.24)$$

$$\mathbf{P}^{E(a)} = \frac{\mu_0}{4\pi} \int_{R^3 \setminus B(a)} d^3\mathbf{s}\, \boldsymbol{\nabla}_{\mathbf{r}} \frac{1}{|\mathbf{r}-\mathbf{s}|} \times \Lambda\tilde{q}. \qquad (5.3.25)$$

Here $B(a)$ = ball of radius a centered on $\mathbf{0}$. The discussion of (5.3.17) shows that the two solenoidal fields $\mathbf{P}^{I(a)}$ and $\mathbf{P}^{E(a)}$ satisfy

$$
\begin{aligned}
\boldsymbol{\nabla} \times \mathbf{P}^{I(a)} &= \mu_0\Lambda\tilde{q} && \text{inside } S(a) \\
&= \mathbf{0} && \text{outside } S(a) \\
\boldsymbol{\nabla} \times \mathbf{P}^{E(a)} &= \mathbf{0} && \text{inside } S(a) \\
&= \mu_0\Lambda\tilde{q} && \text{outside } S(a).
\end{aligned}
$$

Thus $\mathbf{P}^{I(a)}$ and $\mathbf{P}^{E(a)}$ are both poloidal fields since their curls are evidently toroidal. Their poloidal scalars satisfy

$$
\left.
\begin{aligned}
\nabla^2 p^{I(a)} &= -\mu_0\tilde{q} && \text{inside } S(a) \\
\nabla^2 p^{I(a)} &= 0 && \text{outside } S(a) \\
p^{I(a)} \text{ and } \partial_r p^{I(a)} && & \text{continuous at } S(a)
\end{aligned}
\right\} \qquad (5.3.26)
$$

(the continuity conditions are obtained by observing that there are no surface currents in (5.3.24, 5.3.25), so $\mathbf{P}^{I(a)}$ and $\mathbf{P}^{E(a)}$ are continuous). Similarly

$$
\left.
\begin{aligned}
\nabla^2 p^{E(a)} &= 0 && \text{inside } S(a) \\
\nabla^2 p^{E(a)} &= -\mu_0\tilde{q} && \text{outside } S(a) \\
p^{E(a)} \text{ and } \partial_r p^{E(a)} && & \text{continuous at } S(a)
\end{aligned}
\right\} \qquad (5.3.27)
$$

also, clearly,

$$\boldsymbol{\nabla} \times (\mathbf{P}^{I(A)} + \mathbf{P}^{E(a)}) = \mu_0\Lambda\tilde{q} \qquad \text{everywhere,}$$

so if
$$\mathbf{B} = \mathbf{P} + \mathbf{T} \tag{5.3.28}$$

then
$$\mathbf{P} = \mathbf{P}^{I(a)} + \mathbf{P}^{E(a)}. \tag{5.3.29}$$

At least we have succeeded in meaningfully *defining* what we mean by the part of the poloidal magnetic field produced by currents inside $S(a)$ and the part produced by currents outside.

5.3.4 The Generalization of Gauss' Resolution

What can we say about the toroidal field $\mathbf{T}$? By (5.3.7), on $S(b)$

$$\mathbf{T} = \mu_0 \mathbf{\Lambda}[\nabla_1^{-2}(rJ_r)] = \mu_0 \mathbf{\Lambda}\tilde{p} \tag{5.3.30}$$

where $\tilde{p}$ is the poloidal current scalar. What (5.3.30) says is that if two poloidal currents have the same scalar on $S(a)$, they will produce the same toroidal magnetic field on $S(a)$. The toroidal magnetic field on $S(a)$ is produced by the radial current crossing $S(a)$, not by the currents inside or outside $S(a)$. Thus the generalization of Gauss' resolution is

$$\mathbf{B} = \mathbf{P}^{I(a)} + \mathbf{P}^{E(a)} + \mathbf{T}^{S(a)} \qquad \text{on } S(a). \tag{5.3.31}$$

That is, on $S(a)$ an arbitrary magnetic field is the sum of $\mathbf{P}^{I(a)}$, the poloidal field produced by toroidal currents inside $S(a)$, of $\mathbf{P}^{E(a)}$, the poloidal field produced by toroidal currents outside $S(a)$, and of $\mathbf{T}^{S(a)}$, the toroidal field produced by the poloidal current density flowing across $S(a)$.

There remains the question, If we know $\mathbf{B}$ on $S(a)$, how can we find $\mathbf{P}^{I(a)}$, $\mathbf{P}^{E(a)}$, and $\mathbf{T}^{S(a)}$ on $S(a)$? Let $\{\beta_\ell^{-\ell}, \ldots, \beta_\ell^\ell\}$ be an orthonormal basis for $\mathcal{H}_\ell$ and define

$$\mathbf{b}_\ell^m(r\hat{\mathbf{r}}) = -a\nabla\left[\left(\frac{r}{a}\right)^\ell \beta_\ell^m(\hat{\mathbf{r}})\right]$$

$$= -\left(\frac{r}{a}\right)^{\ell-1}[\hat{\mathbf{r}}\ell\beta_\ell^m + \nabla_1\beta_\ell^m]$$

$$\mathbf{B}_\ell^m(r\hat{\mathbf{r}}) = -a\nabla\left[\left(\frac{a}{r}\right)^{\ell+1} \beta_\ell^m(\hat{\mathbf{r}})\right]$$

$$= +\left(\frac{a}{r}\right)^{\ell-1}[\hat{\mathbf{r}}(\ell+1)\beta_\ell^m - \nabla_1\beta_\ell^m].$$

These are of course external and internal magnetic fields associated with a particular basis element according to the Gauss separation. If we define the inner product $\langle f | g \rangle = \langle \bar{f} g \rangle_{S(a)}$, then the theorem at the end of section 4.3 shows that

$$\left.\begin{aligned}
\langle \mathbf{b}_\ell^m | \mathbf{B}_L^M \rangle &= 0 \\
\langle \mathbf{b}_\ell^m | \mathbf{b}_L^M \rangle &= \delta_{\ell L} \delta_{mM} \ell (2\ell + 1) \\
\langle \mathbf{B}_\ell^m | \mathbf{B}_L^M \rangle &= \delta_{\ell L} \delta_{mM} (\ell + 1)(2\ell + 1)
\end{aligned}\right\}. \tag{5.3.32}$$

Also define

$$\mathbf{T}_\ell^m(a\hat{\mathbf{r}}) = \mathbf{\Lambda} \beta_\ell^m(\hat{\mathbf{r}}).$$

Now, because $\langle \boldsymbol{\nabla}_1 f \cdot \mathbf{\Lambda} g \rangle_{S(a)} = -\langle f \boldsymbol{\nabla}_1 \cdot \mathbf{\Lambda} g \rangle_{S(a)} = 0$, we also have

$$\langle \mathbf{T}_\ell^m | \mathbf{b}_L^M \rangle = \langle \mathbf{T}_\ell^m | \mathbf{B}_L^M \rangle = 0. \tag{5.3.33}$$

Also,

$$\langle \mathbf{\Lambda} f \cdot \mathbf{\Lambda} g \rangle_{S(a)} = -\langle f \mathbf{\Lambda} \cdot \mathbf{\Lambda} g \rangle_{S(a)} = -\langle f \nabla_1^2 g \rangle$$

and

$$-\nabla_1^2 \beta_\ell^m = \ell(\ell + 1)\beta_\ell^m,$$

so

$$\langle \mathbf{T}_\ell^m | \mathbf{T}_L^M \rangle = \delta_{\ell L} \delta_{mM} \ell(\ell + 1). \tag{5.3.34}$$

By (5.3.26), there are coefficients $g_\ell^m(a)$ such that for $r > a$

$$p^{I(a)}(r\hat{\mathbf{r}}) = a \sum_{\ell=1}^{\infty} \frac{1}{\ell} \left(\frac{a}{r}\right)^{\ell+1} \sum_{m=-\ell}^{\ell} g_\ell^m(a) \beta_\ell^m(\hat{\mathbf{r}}). \tag{5.3.35}$$

Also, for $r > a$, $\mathbf{P}^{I(a)}$ has a magnetic scalar potential $\psi^{I(a)} = \partial_r r p^{I(a)}$, so

$$\psi^{I(a)}(r\hat{\mathbf{r}}) = a \sum_{\ell=1}^{\infty} \sum_{m=-\ell}^{\ell} g_\ell^m(a) \left(\frac{a}{r}\right)^{\ell+1} \beta_\ell^m(\hat{\mathbf{r}}).$$

Thus

$$\mathbf{P}^{I(a)}(r\hat{\mathbf{r}}) = \sum_{\ell=1}^{\infty} \left(\frac{a}{r}\right)^{\ell+2} \sum_{m=-\ell}^{\ell} g_\ell^m(a)\mathbf{B}_\ell^m(a\hat{\mathbf{r}}) \quad \text{if} \quad r > a. \quad (5.3.36)$$

Similarly, by (5.3.27), there are coefficients $k_\ell^m(a)$ such that

$$p^{E(a)}(r\hat{\mathbf{r}}) = -a\sum_{\ell=1}^{\infty} \frac{1}{(\ell+1)} \left(\frac{r}{a}\right)^{\ell} \sum_{m=-\ell}^{\ell} k_\ell^m(a)\beta_\ell^m(\hat{\mathbf{r}}).$$

And for $r < a$, $\mathbf{P}^{E(a)}$ has a magnetic scalar potential $\psi^{E(a)} = \partial_r r p^{E(a)}$, with

$$\psi^{E(a)}(r\hat{\mathbf{r}}) = a\sum_{\ell=1}^{\infty} \left(\frac{r}{a}\right)^{\ell} \sum_{m=-\ell}^{\ell} k_\ell^m(a)\beta_\ell^m(\hat{\mathbf{r}}).$$

Thus

$$\mathbf{P}^{E(a)}(r\hat{\mathbf{r}}) = \sum_{\ell=1}^{\infty} \left(\frac{r}{a}\right)^{\ell-1} \sum_{m=-\ell}^{\ell} k_\ell^m(a)\mathbf{b}_\ell^m(a\hat{\mathbf{r}}) \quad \text{if} \quad r < a. \quad (5.3.37)$$

Finally, on $S(a)$ we can expand the toroidal magnetic scalar q in spherical harmonics as

$$q(a\hat{\mathbf{r}}) = \sum_{\ell=1}^{\infty} \sum_{m=-\ell}^{\ell} q_\ell^m(a)\beta_\ell^m(\hat{\mathbf{r}}).$$

Thus

$$\mathbf{T}(a\hat{\mathbf{r}}) = \sum_{\ell=1}^{\infty} \sum_{m=-\ell}^{\ell} q_\ell^m(a)\mathbf{T}_\ell^m(\hat{\mathbf{r}}) \quad \text{on} \quad S(a). \quad (5.3.38)$$

If $\mathbf{J}$ is piecewise continuous, (5.3.36, 5.3.37) remain true in the limit $r \to a$. Therefore, on $S(a)$,

$$\mathbf{B}(a\hat{\mathbf{r}}) = \sum_{\ell=1}^{\infty} \sum_{m=-\ell}^{\ell} [g_\ell^m(a)\mathbf{B}_\ell^m(a\hat{\mathbf{r}}) + k_\ell^m(a)\mathbf{b}_\ell^m(a\hat{\mathbf{r}}) + q_\ell^m(a)\mathbf{T}_\ell^m(\hat{\mathbf{r}})].$$

Now (5.3.32, 5.3.33, 5.3.34) give the coefficients $g_\ell^m(a)$, $k_\ell^m(a)$, $q_\ell^m(a)$, namely,

$$\left.\begin{aligned}
(\ell+1)(2\ell+1)g_\ell^m(a) &= \langle \mathbf{B}_\ell^m | \mathbf{B} \rangle = \langle \overline{\mathbf{B}_\ell^m} \cdot \mathbf{B} \rangle_{S(a)} \\
\ell(2\ell+1)k_\ell^m(a) &= \langle \mathbf{b}_\ell^m | \mathbf{B} \rangle = \langle \overline{\mathbf{b}_\ell^m} \cdot \mathbf{B} \rangle_{S(a)} \\
\ell(\ell+1)q_\ell^m(a) &= \langle \mathbf{T}_\ell^m | \mathbf{B} \rangle = \langle \overline{\mathbf{T}_\ell^m} \cdot \mathbf{B} \rangle_{S(a)}
\end{aligned}\right\}. \qquad (5.3.39)$$

These coefficients can be substituted in (5.3.36, 5.3.37, 5.3.38) to give the internally produced poloidal field $\mathbf{P}^{I(a)}$, the externally produced poloidal field $\mathbf{P}^{E(a)}$, and the toroidal field $\mathbf{T}$ on $S(a)$.

5.4 Induction in the Mantle and the Core

From now on in this chapter, a will be the radius of the core and $S(a)$ will be the core–mantle boundary; b will be the radius of the earth and $S(b)$ will be the earth's surface. The shell $S(a,b)$ will be the mantle. As we saw in Chapter 1, the external currents in the magnetosphere and ionosphere vary with time. The resulting variations in $\mathbf{B}^{E(b)}$ induce currents in the conducting mantle, whose conductivity σ increases with depth, from $\sim 10^{-3}$ to 10^{-4} S/m at the bottom of the crust to ~ 10 to 10^3 S/m at the core–mantle boundary. (Recall from Chapter 2 that the SI unit of conductance is the siemens, abbreviation S; $1\,\text{S} = 1\,\Omega^{-1}$.) Those currents produce a time-varying component in $\mathbf{B}^{I(b)}$ and the observed relationship between the cause, $\partial_t \mathbf{B}^{E(b)}$, and the effect, $\mathbf{B}^{I(b)}$, gives information about the variation of σ with r. In fact this is how the electrical conductivity profile in the mantle has been estimated; see subsection 5.4.3. First we derive the equations governing the propagation of electromagnetic fields in the mantle and then reduce these to differential equations for the toroidal and poloidal scalars of the magnetic field. The two kinds of field have separate equations and behave completely differently from each other. Both kinds of field have been used to explore the electrical conductivity in the mantle, and we discuss these efforts briefly. Then we apply our equations to the question of the free decay of magnetic fields in the core.

5.4.1 Equations for the Mie Scalars

We assume $\nabla_1 \sigma = 0$, i.e., σ depends only on r. We define the *magnetic diffusivity* by

$$\eta = 1/(\mu_0 \sigma). \qquad (5.4.1)$$

Note that the dimensions of η are m^2/s, which is why it is called a diffusivity. We adopt the pre-Maxwell approximation,

$$\nabla \times \mathbf{B} = \mu_0 \mathbf{J}. \tag{5.4.2}$$

We neglect magnetization densities $\mathbf{M}$, since $\partial_t \mathbf{M} = \mathbf{0}$ to a good approximation, and we accept Ohm's law,

$$\mathbf{J} = \sigma \mathbf{E}. \tag{5.4.3}$$

The remaining Maxwell equations are

$$\nabla \cdot \mathbf{B} = 0 \tag{5.4.4}$$

$$\nabla \times \mathbf{E} = -\partial_t \mathbf{B} \tag{5.4.5}$$

and

$$\epsilon_0 \nabla \cdot \mathbf{E} = \rho \tag{5.4.6}$$

where ρ is the charge density per unit volume.

The order of solution of these equations is as follows: from (5.4.3), (5.4.2), and (5.4.1),

$$\mathbf{E} = \eta \nabla \times \mathbf{B}. \tag{5.4.7}$$

Then from (5.4.5)

$$\partial_t \mathbf{B} = -\nabla \times \mathbf{E} = -\nabla \times \eta(\nabla \times \mathbf{B}). \tag{5.4.8}$$

We solve (5.4.8) for $\mathbf{B}$, find $\mathbf{E}$ from (5.4.7), and find ρ from (5.4.6). To carry out this program, we use the fact that $\mathbf{B}$ is solenoidal in $S(a,b)$, so that it has the Mie representation

$$\mathbf{B} = \nabla \times \mathbf{\Lambda} p + \mathbf{\Lambda} q. \tag{5.4.9}$$

Then

$$\nabla \times \mathbf{B} = \nabla \times \mathbf{\Lambda} q - \mathbf{\Lambda} \nabla^2 p.$$

It is convenient now to switch to the Helmholtz form of the Mie representation:

$$\nabla \times \mathbf{B} = \hat{\mathbf{r}} \frac{1}{r} \nabla_1^2 q - \nabla_1 \frac{1}{r} \partial_r r q - \mathbf{\Lambda} \nabla^2 p.$$

We are assuming that $\mathbf{\nabla}_1 \eta = 0$. Hence η^M commutes with $\mathbf{\nabla}_1$ and $\mathbf{\Lambda}$, so from (5.4.7)

$$\mathbf{E} = \hat{\mathbf{r}} \left(\frac{\eta}{r} \nabla_1^2 q \right) - \mathbf{\nabla}_1 \left(\frac{\eta}{r} \partial_r r q \right) - \mathbf{\Lambda}(\eta \nabla^2 p). \qquad (5.4.10)$$

Now recall that

$$\mathbf{\nabla} \times \hat{\mathbf{r}} f = -\mathbf{\Lambda} \frac{f}{r}$$

$$\mathbf{\nabla} \times \mathbf{\nabla}_1 g = \mathbf{\Lambda} \frac{\partial_r r g}{r}.$$

Therefore

$$-\mathbf{\nabla} \times \mathbf{E} = +\mathbf{\nabla} \times \mathbf{\Lambda}(\eta \nabla^2 p) + \mathbf{\Lambda} \left(\frac{\eta}{r^2} \nabla_1^2 q + \frac{1}{r} \partial_r \eta \partial_r r q \right). \qquad (5.4.11)$$

But from (5.4.9),

$$\partial_t \mathbf{B} = \mathbf{\nabla} \times \mathbf{\Lambda}(\partial_t p) + \mathbf{\Lambda}(\partial_t q). \qquad (5.4.12)$$

By (5.4.8), equations (5.4.11) and (5.4.12) are two Mie representations of the same vector field. To be sure of this, we must verify that the scalars in (5.4.11) and (5.4.12) average to 0 on each $S(r)$. This is obvious for (5.4.11) and for $\eta/r \nabla_1^2 q = \mathbf{\nabla}_1 \cdot \eta/r \mathbf{\nabla}_1 q$ in (5.4.10). To finish the argument, we note that if $\langle q \rangle_{S(r)} = 0$ for $a < r < b$, then

$$\frac{1}{4\pi} \int_{S(1)} d^2\hat{\mathbf{r}} \, q(r\hat{\mathbf{r}}) = 0 \quad \text{for} \quad a < r < b.$$

Differentiating with respect to r gives

$$\frac{1}{4\pi} \int_{S(1)} d^2\hat{\mathbf{r}} \, \partial_r q(r\hat{\mathbf{r}}) = 0 \quad \text{for} \quad a < r < b.$$

That is, if $\langle q \rangle_{S(r)} = 0$ in an open interval of r, then $\langle \partial_r q \rangle_{S(r)} = 0$ in that interval. Hence $\langle 1/r \partial_r \eta \partial_r r q \rangle_{S(r)} = 0$ in that interval, and

$$\langle \eta \nabla^2 p \rangle_{S(r)} = \frac{\eta}{r^2} \langle r \partial_r (r \partial_r + 1) p + \nabla_1^2 p \rangle_{S(r)} = 0.$$

But the Mie scalars of a solenoidal field are unique. Therefore equating the poloidal scalars of (5.4.11) and (5.4.12), we find

$$\partial_t p = \eta \nabla^2 p \quad \text{in} \quad S(a,b) \qquad (5.4.13)$$

and equating the toroidal scalars:

$$\partial_t q = \frac{1}{r}\partial_r \eta \partial_r rq + \frac{\eta}{r^2}\nabla_1^2 q \quad \text{in} \quad S(a,b). \qquad (5.4.14)$$

Both these equations behave mathematically like the equation for heat conduction or diffusion. To specify a solution for either equation, we need one boundary condition at $S(a)$, the core–mantle boundary; one boundary condition at $S(b)$, the earth's surface; and either an initial condition on all of $S(a,b)$ or the assumption that we can isolate a single frequency $e^{i\omega t}$ in the solution.

5.4.2 Application of Boundary Conditions: Toroidal Field

On $S(a)$, we need to know how the core behaves. We have not studied it yet, so we simply assume that it somehow supplies values of B_r and $\mathbf{E}_S$ just below the core–mantle boundary. Since these quantities are continuous across the core–mantle boundary, we know them just above it if we know the core behavior. But throughout $S(a,b)$, we have $rB_r = \nabla_1^2 p$ from (5.4.9). Thus

$$p = \nabla_1^{-2} rB_r \quad \text{in} \quad S(a,b). \qquad (5.4.15)$$

Also, from (5.4.10) in $S(a,b)$

$$\mathbf{E}_S = -\nabla_1\left(\frac{\eta\partial_r rq}{r}\right) - \mathbf{\Lambda}(\eta\nabla^2 p),$$

so

$$\nabla_1 \cdot \mathbf{E}_S = -\nabla_1^2\left(\frac{\eta\partial_r rq}{r}\right)$$

$$\mathbf{\Lambda} \cdot \mathbf{E}_S = -\nabla_1^2(\eta\nabla^2 p).$$

Thus, inverting the surface Laplacian

$$\eta\partial_r(rq) = -r\nabla_1^{-2}(\nabla_1 \cdot \mathbf{E}_S) \quad \text{in} \quad S(a,b) \qquad (5.4.16)$$

and from (5.4.13)

$$\partial_t p = -\nabla_1^{-2}(\Lambda \cdot \mathbf{E}_S) \quad \text{in} \quad S(a,b). \qquad (5.4.17)$$

Equation (5.4.15) gives a boundary condition for p on $S(a)$, and (5.4.16) gives a boundary condition for q there. Equation (5.4.17) is not new information about p, since (5.4.15) gives us p at all t. But (5.4.17) does show that the core cannot provide B_r and $\mathbf{E}_S$ independently of each other. Combining (5.4.13, 5.4.15, 5.4.17) we see that they are related by

$$\nabla_1^{-2}(r\partial_t B_r + \Lambda \cdot \mathbf{E}_S) = 0,$$

so

$$\partial_t(rB_r) + \Lambda \cdot \mathbf{E}_S = 0. \qquad (5.4.18)$$

This equation is simply the result of dotting $\mathbf{r}$ into (5.4.5). That is, (5.4.18) says the core cannot violate Maxwell's equations.

For the boundary condition at $S(b)$, the earth's surface, we note that just above $S(b)$, $q = 0$. The atmosphere carries no current, so $\mathbf{B}$ there has no toroidal part. There are no surface currents, so $\mathbf{B}$ is continuous, and hence q is continuous. Therefore

$$q = 0 \quad \text{on} \quad S(b). \qquad (5.4.19)$$

This means that the behavior of q in $S(a,b)$ is completely determined by its initial conditions and what the core does to the core–mantle boundary. The toroidal field in the mantle is unaffected by current sources outside $S(b)$, the earth's surface. If we could measure $\mathbf{E}_S$ just below $S(b)$, and if $\nabla_1\eta$ really were $\mathbf{0}$, which of course it isn't, we might be able to study the core's $\mathbf{E}_S$ directly. Even if $\nabla_1\eta = \mathbf{0}$, however, there are serious problems. If σ has a minimum in the upper mantle, as seems likely, then the mantle above this minimum shorts out the electric field from the core and prevents its observation (Backus, 1982). Nonetheless, attempts have been made to detect the toroidal field in the mantle by measuring very slowly varying electric fields in long transoceanic cables (Lanzerotti et al., 1985).

5.4.3 Application of Boundary Conditions: Magnetic Sounding

For the poloidal field, things look better. The boundary condition at $S(b)$ is that p and $\partial_r p$ should be continuous, and so should fit onto a field that dies away as $r \to \infty$ and a known field with external sources. We can expand (5.4.13) in spherical harmonics. We write

$$p(\mathbf{r}, t) = \sum_{\ell=1}^{\infty} \sum_{m=-\ell}^{\ell} p_\ell^m(r, t) \beta_\ell^m(\hat{\mathbf{r}})$$

where $\{\beta_\ell^{-\ell}, \ldots, \beta_\ell^\ell\}$ is an orthonormal basis for $\mathcal{H}_\ell$. Then (5.4.13) implies

$$\partial_t p_\ell^m(r, t) = \frac{\eta(r)}{r^2}[r\partial_r(r\partial_r + 1) - \ell(\ell+1)]p_\ell^m(r, t) \qquad (5.4.20)$$

in $a < r < b$. In $b < r < c$ where c is the radius of the bottom of ionosphere, we have from (5.3.15) with b replacing a:

$$\frac{1}{b}p_\ell^m(r, t) = \frac{1}{\ell}g_\ell^m(t)\left(\frac{b}{r}\right)^{\ell+1} - \frac{1}{(\ell+1)}k_\ell^m(t)\left(\frac{r}{b}\right)^\ell,$$

so

$$\partial_r p_\ell^m(r, t) = -\frac{\ell+1}{\ell}g_\ell^m(t)\left(\frac{b}{r}\right)^{\ell+2} - \frac{\ell}{\ell+1}k_\ell^m(t)\left(\frac{r}{b}\right)^{\ell-1}.$$

Thus on $S(b)$,

$$\frac{1}{b}p_\ell^m(b, t) = \frac{1}{\ell}g_\ell^m(t) - \frac{1}{(\ell+1)}k_\ell^m(t) \qquad (5.4.21)$$

$$\partial_r p_\ell^m(b, t) = -\frac{\ell+1}{\ell}g_\ell^m(t) - \frac{\ell}{(\ell+1)}k_\ell^m(t). \qquad (5.4.22)$$

From measurements on $S(b)$, we can find both $g_\ell^m(t)$ and $k_\ell^m(t)$, so we know both $p_\ell^m(b, t)$ and $\partial_r p_\ell^m(b, t)$. But if we know the former we can compute the latter for any $\eta(r)$. Hence we can try to use the knowledge of $p_\ell^m(b, t)$ and $\partial_r p_\ell^m(b, t)$ to investigate $\eta(r)$. This is called *magnetic sounding*.

In practice, a number of further simplifications are made in the sounding problem. First, it is invariably the case that the general time variation of the fields is replaced by simple periodic behavior $e^{i\omega t}$. To

study long time series of magnetic field measurements, the methods
of spectral analysis (see Priestley, 1981) are applied, representing the
observed complicated function of time by a linear superposition of
elementary sine waves, each with different frequency: we are in effect
Fourier transforming the data. We set $g_\ell^m(t) = \hat{g}_\ell^m(\omega)e^{i\omega t}$, and so
on, for all the variables that depend on time. It is then possible to
estimate at each frequency the complex ratio of internal to external
Gauss coefficients

$$\hat{Q}_\ell^m(\omega) = \frac{\hat{g}_\ell^m(\omega)}{\hat{k}_\ell^m(\omega)}, \tag{5.4.23}$$

which contains the amplitude and phase of the internal signal at
frequency ω relative to the external source field at that frequency.
Dividing (5.4.21) and (5.4.22), we see

$$\frac{b\partial_r \hat{p}_\ell^m}{\hat{p}_\ell^m} = -\frac{(\ell+1)^2 \hat{Q}_\ell^m + \ell^2}{(\ell+1)\hat{Q}_\ell^m - \ell}. \tag{5.4.24}$$

In practice, most of the magnetic sources of importance for sounding
are confined to a single degree $\ell = 1$ corresponding to electric currents
flowing in the radiation belts, called the *ring current*. We mention
now that the frequency of variations of this kind range from several
hertz to 30 nanohertz or less; this corresponds roughly to periods of a
second to one year. If we orient our coordinate axes so that $\hat{\mathbf{z}}$ lies along
the dipole axis of the main field, only the $m = 0$ term contributes in
the natural basis, so that we can direct our attention to the single
harmonic $\ell = 1$, $m = 0$. The external magnetic field is uniform, while
the internal one is dipolar.

With all these simplifications we may drop the sub- and super-
scripts; since we now have $p(r, t) = \hat{p}e^{i\omega t}$, equation (5.4.20) becomes

$$\partial_r^2 \hat{p} + \frac{2}{r^2}\partial_r \hat{p} - \left(\frac{2}{r^2} + \frac{i\omega}{\eta}\right)\hat{p} = 0. \tag{5.4.25}$$

To make further progress we pretend for the moment the mantle is a
uniform conductor of conductivity 1 S/m, a value it almost certainly
attains below 1000 km depth; then $\eta = 8 \times 10^5$ m^2/s. With constant
η we can solve (5.4.25) exactly: the general solution is

$$\hat{p}(r) = Ah_1^{(1)}(kr) + Bh_1^{(2)}(kr) \tag{5.4.26}$$

where $k = (1 + i)\sqrt{\omega/2\eta}$, and the functions, called *spherical Hankel functions*, are given by

$$h_1^{(1)}(z) = -\frac{e^{iz}}{z}\left(1 + \frac{i}{z}\right)$$

$$h_1^{(2)}(z) = -\frac{e^{-iz}}{z}\left(1 - \frac{i}{z}\right)$$

(see Abramowitz and Stegun, 1965, p. 439). The important thing to notice here is that the term with coefficient A in (5.4.26) decays exponentially as r increases, while the other solution grows exponentially with increasing r. The first term corresponds to magnetic fields with origin in the core diffusing to the surface; for these signals, we should set $\hat{Q}_\ell^m$ to be zero in (5.4.24) because these fields have no external sources, and use this boundary condition in (5.4.25). We will solve a similar kind of problem in detail in the next section. These core fields *are not* the same as those discussed in the previous section — those were toroidal and could not be seen on the surface even in principle. However, if we ask by what factor such a poloidal field at a frequency of 1 Hz is attenuated as it passes through the mantle, we find the poloidal scalar reduced by 10^{-2500}, and this translates into a roughly similar effect in the field itself. But at frequencies corresponding to a year, the attenuation factor is only 0.35. We must not take the simplified conductivity model too seriously, but it suggests that above some frequency, generally accepted to be around one cycle per year, magnetic variations caused by effects at the core are invisible at the surface because of attenuation. At periods of 10 years and more such variations are very obvious in the magnetic observatory records; see Chapter 1 and the discussion of westward drift of the nondipole field and the geomagnetic jerk. For much more on the outward diffusion of fields through the mantle, see Backus (1983).

If we restrict ourselves to the higher-frequency variations, the value of $\hat{Q}_\ell^m$ in (5.4.24) is controlled by the profile of mantle conductivity, as given by the ratio on the left. Because of the attenuation downward-traveling signals receive — as evidenced by the second term in (5.4.26) — one may solve the differential equation (5.4.25) with the boundary condition $\hat{p}(a) = 0$, and for any given electrical conductivity profile, $\sigma(r)$, compute the ratio $b\partial_r\hat{p}_\ell^m/\hat{p}_\ell^m$ at the surface $r = b$ as a function of frequency. This can be compared with the estimates of $\hat{Q}_\ell^m$ obtained from observatory data, and adjustments can then be made to $\sigma(r)$ to bring it into agreement with observation. The problem of finding σ from such measurements is a nonlinear geophysical inverse problem that has received a great deal of attention and is relatively well understood. It has been shown, for example, that if the quantity $\hat{Q}_\ell^m$ is known for a single harmonic and all frequencies, the corresponding conductivity profile in the earth is unique. We cannot devote more space to these questions here but refer the interested reader to papers by Parker (1970, 1980) and Weidelt (1972).

5.4.4 Free Decay of Fields in the Core

Another application we can make of (5.4.13) and (5.4.14) is to the question whether the main geomagnetic field itself might be a fossil field, produced by currents of electricity in the core that were emplaced somehow when the earth formed. In this discussion, the core is so much better a conductor than the mantle that we take $\sigma = 0$ in the mantle. For simplicity we take σ to be constant in the core, and we use $\sigma = 3 \times 10^5$ S/m. Then (5.4.14) becomes

$$\partial_t q = \eta \nabla^2 q. \tag{5.4.27}$$

We want to solve (5.4.27) subject to the boundary condition

$$q = 0 \quad \text{on} \quad S(a). \tag{5.4.28}$$

We also want to solve (5.4.13)

$$[p]_-^+ = [\partial_r p]_-^+ = 0 \quad \text{on} \quad S(a), \tag{5.4.29}$$

subject to the conditions

$$\nabla^2 p = 0, \quad p \to 0 \quad \text{as} \quad r \to \infty \quad \text{if} \quad r > a. \tag{5.4.30}$$

We begin with the toroidal problem (5.4.27). We expand q in spherical harmonics $\beta_\ell^m(\hat{\mathbf{r}})$ as

$$q(\mathbf{r}, t) = \sum_{\ell=1}^{\infty} \sum_{m=-\ell}^{\ell} q_\ell^m(r, t) \beta_\ell^m(\hat{\mathbf{r}}). \tag{5.4.31}$$

Then (5.4.27) implies

$$\left[\partial_r^2 + \frac{2}{r}\partial_r - \frac{\ell(\ell+1)}{r^2}\right] q_\ell^m(r,t) = \frac{1}{\eta}\partial_t q_\ell^m(r,t) \qquad (5.4.32)$$

and (5.4.28) implies

$$q_\ell^m(a,t) = 0. \qquad (5.4.33)$$

We solve this system like a heat flow problem. We begin by seeking particular solutions that decay exponentially with time,

$$q_\ell^m(r,t) = q_\ell^m(r)e^{-t/T} \qquad q_\ell^m(a) = 0 \qquad (5.4.34)$$

where T is a constant, the mean life of the solution. We define

$$\kappa^2 = \frac{1}{\eta T} = \frac{\mu_0 \sigma}{T}. \qquad (5.4.35)$$

Then $g_\ell^m(r)$ must satisfy

$$\left[\partial_r^2 + \frac{2}{r}\partial_r + \kappa^2 - \frac{\ell(\ell+1)}{r^2}\right] q_\ell^m(r) = 0 \qquad (5.4.36)$$

$$q_\ell^m(a) = 0. \qquad (5.4.37)$$

The most general solution of (5.4.36) is

$$q_\ell^m(r) = A j_\ell(\kappa r) + B n_\ell(\kappa r) \qquad (5.4.38)$$

where j_ℓ and n_ℓ are the *spherical Bessel functions* (see Abramowitz and Stegun, 1965, p. 437). The function $n_\ell(z)$ behaves like $z^{-\ell-1}$ as $z \to 0$, so we must take $b = 0$ in (5.4.38). Then (5.4.37) requires

$$j_\ell(\kappa a) = 0.$$

The function $j_\ell(z)$ behaves like z^ℓ near $z = 0$, and then oscillates, behaving like $\cos(z - \alpha_\ell)/z$ as $z \to \infty$. It has infinitely many zeros, $0 < z_{\ell,1} < z_{\ell,2} < z_{\ell,3} < \dots$. The zeros $z_{\ell,n}$ increase with both ℓ and n. For each $z_{\ell,n}$, we get a value of κ, $\kappa_{\ell,n}$, and a value of T, $T_{\ell,n}$,

$$\kappa_{\ell,n} = z_{\ell,n}/a$$
$$T_{\ell,n} = \mu_0 \sigma a^2 / z_{\ell,n}^2, \qquad (5.4.39)$$

such that

$$q_\ell(r,t) = j_\ell(\kappa_{\ell,n}r)e^{-t/T_{\ell,n}}$$

solves (5.4.32) and (5.4.33). Since those equations are linear, no matter how we choose constants $c_{\ell,n}^m$, the function

$$q_\ell^m(r,t) = \sum_{n=1}^{\infty} c_{\ell,n}^m j_\ell(\kappa_{\ell,n}r)e^{-t/T_{\ell,n}} \qquad (5.4.40)$$

solves (5.4.32) and (5.4.33). Since (5.4.36, 5.4.37) form a Sturm–Liouville problem (Birkhoff and Rota, 1989), we can always choose the $c_{\ell,n}^m$ so that at $t = 0$ the function (5.4.39) takes the required initial value for the $q_\ell^m(r,0)$ in (5.4.31). Hence we have the solution of (5.4.27, 5.4.28) with an arbitrary initial $q(r,0)$. Clearly it decays with time, and from (5.4.40) its longest-lived part has mean life $T_{1,1}$. Since $z_{1,1} = 4.49341\ldots$, $\sigma = 3 \times 10^5$ S/m, $a = 3.48 \times 10^6$ m, equation (5.4.39) gives

$$T_{1,1}^{TOR} = 2.3 \times 10^{11} \text{ s} = 7200 \text{ years.} \qquad (5.4.41)$$

In the poloidal problem, we expand p in spherical harmonics as in (5.4.31), obtaining

$$p(\mathbf{r},t) = \sum_{\ell=1}^{\infty} \sum_{m=-\ell}^{\ell} p_\ell^m(r,t)\beta_\ell^m(\hat{\mathbf{r}}). \qquad (5.4.42)$$

Then $\partial_t p = \eta\nabla^2 p$ implies

$$\left[\partial_r^2 + \frac{2}{r}\partial_r - \frac{\ell(\ell+1)}{r^2}\right]p_\ell^m(r,t) = \frac{1}{\eta}\partial_t p_\ell^m(r,t). \qquad (5.4.43)$$

Conditions (5.4.30) imply

$$p_\ell^m(r,t) = p_\ell^m(a,t)(a/r)^{\ell+1} \quad \text{if} \quad r > a$$

so

$$r\partial_r p_\ell^m(r,t) + (\ell+1)p_\ell^m(r,t) = 0 \quad \text{at} \quad r = a. \qquad (5.4.44)$$

Since p and $r\partial_r p$ are continuous across $r = a$, the boundary condition (5.4.44) applies to (5.4.43). Again we seek solutions that decay exponentially with time,

$$p_\ell^m(r,t) = p_\ell^m(r)e^{-t/T}.$$

Again we have

$$p_\ell^m(r,t) = Aj_\ell(\kappa r)e^{-t/T}$$

but now $z = \kappa a$ must satisfy (5.4.44), which is

$$z\partial_z j_\ell(z) + (\ell + 1)j_\ell(z) = 0.$$

Fortunately, spherical Bessel functions satisfy the identity

$$z\partial_z j_\ell(z) + (\ell + 1)j_\ell(z) = zj_{\ell-1}(z)$$

(Abramowitz and Stegun, 1965), so we must take κ such that κa is a zero of $j_{\ell-1}(z)$. That is,

$$\kappa_{\ell,n}^{POL} = z_{\ell-1,n}/a.$$

If we then set

$$T_{\ell,n}^{POL} = \frac{\mu_0\sigma}{\kappa_{\ell,n}^2} = \frac{\mu_0\sigma a^2}{z_{\ell-1,n}^2} \tag{5.4.45}$$

we have a solution of (5.4.43, 5.4.44),

$$p_\ell(r,t) = j_\ell(\kappa_{\ell,n}^{POL}r)e^{-t/T_{\ell,n}^{POL}}.$$

The general solution of (5.4.43) and (5.4.44) is then

$$p_\ell^m(r,t) = \sum_{\ell=1}^{\infty}\sum_{m=-\ell}^{\ell} c_{\ell,n}^n j_\ell(\kappa_{\ell,n}^{POL}r)e^{-t/T_{\ell,n}^{POL}}.$$

The longest-lived mode is $T_{1,1}^{POL}$. Since $z_{0,1} = \pi$, equation (5.4.45) together with the other numerical values shows

$$T_{1,1}^{POL} = 4.6 \times 10^{11} \text{ s} = 15{,}000 \text{ years.}$$

We note that this most slowly decaying mode of the magnetic field has degree $\ell = 1$; externally it is a dipole field. Furthermore, the present-day decline of the main dipole has a time constant of about 1000 years, which implies considerably faster decay than the rate we have computed for a solid conductor. Provided that the value of σ we have used is not seriously in error, we must conclude that fluid motions in the core today are acting to withdraw energy from the dipole field.

5.5 Ohmic Heating in the Core

In this section we show that it is possible to relate the Ohmic heat generation by electric current in the core to the spherical harmonic coefficients of the external field. Of course, all that one can say is that the Ohmic losses must be larger than some minimum amount, because there are current systems (those generating toroidal **B**, for example) that have no external expression. We use the Mie representation to derive the bound, first given by Gubbins (1975). As before, we assume the electrical conductivity is uniform throughout the core K, which as in the previous section is a sphere of radius a. Then the objective is to find the smallest value of

$$F = \int_K d^3\mathbf{r} \, |\mu_0 \mathbf{J}|^2$$

when the internal Gauss coefficients are known to be g_1^{-1}, g_1^0, g_1^1, g_2^{-2}, g_2^{-1}, g_2^0, g_2^1, g_2^1, $\ldots$, relative to an orthonormal basis for $\mathcal{H}_\ell$ of $\{\beta_\ell^{-\ell}, \ldots, \beta_\ell^\ell\}$. Inside the core we write as usual

$$\mathbf{B} = \boldsymbol{\nabla} \times \boldsymbol{\Lambda} p + \boldsymbol{\Lambda} q.$$

Then just as in (5.3.3),

$$\mu_0 \mathbf{J} = \boldsymbol{\nabla} \times \mathbf{B} = \boldsymbol{\nabla} \times q - \boldsymbol{\Lambda} \nabla^2 p.$$

Thus

$$F = \int_K d^3\mathbf{r} \, |\boldsymbol{\nabla} \times q - \boldsymbol{\Lambda} \nabla^2 p|^2$$

$$= \int_K d^3\mathbf{r} \, |\boldsymbol{\nabla} \times q|^2 + \int_K d^3\mathbf{r} \, |\boldsymbol{\Lambda} \nabla^2 p|^2. \qquad (5.5.1)$$

This follows because the integral of the cross term vanishes, which is seen as follows:

$$\boldsymbol{\Lambda}\nabla^2 p \cdot \boldsymbol{\nabla} \times \boldsymbol{\Lambda} q = \boldsymbol{\Lambda}\nabla^2 p \cdot (\frac{\hat{\mathbf{r}}}{r}\nabla_1^2 q - \boldsymbol{\nabla}_1 \frac{1}{r}\partial_r r q)$$

$$= -\boldsymbol{\Lambda}\nabla^2 p \cdot \boldsymbol{\nabla}_1 \frac{1}{r}\partial_r r q,$$

and it will be recalled from (5.1.19) that the surface curl and surface grad are orthogonal when averaged over any spherical surface $S(r)$.

Since the toroidal part of $\mathbf{B}$ is invisible at the surface $S(a)$, the corresponding term in (5.5.1), namely $\int_K d^3\mathbf{r} \, |\boldsymbol{\nabla} \times q|^2$, can take on any value without affecting the size of the Gauss coefficients. All we can say is that

$$F \geq \min \int_K d^3\mathbf{r} \, |\boldsymbol{\Lambda}\nabla^2 p|^2,$$

the minimum being subjected to the constraint that the field outside the core is given by

$$\mathbf{B} = \boldsymbol{\nabla} \times \boldsymbol{\Lambda} p$$

with

$$p = -a\sum_{\ell=1}^{\infty}\sum_{m=-\ell}^{\ell} \frac{g_\ell^m}{\ell}\left(\frac{a}{r}\right)^{\ell+1}\beta_\ell^m(\hat{\mathbf{r}}), \quad r \geq a, \qquad (5.5.2)$$

which is (5.3.15) when there is no field of external origin. Since there are no surface currents on $S(a)$, the poloidal scalar p is continuously differentiable at the surface.

We find what p minimizes $\int_K d^3\mathbf{r} \, |\boldsymbol{\Lambda}\nabla^2 p|^2$ by performing a small perturbation on p in the fashion of calculus of variations:

$$\delta \int_K d^3\mathbf{r} \, \boldsymbol{\Lambda}\nabla^2 p \cdot \boldsymbol{\Lambda}\nabla^2 p = 2\int_K d^3\mathbf{r} \, \boldsymbol{\Lambda}\nabla^2 p \cdot \boldsymbol{\Lambda}\nabla^2 \delta p$$

$$= -2\int_K d^3\mathbf{r} \, \Lambda^2\nabla^2 p \, \nabla^2\delta p$$

$$= -2\int_K d^3\mathbf{r} \, (\nabla^2\Lambda^2\nabla^2 p)\delta p$$

$$+ 2\int_K d^3\mathbf{r} \, \boldsymbol{\nabla} \cdot [\boldsymbol{\nabla}\delta p(\Lambda^2\nabla^2 p) - \delta p\boldsymbol{\nabla}\Lambda^2\nabla^2 p]$$

$$= -2 \int_K d^3\mathbf{r} \ (\nabla^2 \Lambda^2 \nabla^2 p) \delta p$$

$$+ 2 \int_{S(a)} d^2\mathbf{r} \ [(\partial_r \delta p)(\Lambda^2 \nabla^2 p) - \delta p(\partial_r \Lambda^2 \nabla^2 p)].$$

At the minimizing p this quantity must vanish for all small perturbations δp. Since $\delta p = \partial_r \delta p = 0$ on $S(a)$, the surface integral vanishes without constraining p. For the perturbation to vanish for arbitrary small δp in the volume integral, we must have $\nabla^2 \Lambda^2 \nabla^2 p = 0$, or on account of the commutivity of ∇^2 and Λ^2 we find $\Lambda^2 \nabla^4 p = 0$. Recall that $\Lambda^2 = \nabla_1^2$ and that the only solution to $\nabla_1^2 f = 0$ on each spherical shell is $f = 0$, when f averages to zero on each $S(r)$. Therefore

$$\nabla^4 p = 0, \qquad 0 \le r \le a.$$

This is called the *biharmonic equation*. We solve it in the same way that we solved Laplace's equation in Chapter 4, by writing

$$p(r\hat{\mathbf{r}}) = a \sum_{\ell=1}^{\infty} \sum_{m=-\ell}^{\ell} f_\ell^m(r)\beta_\ell^m(\hat{\mathbf{r}}).$$

We have dropped the $\ell = 0$ term so that p averages to zero over spherical surfaces. Upon substitution of this expansion into the biharmonic equation, we obtain the following fourth-order ordinary differential equation for each function f_ℓ^m:

$$\left[\frac{1}{r}\partial_r^2 r - \frac{\ell(\ell+1)}{r^2}\right]\left[\frac{1}{r}\partial_r^2 r - \frac{\ell(\ell+1)}{r^2}\right] f_\ell^m = 0.$$

The general solution to this equation can be found by the standard methods of power-series development (Birkhoff and Rota, 1989) to be

$$f_\ell^m = c_1 r^\ell + c_2 r^{\ell+2} + c_3 r^{-(\ell+1)} + c_4 r^{-(\ell-1)}$$

where c_k are constants. We may set c_3 and c_4 to zero because we want p to be differentiable at $r = 0$ and these terms are singular at the origin. Therefore the solution for p can be written

$$p = a \sum_{\ell=1}^{\infty} \sum_{m=-\ell}^{\ell} \left[A_\ell^m \left(\frac{r}{a}\right)^\ell + B_\ell^m \left(\frac{r}{a}\right)^{\ell+2}\right] \beta_\ell^m(\hat{\mathbf{r}}), \qquad 0 \le r \le a.$$

$$(5.5.3)$$

Now we match this equation with (5.5.2) at $r = a$; since the derivatives are continuous, we can also match the respective derivatives:

$$r\partial_r p = a \sum_{\ell=1}^{\infty} \sum_{m=-\ell}^{\ell} \left[\ell A_\ell^m \left(\frac{r}{a}\right)^\ell + (2\ell+1)B_\ell^m \left(\frac{r}{a}\right)^{\ell+2} \right] \beta_\ell^m(\hat{\mathbf{r}}),$$

$$0 \le r \le a \quad (5.5.4)$$

$$= a \sum_{\ell=1}^{\infty} \sum_{m=-\ell}^{\ell} \frac{\ell+1}{\ell} g_\ell^m \left(\frac{a}{r}\right)^{\ell+1} \beta_\ell^m(\hat{\mathbf{r}}), \quad r \ge a. \quad (5.5.5)$$

Because of the orthogonality of the β_ℓ^m, we may equate the individual coefficients in the two pairs of equations: from matching (5.5.1) and (5.5.3) and then matching (5.5.4) and (5.5.5) we find

$$A_\ell^m + B_\ell^m = -\frac{g_\ell^m}{l}$$

$$\ell A_\ell^m + (\ell+2)B_\ell^m = \frac{\ell+1}{\ell} g_\ell^m.$$

Solving this pair gives us

$$A_\ell^m = -\frac{2\ell+3}{2\ell} g_\ell^m, \quad B_\ell^m = \frac{2\ell+1}{2\ell} g_\ell^m. \quad (5.5.6)$$

Finally we wish to compute the integral. First we note that

$$F_{\min} = \int_K d^3\mathbf{r} \, |\mathbf{\Lambda}\nabla^2 p|^2 = -\int_K d^3\mathbf{r} \, \nabla^2 p \, \Lambda^2\nabla^2 p. \quad (5.5.7)$$

Substituting (5.5.6) into (5.5.3) and performing some straightforward algebra yields

$$\nabla^2 p = \frac{1}{a} \sum_{\ell=1}^{\infty} \sum_{m=-\ell}^{\ell} \frac{(2\ell+1)(2\ell+3)}{\ell} g_\ell^m \left(\frac{r}{a}\right)^\ell \beta_\ell^m(\hat{\mathbf{r}})$$

$$\Lambda^2\nabla^2 p = -\frac{1}{a} \sum_{\ell=1}^{\infty} \sum_{m=-\ell}^{\ell} (\ell+1)(2\ell+1)(2\ell+3)g_\ell^m \left(\frac{r}{a}\right)^\ell \beta_\ell^m(\hat{\mathbf{r}})$$

for $0 \le r \le a$. Inserting these into (5.5.7) and continuing,

$$
F_{\min} = \int_0^a dr \, r^2 \int_{S(1)} d^2\hat{\mathbf{r}} \, (\nabla^2 p)(-\Lambda^2 \nabla^2 p)
$$

$$
= \int_0^a dr \, r^2 \, \frac{1}{a^2} \sum_{\ell=1}^{\infty} \sum_{m=-\ell}^{\ell} \frac{(\ell+1)(2\ell+1)^2(2\ell+3)^2}{\ell} |g_\ell^m|^2 \left(\frac{r}{a}\right)^{2\ell}
$$

$$
= a \sum_{\ell=1}^{\infty} \sum_{m=-\ell}^{\ell} \frac{(\ell+1)(2\ell+1)^2(2\ell+3)}{\ell} |g_\ell^m|^2.
$$

Therefore finally we have the following bound on Ohmic heat generation in the core:

$$
\frac{1}{\sigma} \int_K d^3\mathbf{r} \, |\mathbf{J}|^2 \ge \frac{a}{\mu_0^2 \sigma} \sum_{\ell=1}^{\infty} \sum_{m=-\ell}^{\ell} \frac{(\ell+1)(2\ell+1)^2(2\ell+3)}{\ell} |g_\ell^m|^2.
$$

Expressed in terms of Schmidt-normalized Gauss coefficients $\breve{g}_\ell^m = \sqrt{2\ell+1}\, g_\ell^m$, this is

$$
\frac{1}{\sigma} \int_K d^3\mathbf{r} \, |\mathbf{J}|^2 \ge \frac{a}{\mu_0^2 \sigma} \sum_{\ell=1}^{\infty} \sum_{m=-\ell}^{\ell} \frac{(\ell+1)(2\ell+1)(2\ell+3)}{\ell} |\breve{g}_\ell^m|^2.
$$

Exercises

1. Let

$$
R(x, \mu) = \sum_{l=1}^{\infty} \frac{x^l}{l} P_l(\mu)
$$

$$
S(x, \mu) = \sum_{l=1}^{\infty} \frac{x^l}{l+1} P_l(\mu)
$$

$$
K(\mu) = \sum_{l=1}^{\infty} \frac{(2l+1)}{l(l+1)} P_l(\mu).
$$

Recall from (5.1.9, 5.1.11) that K is the averaging function that solves the surface form of Poisson's equation.
(a) Show that $K(\mu) = R(1,\mu) + S(1,\mu)$.
(b) Show that

$$x\partial_x[R(x,\mu)] = \partial_x[xS(x,\mu)] = (1 + x^2 - 2\mu x)^{-1/2} - 1.$$

(c) Integrate the equations in (b) with the appropriate initial conditions so as to find $R(x,\mu)$ and $S(x,\mu)$ for $0 \le x \le 1$. Hence find $K(\mu)$.

2. Decide whether each of the following statements about a solenoidal vector field in $S(a,b)$ is true or false. Prove your conclusions.
   (a) If $\mathbf{v}$ is toroidal, $\boldsymbol{\nabla} \times \mathbf{v}$ is poloidal.
   (b) If $\boldsymbol{\nabla} \times \mathbf{v}$ is poloidal, $\mathbf{v}$ is toroidal.

3. If $\mathbf{B}$ is the magnetic field of a point dipole with dipole moment $\mathbf{m}$ placed at the origin, we already had two ways to write $\mathbf{B}$: $\mathbf{B} = -\boldsymbol{\nabla}\psi$ with $\psi = -(\mu_0/4\pi)\mathbf{m} \cdot \boldsymbol{\nabla}(1/r)$ and $\mathbf{B} = \boldsymbol{\nabla} \times \mathbf{A}$, with $\mathbf{A} = -(\mu_0/4\pi)\mathbf{m} \times \boldsymbol{\nabla}(1/r)$. Now we know there is a third representation. Since $\boldsymbol{\nabla} \cdot \mathbf{B} = 0$ in $r > 0$, there are unique scalars p and q in $r > 0$ such that $\langle p \rangle_{S(r)} = \langle q \rangle_{S(r)} = 0$ and $\mathbf{B} = \boldsymbol{\nabla} \times \boldsymbol{\Lambda}p + \boldsymbol{\Lambda}q$. Find p and q and prove that you have done so.

6

HYDROMAGNETICS OF THE CORE

In this chapter we study how the fluid outer core generates the geomagnetic field $\mathbf{B}$. The core is molten, and very likely mostly iron, and liquid iron cannot be permanently magnetized. However, the liquid moves, and it is an electrical conductor. Moving conductors can generate magnetic fields, and in fact form the basis for most commercial dynamos. The current in these dynamos, however, is confined to a topologically complicated path by insulated wire windings. The simplest such dynamo, and the easiest to understand in detail, is the disk dynamo of Bullard (1955); we describe the behavior of this model dynamo in some detail in the next section.

Of course the earth's outer core is completely unlike the disk dynamo. So in the following section we consider the equations governing fluid motion and its interaction with a magnetic field. The normal form of Ohm's law needs modification to account for the fact that the conductor is moving, and we derive the well-known adaptation by appealing to special relativity. Next we find the conventional form of the so-called dynamo equations by writing the electrical contributions to the system in terms of the magnetic field $\mathbf{B}$ alone. Two limiting cases can be understood completely. The first is the almost trivial situation of a stationary conductor; we have already treated some aspects of this problem in the previous chapter. We conclude that without motion, the field in the core would vanish in several thousand years. More interesting is the situation in which the electrical conductivity is extremely high, or in the limit, a perfect conductor. To handle this case we need to introduce the Lagrangian description of a deforming body, in which the evolution of various properties (principally the magnetic field) is considered from the perspective of a particle moving with the flow. This approach enables us to show there is a remarkable connection between the magnetic field lines and the fluid particles: they move together. This behavior leads to some conservation laws — in particular, the frozen-flux condition, which can be exploited to learn about the fluid velocity at the top of the core, provided of course that the conductivity there is great enough and the time of observation short enough to justify the perfect-conductor approximation. While the core may appear to act as a perfect conductor over periods of several decades, on scales of thousands of years, diffusion of the magnetic field must be important; the geomagnetic field has existed for much longer than this, and therefore to understand the longevity of the field we cannot ignore the diffusion terms in the equations.

In any attempt to understand the operating principles of something as complex and inaccessible as the generator of the geomagnetic field, it is sensible to study simple models first. The discussion turns in section 6.5 to the simplest class, the kinematic dynamos: here the fluid flow is assumed to be specified and is presumed to be unaffected by the forces arising from the presence of electric currents and magnetic fields. At first encounter, it seems natural to impose symmetries on the flow to help in the calculations; one might expect axial symmetry to be a good starting point, since the geomagnetic field is approximately dipolar. As we will see, this expectation is thwarted by a famous theorem due to Cowling: no dynamo can exist with this particular symmetry. In fact, many kinds of symmetry in the flow geometry can inhibit dynamo action, something that makes the study of dynamos much more difficult than it might at first appear. Simplicity can be taken only so far. We prove two such prohibiting theorems including Cowling's original.

In the early days of dynamo theory the failure of axisymmetric systems to perform as hoped led to a belief that perhaps all dynamos were impossible because of a yet-to-be-discovered general antidynamo theorem; then, of course, the source of the geomagnetic field would have to be sought elsewhere. It was therefore most important to establish mathematically the existence of a working dynamo as a matter of principle, no matter how artificial the motions. We briefly sketch the first two rigorous demonstrations, one involving steady velocities, the other requiring intermittent motions. Once the mathematical possibility of dynamo action had been settled, the next task, still unfinished, is to identify plausible motions and forces that will sustain magnetic fields of a realistic type. Numerical computations have been indispensable for this; we give a very brief sketch of some early results.

The final section deals with one of a number of dynamos based on a notion originally invented in astrophysics: that fluid motions on very small scales can collaboratively generate large-scale magnetic fields, under the proper conditions. An approximate theory can be developed in which the equations are averaged over the small scale. The outer core fluid may well be in a state of hydrodynamic turbulence in which there is considerable kinetic energy on small scales. If the turbulence is not completely isotropic, but exhibits correlations entailing the kind of corkscrew motion that might reasonably be expected in a rotating fluid, then the theory shows that energy is automatically transferred into the magnetic field in certain ranges of wavelength. We describe a famous example of this kind, called the α-effect dynamo.

Our discussion in this chapter represents a small sampling of a large and currently active field. In addition to the α-effect dynamo, there are many more kinematic dynamos based on the approximation of averaging the equations. Moffatt's (1978) book is a good source for more on this topic.

6.1 The Bullard Disk Dynamo

As depicted in Figure 6.1.1, the Bullard disk dynamo consists of two rigid pieces of copper, one an axle with a disk welded to it, and the other a wire that makes sliding contact with the edge of the disk at point C' and with the axle at C. The wire is held fixed and the disk and axle are rotated with angular velocity $\mathbf{\Omega} = \Omega\hat{\mathbf{z}}$. If there is a current I in the wire, it will produce a magnetic field $\mathbf{B}$ in the disk. The copper in the disk moves with velocity $\mathbf{u} = \Omega\varpi\hat{\boldsymbol{\lambda}}$ where $\varpi = $ distance from z axis and $\hat{\boldsymbol{\lambda}} = \hat{\mathbf{z}} \times \hat{\boldsymbol{\varpi}}$. This produces an electric field $\mathbf{u} \times \mathbf{B}$ whose ϖ component is $\hat{\boldsymbol{\varpi}} \cdot \mathbf{u} \times \mathbf{B} = \Omega\varpi(\hat{\boldsymbol{\varpi}} \times \hat{\boldsymbol{\lambda}}) \cdot \mathbf{B} = \Omega\varpi B_z$. The result is an electromotive force $\mathcal{E}$ applied to the wire by the disk via its electric field $\mathbf{E} = \mathbf{u} \times \mathbf{B}$, with $\mathbf{u} = \Omega\varpi\hat{\boldsymbol{\lambda}}$, given by

$$\mathcal{E} = \int_{\varpi_1}^{\varpi_2} d\varpi \; E_\varpi = \Omega \int_{\varpi_1}^{\varpi_2} d\varpi \; \varpi B_z = \frac{\Omega}{2\pi} \int_{\text{disk}} d^2\mathbf{r} \; B_z = \frac{\Omega}{2\pi}\Phi$$

$$(6.1.1)$$

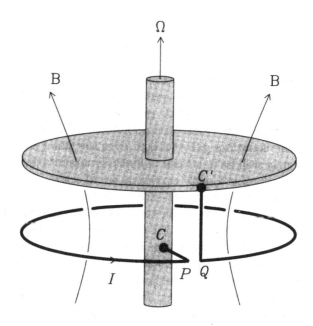

Figure 6.1.1: Diagram of the Bullard disk dynamo: the gray regions are the copper rotor and axle; the loop of wire, which carries the current I and generates the magnetic field **B**, is shown by the thick line; it makes electrical contact at C and C'.

where Φ is the magnetic flux through the disk, positive in the upward direction. If M is the mutual inductance between the wire and the two boundaries of the disk,

$$\Phi = MI$$

where I is the current in the wire. We take $I > 0$ when it flows outward in the disk, so $M > 0$ in the figure. If the wire were replaced by its mirror image, it would have $M < 0$.

Finally, if R is the resistance in the circuit where I flows, and L is the self-inductance of that circuit,

$$\mathcal{E} = L\frac{dI}{dt} + RI.$$

Eliminating Φ and $\mathcal{E}$ from these three equations gives

$$\frac{\Omega M}{2\pi}I = L\frac{dI}{dt} + RI$$

or

$$L\frac{dI}{dt} = \left[\frac{\Omega M}{2\pi} - R\right] I, \qquad (6.1.2)$$

whose solution, if Ω is held constant, is

$$I(t) = I(0)e^{(\Omega M/2\pi - R)t/L}.$$

Thus, if

$$\frac{\Omega M}{R} > 2\pi \qquad (6.1.3)$$

then $I(t)$ will grow exponentially with time. Otherwise, $I(t)$ will decay.

A real disk dynamo will not be able to keep I growing exponentially forever, because it will have only a finite source of energy. The torque in the positive sense about the $\hat{\mathbf{z}}$ axis that $\mathbf{B}$ exerts on $\mathbf{J}$ is $\mathcal{T} = \int_{\mathrm{disk}} d^3\mathbf{r}\, \varpi\hat{\boldsymbol{\lambda}} \cdot (\mathbf{J} \cdot \mathbf{B})$. But

$$\hat{\boldsymbol{\lambda}} \cdot \mathbf{J} \times \mathbf{B} = (\hat{\boldsymbol{\lambda}} \times \mathbf{J}) \cdot \mathbf{B} = (\hat{\boldsymbol{\varpi}} J_z - \hat{\mathbf{z}} J_\varpi) \cdot \mathbf{B} = -J_\varpi B_z$$

on the assumption that $J_z \approx 0$. Then the torque

$$\mathcal{T} = \int_{\mathrm{bottom}}^{\mathrm{top}} dz \int_0^{2\pi} d\lambda \int_{\varpi_1}^{\varpi_2} d\varpi\, \varpi[-\varpi J_\varpi B_z].$$

Since $\partial_\lambda B_z \approx \partial_z B_z \approx 0$, this is approximately

$$\mathcal{T} = -\int_{\varpi_1}^{\varpi_2} d\varpi\, \varpi B_z(\varpi) \int_{\mathrm{bottom}}^{\mathrm{top}} dz \int_0^{2\pi} d\lambda\, \varpi J_\varpi$$

$$= -I \int_{\varpi_1}^{\varpi_2} d\varpi\, \varpi B_z,$$

so

$$\mathcal{T} = -\frac{I}{2\pi}\Phi = -\frac{M}{2\pi}I^2. \qquad (6.1.4)$$

Thus as I increases, the torque required to keep Ω constant rises very rapidly. If G is the external torque that drives the dynamo, and K is the moment of inertia of the axle and disk, then the equation for the angular acceleration of the disk is

$$G - \frac{M}{2\pi}I^2 = K\frac{d\Omega}{dt}. \tag{6.1.5}$$

For a physically realizable dynamo, (6.1.2) and (6.1.5) must both be satisfied. Equation (6.1.2) alone is the *kinematic dynamo* equation. It shows how a given velocity Ω changes the current (and magnetic field). Equation (6.1.5) is the dynamic equation, which describes the forces produced electromagnetically, and how they modify the original driving force.

Equations (6.1.5) and (6.1.2) have two steady solutions when G is constant and $MG > 0$, namely $\Omega_0 = 2\pi R/M$ and $I_0 = \pm(2\pi G/M)^{\frac{1}{2}}$. Rearranging (6.1.2) we find

$$\frac{1}{I}\frac{dI}{dt} = \frac{d\ln I}{dt} = \Omega\frac{M}{2\pi L} - \frac{R}{L}.$$

If we differentiate this with respect to t and substitute from (6.1.5) we have an equation for $\ln I$, which we name y:

$$\frac{d^2y}{dt^2} = \alpha - \beta e^{2y} \tag{6.1.6}$$

where the constants are $\alpha = MG/2\pi LK$ and $\beta = M^2/4\pi^2 LK$. If the departures from equilibrium are small so that y is small we can expand the right side of (6.1.6) in a Taylor series and then we see at once that the system oscillates about the equilibrium with a period of

$$T = 2\pi\sqrt{KL\pi/MG}.$$

Equation (6.1.6) is an example of a second-order autonomous system, and although it is nonlinear, the theory of such systems is well understood (see Coddington and Levinson, 1955, chap. 15). The character of the solution depends on only one parameter, despite appearances to the contrary: if we scale time so that $s = t\alpha^{1/2}$ and set $w = y - \ln\sqrt{\beta/\alpha}$ then the equation becomes

$$\frac{d^2w}{ds^2} = 1 - e^{2w},$$

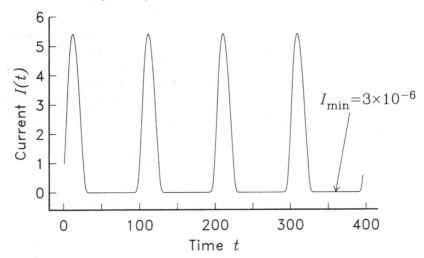

Figure 6.1.2: A typical large-amplitude system for the Bullard disk dynamo when the driving torque is held constant.

and so the essential behavior depends only on the initial conditions. It can be shown that the general solution for $w(t)$ is periodic with finite amplitude, and hence the sign of I never changes. Thus the initial conditions set the phase of the solution, and the amplitude; the amplitude is the only parameter governing the general behavior of the system. For large-amplitude oscillations, the current stays very close to zero for a long time, then rises to a peak and collapses again, the cycle repeating indefinitely; a numerical solution for such a system is shown in Figure 6.1.2.

It is quite certain from paleomagnetic records that the behavior of the earth's dipole moment does not in the least resemble the curve in Figure 6.1.2. Yet a small change in the physics of the disk dynamo completely alters its behavior: if a small inductance is inserted connecting points P and Q of Figure 6.1.1, the solution is no longer periodic – it can be chaotic, and show reversals; see Robbins (1976).

6.2 Hydromagnetics in an Ohmic Conductor

The earth's core is, of course, much more complicated than Bullard's disk dynamo. It is not completely understood, but much progress has been made. We begin with the governing equations. The analog of (6.1.5) is a fairly standard piece of continuum mechanics; readers unfamiliar with this subject should consult a text such as Malvern (1969). If **u** is the velocity of the core fluid, $\breve{\rho}$ is its density, **g** is gravity (including centrifugal force), and $\boldsymbol{\Omega}$ is the angular velocity of the reference frame in which the core is viewed (usually a frame fixed with respect to the mantle), then the momentum equation for the fluid is

$$\breve{\rho}(\mathbf{a} + 2\boldsymbol{\Omega} \times \mathbf{u}) = -\boldsymbol{\nabla}p + \breve{\rho}\mathbf{g} + \mathbf{J} \times \mathbf{B} + \mathbf{f}_{\text{visc}}. \qquad (6.2.1)$$

Here **a** is the acceleration of a fluid particle, as seen in the observer's reference frame, so

$$\mathbf{a} = \partial_t\mathbf{u} + (\mathbf{u} \cdot \boldsymbol{\nabla})\mathbf{u} = D_t\mathbf{u}.$$

This is the definition of D_t, which is sometimes called the *material time derivative*. We will discuss such derivatives in more detail in subsection 6.2.4. On the right we have the forcing terms; first is the gradient of the pressure in the fluid. Next is the *Lorentz force*, which gives the force exerted on an electric current by a magnetic field. Since **J** is the electric current density

$$\mathbf{J} \times \mathbf{B} = \frac{1}{\mu_0}(\boldsymbol{\nabla} \times \mathbf{B}) \times \mathbf{B}.$$

The last term on the right in (6.2.1) is $\mathbf{f}_{\text{visc}}$, the viscous force. By definition in a so-called Newtonian fluid, its j-th Cartesian component is $f_j = \partial_i V_{ij}$ where

$$V_{ij} = (\kappa - \tfrac{2}{3}\zeta)(\boldsymbol{\nabla} \cdot \mathbf{u})\delta_{ij} + \zeta(\partial_i u_j + \partial_j u_i)$$

where V_{ij} are the Cartesian components of the viscous stress tensor, κ is the bulk viscosity of the fluid, and ζ is its shear viscosity. Both κ and ζ are very poorly known for the fluid of the core. Estimates of ζ, based on various semiempirical theories, range from 10^{-3} to 10^2 kg/m s, and upper bounds from the damping of seismic waves are much higher. The value of κ is even less certain.

To use (6.2.1), we also need the equation of mass conservation,

$$\partial_t\breve{\rho} + \boldsymbol{\nabla} \cdot (\breve{\rho}\mathbf{u}) = 0, \qquad (6.2.2)$$

often called the *continuity equation*. From this equation it is clear that the bulk viscosity would be unimportant for a liquid that was incompressible, i.e., one with $\breve{\rho}$ = constant; but a material parcel moving from the bottom of the outer core to the top experiences a change in density of more than 20 percent.

6.2.1 Ohm's Law for a Moving Conductor

It remains to find the analog of (6.1.2) in the core. The discussion is like the discussion of induction in the mantle in section 5.4, except that Ohm's law is no longer so obvious, since we are looking at a *moving* conductor, the core fluid. Ohm's law is an empirical observation about materials, made in a laboratory in which the conductor is at rest. We must understand how the results of this experiment will appear to an observer not at rest relative to the conductor.

This leads us to the foundations of physics. To make any physical observations, an observer establishes a reference frame, a way of measuring the time and spatial coordinates of events. We will assume that she chooses an origin and three orthogonal axes, and uses them to establish a Cartesian frame. All this means we neglect the curvature of space that would be produced by nearby large masses and would make it impossible to describe events in a Cartesian frame.

We also assume that the observer's reference frame is unaccelerated or *inertial*. That is, she is able to assign masses to bodies by observing their collisions with a standard mass, and then the total force acting on a body (i.e., its mass times its acceleration) is the gravitational force computed from the other masses via Newton's law of gravity, plus an electromagnetic force $\mathbf{f}^{EM}$. There are no Coriolis or centrifugal forces, and we neglect weak and nuclear forces by assuming that the particles never approach within nuclear distances. The force $\mathbf{f}^{EM}$ is produced by other bodies, and this can be checked by removing them. We assert that $\mathbf{f}^{EM}$ has the following properties:

(1) If two bodies are at rest in the frame and are at positions $\mathbf{r}_i$ and $\mathbf{r}_j$, and $|\mathbf{r}_i - \mathbf{r}_j|$ is much greater than the sizes of the bodies, then the force $\mathbf{f}_{ij}^{EM}$ exerted by the ith body on the jth is

$$\mathbf{f}_{ij}^{EM} = C_{ij}\frac{\mathbf{r}_i - \mathbf{r}_j}{|\mathbf{r}_i - \mathbf{r}_j|^3} \qquad (6.2.3)$$

where the constant C_{ij} depends only on the two bodies and not on their positions or velocity.

(2) The observer finds that she can assign a number q_i to the ith body (obviously its electrical charge) such that for all pairs of bodies, $C_{ij}/q_i q_j$ has the same value. This value is positive, and we will write it $(4\pi\epsilon_0)^{-1}$, so

$$C_{ij} = \frac{q_i q_j}{4\pi\epsilon_0}. \qquad (6.2.4)$$

The observer has some freedom in assigning the charges. For any constant k, positive or negative, she can use kq_i for the ith body as long as she replaces ϵ_0 by $k^2\epsilon_0$. Thus choosing ϵ_0 is equivalent to choosing a unit of charge. For $\epsilon_0 = (4\pi)^{-1}$, the unit of charge is called the statcoulomb. For $\epsilon_0 = (\mu_0 c^2)^{-1}$, where $\mu_0 = 4\pi \times 10^{-7}$ and c is the speed of light in meters per second, the unit of charge is called the coulomb, and this is the SI unit (see section 2.3).

(3) The observer finds that she can assign to each point in space and each time a pair of vectors $\mathbf{E}(\mathbf{r}, t)$ and $\mathbf{B}(\mathbf{r}, t)$ such that if a small body with charge q is near $\mathbf{r}$ at time t and the body's velocity then is $\mathbf{u}$, the nongravitational force on the body is

$$\mathbf{f}^{EM} = q[\mathbf{E}(\mathbf{r}, t) + \mathbf{u} \times \mathbf{B}(\mathbf{r}, t)]. \qquad (6.2.5)$$

$\mathbf{E}$ and $\mathbf{B}$ are the electric and magnetic fields seen by the observer at $\mathbf{r}, t$. Observing only three particles with linearly independent velocities near $\mathbf{r}, t$ enables the observer to measure both $\mathbf{E}$ and $\mathbf{B}$. For simplicity, suppose the velocities are $\hat{\mathbf{x}}, \hat{\mathbf{y}}$, and $\hat{\mathbf{z}}$. The observer measures $\mathbf{f}^{EM}$ and knows q for each particle, so she can find $\mathbf{E} + \mathbf{u} \times \mathbf{B}$ for each. But $\mathbf{u} \cdot (\mathbf{E} + \mathbf{u} \times \mathbf{B}) = \mathbf{u} \cdot \mathbf{E}$, so she knows $\mathbf{u} \cdot \mathbf{E}$ for each particle. That is, she knows $\hat{\mathbf{x}} \cdot \mathbf{E}$, $\hat{\mathbf{y}} \cdot \mathbf{E}$, and $\hat{\mathbf{z}} \cdot \mathbf{E}$. Then she can find $\mathbf{u} \times \mathbf{B}$ for each particle. But $\hat{\mathbf{x}} \times \mathbf{B} = -\hat{\mathbf{y}}B_z + \hat{\mathbf{z}}B_y$ and $\hat{\mathbf{y}} \times \mathbf{B} = \hat{\mathbf{x}}B_z - \hat{\mathbf{z}}B_x$, so she knows B_x, B_y, B_z.

Now consider a second observer moving at constant velocity $\mathbf{v}$ relative to the first observer. Suppose $|\mathbf{v}|$ is much less than c, so both observers can use the same time t. Suppose the second observer chooses his coordinate axes parallel to those of the first observer, and chooses his origin so that at time $t = 0$ the two observers have the same origin. If the first observer sees an event at position $\mathbf{r}$ and time t, the second observer will see the same event at position $\mathbf{r}'$ at time t' where

$$t' = t \qquad (6.2.6)$$

$$\mathbf{r}' = \mathbf{r} - \mathbf{v}t. \tag{6.2.7}$$

The relativistic Lorentz transformation has been replaced by the Galilean transformation because $|\mathbf{v}| \ll c$.

If the first observer claims that a small body has velocity $\mathbf{u}(t)$ at time t, the second observer will claim that it has velocity $\mathbf{u}'(t)$, where

$$\mathbf{u}'(t) = \mathbf{u}(t) - \mathbf{v}. \tag{6.2.8}$$

The acceleration $\mathbf{a}$ of a body is $\mathbf{a} = d\mathbf{u}/dt$. Since $d\mathbf{v}/dt = 0$, (6.2.8) implies

$$\mathbf{a}'(t) = \mathbf{a}(t), \tag{6.2.9}$$

the two observers will agree on the acceleration of the body. If the two observers adopt the same mass unit, they will agree on the masses of bodies

$$m' = m \tag{6.2.10}$$

and hence on the value of $m\mathbf{a}$, the total force on a body. They will also agree on the gravitational force exerted on the body by the surrounding masses, so they will agree on the nongravitational force. For the first observer this is (6.2.5). For the second observer let us call it $(\mathbf{f}^{NG})'$.

Suppose the second observer decides to repeat the experiments that led the first observer to (6.2.5). He can simply watch the first observer at work and describe the first observer's results in his own terms. He will agree with the first observer about observations (1) and (2). If he chooses the same value of ϵ_0 as the first observer, and agrees with the first observer that electrons are negative, he will assign the same charge to each body as does the first observer. Now the second observer will write the first observer's observation (6.2.5) as

$$(\mathbf{f}^{NG})' = q'[\mathbf{E} + (\mathbf{u}' + \mathbf{v}) \times \mathbf{B}] \tag{6.2.11}$$

where $\mathbf{E}$ and $\mathbf{B}$ are the electric and magnetic fields observed by the first observer. But (6.2.11) has the same form as (6.2.5). It is an electromagnetic force, with

$$\mathbf{E}' = \mathbf{E} + \mathbf{v} \times \mathbf{B} \tag{6.2.12}$$
$$\mathbf{B}' = \mathbf{B}. \tag{6.2.13}$$

Note that the primed observer has velocity **v** as seen by the unprimed observer in these two equations. Thus the moving observer will see the nongravitational force to be electromagnetic, but he will disagree with the first (stationary) observer about the electric field. He will agree about the magnetic field.

Equations (6.2.12, 6.2.13) are the Galilean transformation law for the electromagnetic field. This law is a prerelativistic approximation to the correct transformation, the Lorentz transformation. The Lorentz transformation differs from (6.2.6, 6.2.7) by terms of order $|\mathbf{v}|^2/c^2$. The true Lorentz form of (6.2.8) is (Panofsky and Phillips, 1962, chap. 16)

$$\mathbf{u}' = \mathbf{u} - \mathbf{v}\frac{1 - (\mathbf{u} \cdot \mathbf{v}/cv)^2}{1 - \mathbf{u} \cdot \mathbf{v}/c^2} - \hat{\mathbf{s}}\hat{\mathbf{s}} \cdot \mathbf{u}\left(1 - \frac{1}{\Gamma(1 - \mathbf{u} \cdot \mathbf{v}/c^2)}\right)$$

where $\hat{\mathbf{s}}$ is a unit vector in the plane of **u** and **v** and at right angles to **v** and $\Gamma = (1 - v^2/c^2)^{-1/2}$. The Lorentz version of (6.2.12, 6.2.13) is (Panofsky and Phillips, 1962, chap. 18)

$$\mathbf{B}'^{\|} = \mathbf{B}^{\|} \qquad \mathbf{B}'^{\perp} = \Gamma[\mathbf{B} - \mathbf{v} \times \mathbf{E}/c^2]^{\perp}$$

$$\mathbf{E}'^{\|} = \mathbf{E}^{\|} \qquad \mathbf{E}'^{\perp} = \Gamma[\mathbf{E} + \mathbf{v} \times \mathbf{B}]^{\perp}$$

where $\mathbf{E}^{\|}$ is the part of **E** parallel to **v**, and $\mathbf{E}^{\perp}$ is the part perpendicular to **v**. Clearly (6.2.12) is a good approximation as long as $(v/c)^2 \ll 1$. The approximate validity of (6.2.13) requires that both $(v/c)^2 \ll 1$ and

$$vE/Bc^2 \ll 1 \qquad\qquad (6.2.14)$$

where $v = |\mathbf{v}|$, $E = |\mathbf{E}|$, and $B = |\mathbf{B}|$.

In order to see how Ohm's law looks to a moving observer, it will also be necessary to know how the two observers' charge densities and current densities compare. For simplicity, suppose that all the current is carried by charged particles of charge $+q$ or $-q$. Suppose the unprimed observer sees n^+ positive and n^- negative particles per unit volume, moving with average velocities $\mathbf{u}^+$ and $\mathbf{u}^-$. Then that observer will see a charge density ρ and current density **J** given by

$$\mathbf{J} = (n^+\mathbf{u}^+ - n^-\mathbf{u}^-)q$$

$$\rho = (n^+ - n^-)q.$$

The primed observer will see

$$\mathbf{J}' = (n'^+\mathbf{u}'^+ - n'^-\mathbf{u}'^-)q$$
$$\rho' = (n'^+ - n'^-)q.$$

Since $n'^+ = n^+$, $n'^- = n'$, and the velocities are related by (6.2.8), the result is

$$\rho' = \rho \tag{6.2.15}$$

$$\mathbf{J}' = \mathbf{J} - \rho\mathbf{v}. \tag{6.2.16}$$

The two observers see the same charge density but not the same current density. However, for electromagnetic fields with length scale L and time scale T, $\rho \approx \epsilon_0 E/L$ and $J \approx B/\mu_0 L$, so

$$\frac{\rho v}{J} \approx \frac{\epsilon_0\mu_0 Ev}{B} = \frac{Ev}{Bc^2}. \tag{6.2.17}$$

We already know that we need (6.2.14) in order to use (6.2.13) instead of the relativistic form; but then (6.2.17) justifies the approximation

$$\mathbf{J} = \mathbf{J}'. \tag{6.2.18}$$

Now consider an Ohmic conductor observed at rest in a laboratory. If an electric field $\mathbf{E}^R$ is applied to the conducting material, it responds by conducting a current with density $\mathbf{J}^R$ where

$$\mathbf{J}^R = \sigma\mathbf{E}^R. \tag{6.2.19}$$

The physical mechanism is that $\mathbf{E}^R$ accelerates electrons and ions until they collide with material atoms. Because of these collisions, the electrons and ions do not gain energy on average. All of the power per unit volume, $\mathbf{J}^R \cdot \mathbf{E}^R$, is dissipated as ohmic heat, i.e., it appears as random kinetic energy of the atoms in the material, so that the material becomes warmer. The heating rate is

$$h^R = \mathbf{J}^R \cdot \mathbf{E}^R = \sigma|\mathbf{E}^R|^2 = \frac{|\mathbf{J}^R|^2}{\sigma} \quad \text{W/m}^3. \tag{6.2.20}$$

Now suppose the conductor moves past us with velocity $\mathbf{u}$, and we measure the electric and magnetic fields in it to be $\mathbf{E}$ and $\mathbf{B}$. What current density $\mathbf{J}$ and what ohmic heating rate h will we see in the conductor?

According to (6.2.18), $\mathbf{J}^R = \mathbf{J}$. According to (6.2.12) and (6.2.13)

$$\mathbf{B}^R = \mathbf{B} \quad \text{and} \quad \mathbf{E}^R = \mathbf{E} + \mathbf{u} \times \mathbf{B}.$$

Therefore, expressed in our terms, (6.2.19) is

$$\mathbf{J} = \sigma(\mathbf{E} + \mathbf{u} \times \mathbf{B}). \tag{6.2.21}$$

The ohmic heating rate h is measured in the real frame of the material, and to measure it we would put ourselves in that frame. Therefore we will set it equal to h^R, and it will be, in our terms,

$$h = \mathbf{J} \cdot (\mathbf{E} + \mathbf{u} \times \mathbf{B}) = \sigma|\mathbf{E} + \mathbf{u} \times \mathbf{B}|^2 = \frac{|\mathbf{J}|^2}{\sigma}, \tag{6.2.22}$$

which agrees of course with (6.2.20).

6.2.2 Equations Governing the Geodynamo

Now we can discuss how $\mathbf{E}$ and $\mathbf{B}$ change with time in a moving ohmic conductor. The pre-Maxwell equations and Ohm's law are

$$\left. \begin{aligned} \boldsymbol{\nabla} \cdot \mathbf{B} &= 0 \\ \epsilon_0 \boldsymbol{\nabla} \cdot \mathbf{E} &= \rho \\ \partial_t \mathbf{B} &= -\boldsymbol{\nabla} \times \mathbf{E} \\ \boldsymbol{\nabla} \times \mathbf{B} &= \mu_0 \mathbf{J} \\ \mathbf{J} &= \sigma(\mathbf{E} + \mathbf{u} \times \mathbf{B}) \end{aligned} \right\} \tag{6.2.23}$$

where ρ is the electric charge density. Recall from Chapter 5 the magnetic diffusivity η, defined by

$$\eta = \frac{1}{\mu_0 \sigma}. \tag{6.2.24}$$

Then we note that $\boldsymbol{\nabla} \times \mathbf{B} = \mu_0 \mathbf{J}$ and Ohm's law combine to give

$$\mathbf{E} = -\mathbf{u} \times \mathbf{B} + \eta \boldsymbol{\nabla} \times \mathbf{B}. \tag{6.2.25}$$

Then the induction equation, the third Maxwell equation in (6.2.23), implies

$$\partial_t \mathbf{B} = \mathbf{\nabla} \times (\mathbf{u} \times \mathbf{B} - \eta \mathbf{\nabla} \times \mathbf{B}). \tag{6.2.26}$$

The procedure for solving (6.2.23) when we know $\mathbf{u}$ is to solve (6.2.26) for $\mathbf{B}$, then to find $\mathbf{E}$ from (6.2.25), and then find $\mathbf{J}$ from

$$\mathbf{J} = \frac{1}{\mu_0} \mathbf{\nabla} \times \mathbf{B} = \sigma(\mathbf{E} + \mathbf{u} \times \mathbf{B}) \tag{6.2.27}$$

(either expression gives $\mathbf{J}$); then ρ can be calculated from

$$\rho = \epsilon_0 \mathbf{\nabla} \cdot \mathbf{E} = \epsilon_0 \mathbf{\nabla} \cdot [\eta \mathbf{\nabla} \times \mathbf{B} - \mathbf{u} \times \mathbf{B}]. \tag{6.2.28}$$

Obviously the fundamental problem is to solve (6.2.26). To do so, we need initial conditions on $\mathbf{B}$. If at time $t = 0$, $\mathbf{B}$ satisfies

$$\mathbf{\nabla} \cdot \mathbf{B} = 0 \tag{6.2.29}$$

then this will remain true forever, because any solution $\mathbf{B}$ of (6.2.26) satisfies

$$\partial_t(\mathbf{\nabla} \cdot \mathbf{B}) = 0. \tag{6.2.30}$$

Equation (6.2.26) is the analog of (6.1.2) in the Bullard dynamo system, just as the momentum equation (6.2.1) is the analog of (6.1.5). Equation (6.2.26) is called the *kinematic dynamo equation*, or simply the *dynamo equation*. The kinematic dynamo problem consists in solving (6.2.26) for $\mathbf{B}$ when $\mathbf{u}$ is given. The *full dynamo problem* consists in solving (6.2.26) and (6.2.1) simultaneously for $\mathbf{B}$ and $\mathbf{u}$.

To solve the full system requires accounting for other interactions, for example, heating of the fluid by electric currents, which in turn causes expansion of the material and buoyancy forces. The rate of ohmic heat production is

$$h_{EM} = |\mathbf{J}|^2/\sigma. \tag{6.2.31}$$

This feeds back into the system via the energy equation, which we have not yet written down. That equation is

$$\breve{\rho}(D_t U) + \mathbf{\nabla} \cdot \mathbf{h} = V_{ij} \partial_i u_j + h_{EM} + h_{RAD}. \tag{6.2.32}$$

Here U is the internal energy per kilogram of material,

$$D_t U = \partial_t U + \mathbf{u} \cdot \nabla U$$

where $\mathbf{h}$ is the heat flow by heat conduction, V_{ij} are the Cartesian components of the viscous stress tensor, and h_{RAD} is the radioactive heating rate in W/m^3. The full dynamo problem requires the simultaneous solution of (6.2.1), (6.2.2), (6.2.32), (6.2.26), and the equation of state, which gives U as a function of $\breve{\rho}$ and of S, the entropy per kilogram. From this function, the temperature is obtained as $\theta = \partial U / \partial S$, and hence the heat flow $\mathbf{h}$ can be found from

$$\mathbf{h} = -\chi \nabla \theta$$

where χ is thermal conductivity. The driving force for the whole system is the fact that temperature changes change $\breve{\rho}$ and produce a $\breve{\rho}\mathbf{g}$ in (6.2.1), which cannot be balanced by a hydrostatic pressure gradient. It has been suggested (Braginsky, 1963) that chemical fractionation occurs at the inner core–outer core boundary as the liquid outer core freezes out on the solid inner core because of the gradual cooling of the earth. This chemical fractionation would be another source of buoyancy differences in the $\breve{\rho}\mathbf{g}$ term in (6.2.1), possibly even more important than the thermal term. In view of the complexity of this system, it is not very surprising that no complete solution of the full dynamo problem is known.

6.2.3 The Kinematic Problem: Limiting Case with $\mathbf{u} = 0$

We will try to shed some light on the nature of the solutions $\mathbf{B}$ of the kinematic dynamo equation (6.2.26) when $\mathbf{u}$ is given. In order to simplify matters, we will assume that the magnetic diffusivity η is constant, so (6.2.26) can be written

$$\partial_t \mathbf{B} = \nabla \times (\mathbf{u} \times \mathbf{B}) + \eta \nabla^2 \mathbf{B}. \qquad (6.2.33)$$

When $\mathbf{u} = 0$, (6.2.33) is a heat equation for each of the three Cartesian components of $\mathbf{B}$. If $\mathbf{B}$ is a complex solution, its real part is a real solution of (6.2.33) as long as $\mathbf{u}$ and η are real, so we may consider complex solutions $\mathbf{B}$. With $\mathbf{u} = 0$ suppose

$$\mathbf{B}(\mathbf{r}, 0) = \mathbf{B}_0 e^{i\mathbf{k}\cdot\mathbf{r}} \qquad (6.2.34)$$

where $\mathbf{B}_0$ and $\mathbf{k}$ are constant vectors and $\mathbf{k}$ is real. Equation (6.2.34) represents a plane wave at $t = 0$ with wavevector $\mathbf{k}$. If the medium is infinite,

$$\partial_t \mathbf{B} = \eta \nabla^2 \mathbf{B} \qquad (6.2.35)$$

throughout R^3, so a solution to (6.2.35) can be found by assuming

$$\mathbf{B}(\mathbf{r}, t) = \mathbf{B}(t) e^{i\mathbf{k}\cdot\mathbf{r}}.$$

Then by direct calculation

$$\nabla^2 \mathbf{B} = -k^2 \mathbf{B}$$

and therefore by (6.2.35)

$$\partial_t \mathbf{B}(t) = -k^2 \eta \mathbf{B}(t)$$

and solving this equation gives

$$\mathbf{B}(t) = \mathbf{B}_0 e^{-k^2 \eta t}.$$

Therefore finally

$$\mathbf{B}(\mathbf{r}, t) = \mathbf{B}_0 e^{i\mathbf{k}\cdot\mathbf{r} - k^2 \eta t}. \qquad (6.2.36)$$

Waves with wavenumber k decay exponentially in time without changing shape with a mean life $T = (\eta k^2)^{-1}$. We can use the solution to tell us roughly how any feature with a length scale L in the magnetic field dies away: if we set the scale to be a half wavelength, $k = \pi/L$, this gives a mean life of

$$T = L^2/\pi^2 \eta = \mu_0 \sigma L^2 / \pi^2.$$

Applying this result to the core, we take $\sigma = 3 \times 10^5$ S/m and we use the core radius, 3.48×10^6 m for L; then $T = 4.6 \times 10^{11}$ s $=$ 15,000 years. The plane-wave geometry of (6.2.34) is poorly suited to a sphere like the core, yet remarkably, when we solved this problem carefully in section 5.4.4, we found the identical time scale for the slowest-decaying mode. This inexorable decay means that the earth's field is not a relic supported by eddy currents in a rigid core but is continually renewed by the fluid motion. Suppose we examine smaller-scale features on the core, say, with scale 10^6 m; these have a mean life of 1200 years, which is long enough to suggest that diffusion can be ignored in studies of secular variation over intervals of several decades to a century (more about this in the next two sections).

6.2.4 Eulerian and Lagrangian Descriptions

The other limit in (6.2.26) is to take $\eta = 0$ instead of $\mathbf{u} = 0$. In this case $\sigma = \infty$, and the material is a perfect conductor. In a perfect conductor $\mathbf{E} + \mathbf{u} \times \mathbf{B} = 0$. That is, $\mathbf{E}^R = \mathbf{0}$. The electric field in the local rest frame of the conductor always vanishes.

In this case, a little finesse serves to integrate (6.2.26) exactly. To carry out the integration, we must consider the two ways of describing the fluid motion, *Eulerian* and *Lagrangian*. Then every physical quantity, e.g., fluid pressure, temperature, mass density $\breve{\rho}$, charge density ρ, and magnetic field $\mathbf{B}$, has an Eulerian and a Lagrangian description. The Eulerian description is the one given by an observer fixed in space. To describe f, a physical quantity, in the Eulerian style, we must provide the function f_E that gives the value of f at spatial position $\mathbf{r}$ at time t;

$$f = f_E(\mathbf{r}, t) = \text{value of } f \text{ at spatial position } \mathbf{r} \text{ at time } t. \quad (6.2.37)$$

The Lagrangian description is given from the perspective of an observer moving along with the fluid. The idea is to label each fluid particle by its position $\mathbf{x}$ at time $t = 0$. The Lagrangian description of f is the function f_L that gives the value of f at time t at the particle labeled $\mathbf{x}$ (the particle that was at position $\mathbf{x}$ at time $t = 0$);

$$f = f_L(\mathbf{x}, t) = \text{value of } f \text{ at particle labeled } \mathbf{x} \text{ at time } t. \quad (6.2.38)$$

To specify the fluid motion in the Lagrangian fashion, we simply give the Lagrangian description of particle position, $\mathbf{r}_L$. It is

$$\mathbf{r} = \mathbf{r}_L(\mathbf{x}, t) = \text{position at time } t \text{ of particle labeled } \mathbf{x} \quad (6.2.39)$$

(i.e., of the particle that was at $\mathbf{x}$ at time $t = 0$). The Eulerian description of the fluid motion is the Eulerian description of particle velocity, $\mathbf{u}_E$. It is

$$\mathbf{u} = \mathbf{u}_E(\mathbf{r}, t) = \begin{array}{l} \text{velocity at time } t \text{ of fluid particle} \\ \text{which is at position } \mathbf{r} \text{ at time } t. \end{array} \quad (6.2.40)$$

Note that we write components of Lagrangian vectors thus: $(\mathbf{u}_L)_i = u_i^L$, and so on.

If either the Lagrangian or the Eulerian description of a physical quantity f is known, and if Lagrangian description (6.2.39) of the motion is known, then the other description of f can be calculated. To begin the calculation, we fix t and $\mathbf{r}$ and solve (6.2.39) for $\mathbf{x}$, obtaining

$$\mathbf{x} = \mathbf{x}_E(\mathbf{r}, t) = \begin{array}{l} \text{label (initial position) of particle} \\ \text{that is at position } \mathbf{r} \text{ at time } t. \end{array} \qquad (6.2.41)$$

Then if $\mathbf{x}$ and $\mathbf{r}$ are related by (6.2.39) and (6.2.41),

$$f = f_E(\mathbf{r}, t) = f_L(\mathbf{x}, t). \qquad (6.2.42)$$

Therefore

$$f_E(\mathbf{r}, t) = f_L[\mathbf{x}_E(\mathbf{r}, t), t] \qquad (6.2.43)$$

and

$$f_L(\mathbf{x}, t) = f_E[\mathbf{r}_L(\mathbf{x}, t), t]. \qquad (6.2.44)$$

If the Lagrangian description of the motion is known, the Eulerian description can be found as follows. The velocity at time t of the particle labeled $\mathbf{x}$ is obviously

$$\mathbf{u}_L(\mathbf{x}, t) = \frac{\partial}{\partial t} \mathbf{r}_L(\mathbf{x}, t) = \mathbf{u}_E(\mathbf{r}, t). \qquad (6.2.45)$$

Then we apply (6.2.43) with $f = \mathbf{u}$ to find $\mathbf{u}_E(\mathbf{r}, t)$. Conversely, if the Eulerian description of the motion is known, the Lagrangian description can be found. We can use (6.2.44) to write (6.2.45) as

$$\frac{\partial \mathbf{r}_L(\mathbf{x}, t)}{\partial t} = \mathbf{u}_E[\mathbf{r}_L(\mathbf{x}, t), t]. \qquad (6.2.46)$$

If we fix $\mathbf{x}$ (i.e., focus on one particular particle), then (6.2.46) is just an ordinary differential equation in t for $\mathbf{r}_L(\mathbf{x}, t)$. We know the initial condition.

$$\mathbf{r}_L(\mathbf{x}, 0) = \mathbf{x}, \qquad (6.2.47)$$

so we can integrate (6.2.46) to find $\mathbf{r}_L(\mathbf{x}, t)$.

If f is any physical quantity, then its time rate of change at a particular fluid particle will be written $D_t f$, and its time rate of change at a particular point in space will be written $\partial_t f$. These are both physical quantities, and clearly

$$(D_t f)_L(\mathbf{x}, t) = \frac{\partial}{\partial t} f_L(\mathbf{x}, t), \qquad \mathbf{x} \text{ fixed} \qquad (6.2.48)$$

$$(\partial_t f)_E(\mathbf{r}, t) = \frac{\partial}{\partial t} f_E(\mathbf{r}, t), \qquad \mathbf{r} \text{ fixed.} \qquad (6.2.49)$$

The relation between $D_t f$ and $\partial_t f$ can be obtained by applying the chain rule for t-differentiation to (6.2.44). Whenever $\mathbf{r}$ and $\mathbf{x}$ are related by (6.2.39) or (6.2.41), we have

$$\begin{aligned}
(D_t f)_L(\mathbf{x}, t) &= \frac{\partial}{\partial t} f_L(\mathbf{x}, t) = \frac{\partial}{\partial t} \{f_E[\mathbf{r}_L(\mathbf{x}, t), t]\} \\
&= \frac{\partial}{\partial t} f_E(\mathbf{r}, t) + \frac{\partial r_i^L(\mathbf{x}, t)}{\partial t} \frac{\partial f_E(\mathbf{r}, t)}{\partial r_i} \\
&= \frac{\partial}{\partial t} f_E(\mathbf{r}, t) + u_i^L(\mathbf{x}, t) \frac{\partial f_E(\mathbf{r}, t)}{\partial r_i}.
\end{aligned}$$

Notice the Einstein summation convention is at work. But

$$u_i^L(\mathbf{x}, t) = u_i^E(\mathbf{r}, t)$$
$$(D_t f)_L(\mathbf{x}, t) = (D_t f)_E(\mathbf{r}, t)$$

and

$$\frac{\partial}{\partial t} f_E(\mathbf{r}, t) = (\partial_t f)_E(\mathbf{r}, t).$$

Therefore,

$$(D_t f)_E(\mathbf{r}, t) = (\partial_t f)_E(\mathbf{r}, t) + \mathbf{u}_E(\mathbf{r}, t) \cdot \boldsymbol{\nabla} f_E(\mathbf{r}, t)$$

or

$$D_t f = \partial_t f + \mathbf{u} \cdot \boldsymbol{\nabla} f. \qquad (6.2.50)$$

This is true for any f, so we can summarize it by the operator equation

$$D_t = \partial_t + \mathbf{u} \cdot \boldsymbol{\nabla}. \qquad (6.2.51)$$

The reader may recall that this operator is called the material time derivative; we have already seen it at the beginning of section 6.2.

6.2.5 The Kinematic Problem: Limiting Case with $\eta = 0$

With this machinery on hand, we can return to the dynamo equation with diffusivity $\eta = 0$:

$$\partial_t \mathbf{B} = \boldsymbol{\nabla} \times (\mathbf{u} \times \mathbf{B}). \qquad (6.2.52)$$

We rewrite this, using $\boldsymbol{\nabla} \cdot \mathbf{B} = 0$, and the well-known identity as

$$\partial_t \mathbf{B} = (\mathbf{B} \cdot \boldsymbol{\nabla})\mathbf{u} - (\mathbf{u} \cdot \boldsymbol{\nabla})\mathbf{B} - (\boldsymbol{\nabla} \cdot \mathbf{u})\mathbf{B}.$$

Then, using (6.2.51), we have

$$D_t \mathbf{B} = (\mathbf{B} \cdot \boldsymbol{\nabla})\mathbf{u} - (\boldsymbol{\nabla} \cdot \mathbf{u})\mathbf{B}. \qquad (6.2.53)$$

Our objective in this subsection is to link the motion of the particles of the fluid with the behavior of the magnetic field. We do not assume the core fluid is incompressible, so we will need the continuity equation (6.2.2); this can be written in three equivalent ways:

$$\partial_t \breve{\rho} + \boldsymbol{\nabla}\breve{\rho} \cdot \mathbf{u} + \breve{\rho}\boldsymbol{\nabla} \cdot \mathbf{u} = 0$$

$$D_t \breve{\rho} + \breve{\rho}\boldsymbol{\nabla} \cdot \mathbf{u} = 0$$

$$D_t \ln \breve{\rho} + \boldsymbol{\nabla} \cdot \mathbf{u} = 0$$

where it will be recalled $\breve{\rho}$ is the ordinary material density. Thus

$$\boldsymbol{\nabla} \cdot \mathbf{u} = -D_t \ln \breve{\rho} = D_t \ln \frac{1}{\breve{\rho}} = \breve{\rho} D_t \frac{1}{\breve{\rho}}.$$

Therefore we can write (6.2.53) as

$$D_t \mathbf{B} + \breve{\rho}\left(D_t \frac{1}{\breve{\rho}}\right)\mathbf{B} = (\mathbf{B} \cdot \boldsymbol{\nabla})\mathbf{u}.$$

Dividing by $\breve{\rho}$ gives

$$D_t \left(\frac{\mathbf{B}}{\breve{\rho}}\right) = \left(\frac{\mathbf{B}}{\breve{\rho}} \cdot \boldsymbol{\nabla}\right)\mathbf{u}. \qquad (6.2.54)$$

Still more progress can be made by noting that

$$\mathbf{u} = D_t \mathbf{r} \qquad \text{or} \qquad u_L(\mathbf{x}, t) = \frac{\partial}{\partial t} \mathbf{r}_L(\mathbf{x}, t).$$

Then writing out Cartesian components of (6.2.54) gives

$$D_t \left(\frac{B_i}{\breve{\rho}} \right) = \left[\frac{B_j}{\breve{\rho}} \frac{\partial}{\partial r_j} \right] \frac{\partial}{\partial t} r_i^L(\mathbf{x}, t). \tag{6.2.55}$$

We cannot claim

$$\frac{\partial}{\partial r_j} \frac{\partial}{\partial t} = \frac{\partial}{\partial t} \frac{\partial}{\partial r_j}$$

in (6.2.55) because t and r_j are not independent there. However, t and x_j are independent, so we can write (6.2.55) as

$$D_t \left(\frac{B_i}{\breve{\rho}} \right) = \left[\frac{B_j}{\breve{\rho}} \frac{\partial x_k}{\partial r_j} \frac{\partial}{\partial x_k} \right] \frac{\partial}{\partial t} r_i^L(\mathbf{x}, t)$$

$$= \frac{B_j}{\breve{\rho}} \frac{\partial x_k}{\partial r_j} \frac{\partial}{\partial t} \left(\frac{\partial r_i^L}{\partial x_k}(\mathbf{x}, t) \right)$$

or, with some obvious abbreviations,

$$D_t \left(\frac{B_i}{\breve{\rho}} \right) = \frac{B_j}{\breve{\rho}} \frac{\partial x_k}{\partial r_j} D_t \frac{\partial r_i}{\partial x_k}. \tag{6.2.56}$$

But

$$\frac{\partial x_m}{\partial r_i} \frac{\partial r_i}{\partial x_k} = \frac{\partial x_m}{\partial x_k} = \delta_{mk}.$$

Since δ_{mk} is constant, $D_t \delta_{mk} = 0$, and so

$$\left(D_t \frac{\partial x_m}{\partial r_i} \right) \frac{\partial r_i}{\partial x_k} + \frac{\partial x_m}{\partial r_i} \left(D_t \frac{\partial r_i}{\partial x_k} \right) = 0.$$

Therefore, if we multiply (6.2.56) by $\partial x_m / \partial x_i$, we obtain

$$\frac{\partial x_m}{\partial r_i} D_t \left(\frac{B_i}{\breve{\rho}} \right) = \frac{B_j}{\breve{\rho}} \frac{\partial x_k}{\partial r_j} \frac{\partial r_i}{\partial x_k} D_t \left(\frac{\partial x_m}{\partial r_i} \right).$$

We also have

$$\frac{\partial x_k}{\partial r_j} \frac{\partial r_i}{\partial x_k} = \delta_{ij},$$

(6.2.57)

so

$$\frac{\partial x_m}{\partial r_i} D_t \left(\frac{B_i}{\breve{\rho}} \right) = -\frac{B_j}{\breve{\rho}} \delta_{ij} D_t \left(\frac{\partial x_m}{\partial r_i} \right) = -\frac{B_i}{\breve{\rho}} D_t \left(\frac{\partial x_m}{\partial r_i} \right).$$

Thus, finally,

$$D_t \left[\frac{\partial x_m}{\partial r_i} \frac{B_i}{\breve{\rho}} \right] = 0.$$

(6.2.58)

Equation (6.2.58) is equivalent to (6.2.52). But (6.2.58) says that the value of $(\partial x_m/\partial r_i)\,(B_i/\breve{\rho})$ is the same at a fluid particle labeled $\mathbf{x}$ at all times. At time $t = 0$, the labeling process required (6.2.47), so obviously

$$\mathbf{x}_E(\mathbf{r}, 0) = \mathbf{r}.$$

(6.2.59)

Therefore

$$\frac{\partial x_m}{\partial r_i} = \delta_{im} \quad \text{at} \quad t = 0,$$

and thus

$$\frac{\partial x_m}{\partial r_i} \frac{B_i}{\breve{\rho}} = \frac{B_m}{\breve{\rho}}.$$

Then

$$\frac{\partial x_m}{\partial r_i} \frac{B_i}{\breve{\rho}} = \frac{B_m^L(\mathbf{x}, 0)}{\breve{\rho}_L(\mathbf{x}, 0)} = \frac{B_m(\mathbf{x}, 0)}{\breve{\rho}(\mathbf{x}, 0)}.$$

Multiplying by $\partial r_j/\partial x_m$ and using (6.2.57) gives

$$\frac{B_j^L(\mathbf{x}, t)}{\breve{\rho}_L(\mathbf{x}, t)} = \frac{\partial r_j^L(\mathbf{x}, t)}{\partial x_m} \frac{B_m^L(\mathbf{x}, 0)}{\breve{\rho}_L(\mathbf{x}, 0)}.$$

(6.2.60)

Equation (6.2.60) has a simple geometrical interpretation. Consider two nearby particles, with labels $\mathbf{x}$ and $\tilde{\mathbf{x}}$. At time t their positions are $\mathbf{r}_L(\mathbf{x}, t)$ and $\mathbf{r}_L(\tilde{\mathbf{x}}, t)$ and the vector connecting them is

$$\mathbf{r}_L(\tilde{\mathbf{x}}, t) - \mathbf{r}_L(\mathbf{x}, t) = (\tilde{x}_m - x_m)\frac{\partial \mathbf{r}_L(\mathbf{x}, t)}{\partial x_m} + O|\tilde{\mathbf{x}} - \mathbf{x}|^2.$$

If $|\tilde{\mathbf{x}} - \mathbf{x}|$ is small enough to permit neglecting the quadratic error, then the Cartesian components of this equation are

$$\left[r_j^L(\tilde{\mathbf{x}}, t) - r_j^L(\mathbf{x}, t) \right] = \frac{\partial r_j^L(\mathbf{x}, t)}{\partial x_m} \left[r_m^L(\tilde{\mathbf{x}}, 0) - r_m^L(\mathbf{x}, 0) \right]. \quad (6.2.61)$$

Comparison of (6.2.60) and (6.2.61) tells us that the motion of the fluid deforms the vector $\mathbf{B}/\breve{\rho}$ at a fluid particle $\mathbf{x}$ in exactly the same way as it deforms the fluid. If we imagine the vector $\mathbf{B}/\breve{\rho}$ scaled so that at time $t = 0$ its tail is at particle $\mathbf{x}$ and its head is at a nearby particle $\tilde{\mathbf{x}}$, then its tail and head will continue to occupy those particle positions as the fluid moves.

In an incompressible fluid, $D_t \breve{\rho} = 0$, so $\breve{\rho}$ is constant in t at each fluid particle $\mathbf{x}$, and thus $\breve{\rho}_L$ can be canceled in (6.2.60). The magnetic field itself moves as if its head and tail were fixed in the fluid at two very close particles.

One obvious consequence of this situation is that, whether a fluid is compressible or not, if it is a perfect conductor, then each magnetic line of force always consists of the same fluid particles. More precisely, if a curve moves so as always to consist of the same fluid particles, and if the curve is a magnetic line of force at time $t = 0$, then it is always a magnetic line of force.

For a formal proof, suppose that the curve $\mathbf{x}(\lambda)$ is a magnetic line of force at $t = 0$. Then we can choose the parameter λ so that

$$\frac{d\mathbf{x}(\lambda)}{d\lambda} = \frac{\mathbf{B}_L[\mathbf{x}(\lambda), 0]}{\breve{\rho}_L[\mathbf{x}(\lambda), 0]}.$$

We can regard this as a differential equation defining the shape of a line of force in a given magnetic field $\mathbf{B}$. At $t = 0$, each particle on the curve is identified by a particular value of λ. At a later time t, the particle labeled $\mathbf{x}(\lambda)$ is at position $\mathbf{r}(\lambda)$ where

$$\mathbf{r}(\lambda) = \mathbf{r}_L[\mathbf{x}(\lambda), t].$$

By the chain rule,

$$\frac{dr_i}{d\lambda} = \frac{\partial r_i^L}{\partial x_j} \frac{dx_j}{d\lambda} = \frac{\partial r_i^L}{\partial x_j} \frac{B_j^L[\mathbf{x}(\lambda), 0]}{\breve{\rho}_L[\mathbf{x}(\lambda), 0]}.$$

Then, by (6.2.60),

$$\frac{dr_i}{d\lambda} = \frac{B_i^L[\mathbf{x}(\lambda), t]}{\breve{\rho}_L[\mathbf{x}(\lambda), t]} = \frac{B_i^E[\mathbf{r}(\lambda), t]}{\breve{\rho}_E[\mathbf{r}(\lambda), t]}.$$

Thus if the curve of particles is a line of force at $t = 0$ it remains one at all later times because it continues to obey the differential equation for lines of force. The result is usually stated in the following somewhat imprecise form:

> In a perfectly conducting material, the magnetic lines
> of force are permanently attached to the material,   (6.2.62)
> and move with it.

The magnetic field lines are said to be *frozen* in the fluid.

6.2.6 Frozen-Flux Condition

There is more to (6.2.60) than (6.2.62), because (6.2.62) says nothing about the strength of the field. There is a nonlocal consequence of (6.2.52) or (6.2.60) that is often very helpful in visualizing how $\mathbf{B}$ changes as the fluid moves. Let $\mathbf{A}$ be the magnetic vector potential in all of R^3. Then

$$\mathbf{B} = \mathbf{\nabla} \times \mathbf{A} \quad \text{in} \quad R^3. \qquad (6.2.63)$$

Suppose S_1 and S_2 are two oriented surfaces with the same edge C. Then the flux of $\mathbf{B}$ is the same across S_1 and S_2. By Stokes' Theorem it is $\oint_C d\mathbf{r} \cdot \mathbf{A}$. Therefore, this flux is a property of the edge curve C, and does not depend on which surface S with edge C is used to calculate the flux. We call this flux the magnetic flux $F(C)$ *linking* the edge C and denote it by $F(C)$. It is

$$F(C) = \oint_C d\mathbf{r} \cdot \mathbf{A}. \qquad (6.2.64)$$

Now suppose that C lies inside and moves with a perfect conductor, and always consists of the same material particles. The perfect conductor need not contain any surface S whose edge is C; the conductor can be simply a wire containing C. Since C now changes

with time, we write it $C(t)$. How does $F[C(t)]$ change with time? It
doesn't:

$$\frac{d}{dt}F[C(t)] = 0. \qquad (6.2.65)$$

To prove this, we parametrize $C(0)$ with a parameter λ such that
$0 \leq \lambda \leq 1$ and

$$\mathbf{x}(0) = \mathbf{x}(1). \qquad (6.2.66)$$

We use the same λ to parametrize $C(t)$ by assigning to each particle
on $C(t)$ the value of λ it had on $C(0)$. Thus the parametrization of
$C(t)$ is

$$\mathbf{r}(\lambda, t) = \mathbf{r}_L[\mathbf{x}(\lambda), t] \qquad (6.2.67)$$

where $\mathbf{x}(\lambda)$ is the initial position of the particle whose parameter is
always λ. If f is any physical quantity, then for the particles on $C(t)$
we can regard f as a function of λ and t, namely

$$f(\lambda, t) = f_L[\mathbf{x}(\lambda), t] = f_E[\mathbf{r}(\lambda, t), t]. \qquad (6.2.68)$$

Then clearly

$$\partial_t f(\lambda, t) = (D_t f)(\lambda, t) = (D_t f)_E(\mathbf{r}(\lambda, t), t) \qquad (6.2.69)$$

$$\partial_\lambda f(\lambda, t) = \partial_\lambda \mathbf{r}(\lambda, t) \cdot \boldsymbol{\nabla} f_E(\mathbf{r}(\lambda, t), t)$$
$$= \frac{\partial r_i(\lambda, t)}{\partial \lambda} \frac{\partial f_E}{\partial r_i}(\mathbf{r}(\lambda, t), t). \qquad (6.2.70)$$

Moreover,

$$\partial_t \mathbf{r}(\lambda, t) = (D_t \mathbf{r})(\lambda, t) = \mathbf{u}(\lambda, t). \qquad (6.2.71)$$

Now we can write (6.2.64) as

$$F[C(t)] = \int_0^1 d\lambda \, \partial_\lambda \mathbf{r}(\lambda, t) \cdot \mathbf{A}.$$

Differentiating,

$$\frac{d}{dt}F[C(t)] = \int_0^1 d\lambda \, [\partial_t \, \partial_\lambda \mathbf{r}(\lambda, t) \cdot \mathbf{A} + \partial_\lambda \mathbf{r} \cdot D_t \mathbf{A}]$$
$$= \int_0^1 d\lambda \, [\partial_\lambda \mathbf{u} \cdot \mathbf{A} + \partial_\lambda \mathbf{r} \cdot D_t \mathbf{A}] \qquad (6.2.72)$$

because by (6.2.71)

$$\partial_t \partial_\lambda \mathbf{r} = \partial_\lambda \partial_t \mathbf{r}(\lambda, t) = \partial_\lambda \mathbf{u}(\lambda, t).$$

Since $C(t)$ is in a perfect conductor, there we have

$$\mathbf{E} = -\mathbf{u} \times \mathbf{B} = -\mathbf{u} \times (\nabla \times \mathbf{A}).$$

In components

$$E_i = -\epsilon_{ijk} u_j \epsilon_{k\ell m} \partial_\ell A_m$$
$$= -(\delta_{i\ell}\delta_{jm} - \delta_{im}\delta_{j\ell}) u_j \partial_\ell \partial_m A_m.$$

Therefore

$$E_i = u_j \partial_j A_i - u_j \partial_i A_j. \tag{6.2.73}$$

But everywhere in R^3,

$$\nabla \times \mathbf{E} = -\partial_t \mathbf{B} = -\partial_t \nabla \times \mathbf{A} = -\nabla \times \partial_t \mathbf{A},$$

so $\nabla \times (\mathbf{E} + \partial_t \mathbf{A}) = 0$. Thus there is a scalar field ϕ in R^3 such that $\mathbf{E} + \partial_t \mathbf{A} = -\nabla\phi$, or

$$E_i = -\partial_i \phi - \partial_t A_i. \tag{6.2.74}$$

Combining (6.2.73) and (6.2.74) gives

$$D_t A_i = -\partial_i \phi + u_j \partial_i A_j. \tag{6.2.75}$$

Then

$$\partial_\lambda \mathbf{r} \cdot D_t \mathbf{A} = \partial_\lambda r_i D_t A_i = -(\partial_\lambda r_i)(\partial_i \phi) + u_j(\partial_\lambda r_i)(\partial_i A_j),$$

so

$$\partial_\lambda \mathbf{r} \cdot D_t \mathbf{A} = -\partial_\lambda \phi(\lambda, t) + \mathbf{u} \cdot \partial_\lambda \mathbf{A}(\lambda, t). \tag{6.2.76}$$

Substituting (6.2.76) in (6.2.72) gives

$$\frac{d}{dt} F[C(t)] = \int_0^1 d\lambda \, \partial_\lambda [\mathbf{u} \cdot \mathbf{A} - \phi] = [\mathbf{u} \cdot \mathbf{A} - \phi]_{\lambda=0}^{\lambda=1}. \tag{6.2.77}$$

But because of (6.2.66) and (6.2.67),

$$\mathbf{r}(0, t) = \mathbf{r}(1, t)$$

for all t. Hence (6.2.77) gives (6.2.65), that the flux linking C is unchanging in time.

One consequence of (6.2.65) is that the magnetic flux through a superconducting wire loop cannot change with time. In the quantum theory of superconductors, the wave function of a charge carrier must be coherent around the loop, and can change only by an integral number of wavelengths, which implies that the flux through the loop can change only by an integral multiple of h/q where $h =$ Planck's constant and $q =$ charge of charge carriers. In both the old metallic and the new ceramic superconductors, $q = 2e$ because the charge carriers are bound pairs of electrons. Thus the flux quantum is

$$\frac{h}{2e} = 2.07 \times 10^{-15} \text{ T m}^2 = 2.07 \times 10^{-15} \text{ Wb}.$$

The flux cannot change at all unless there is a "weak link" that permits a part of the loop to fluctuate out of superconductivity occasionally. The counting of integral changes of flux in a loop with such a link is the working principle of modern superconducting magnetometers widely used in paleomagnetism.

Assertion (6.2.65) is true as long as $C(t)$ always lies in a perfect conductor and moves with it, even if $C(t)$ is not the edge of any surface lying in the conductor. However, if such a surface does exist, a special case of (6.2.65) is as follows:

> The magnetic flux across a surface is constant in time
> if that surface lies in a perfectly conducting material $\Big\}$ (6.2.78)
> and always consists of the same material particles.

The invariance in time of the flux integral over a surface consisting of the same fluid particles is called the *frozen-flux condition*.

This condition has important applications to the core, as we will see in section 6.4. As a brief illustration, suppose the sun were to collapse from its present radius, $a = 7 \times 10^5$ km, to that of a neutron star, say, $a = 10$ km. The average field intensity out of the sun's northern hemisphere is about 10^{-4} T, or a little more than the average

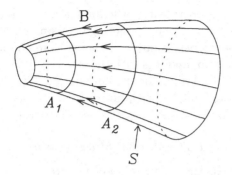

Figure 6.2.1: A flux tube. The walls of the tube contain the magnetic field lines. Two surfaces A_1 and A_2 cut the tube.

field at the earth's surface. By (6.2.78), Ba^2 would remain constant during the collapse, so after collapse $B \approx 5 \times 10^5$ T. Pulsars, whose magnetic fields can be estimated, are believed to be neutron stars, and do appear to have magnetic fields in the megatesla range.

A *flux tube* (Figure 6.2.1) is a long, thin volume consisting of magnetic lines of force. If A_1 and A_2 are any two surfaces cutting the tube, and $\hat{n}_1$ and $\hat{n}_2$ are the unit normals to A_1 and A_2 in the direction of **B** along the tube,

$$\int_{A_1} d^2\mathbf{r} \, \hat{n}_1 \cdot \mathbf{B} = \int_{A_2} d^2\mathbf{r} \, \hat{n}_2 \cdot \mathbf{B} \qquad (6.2.79)$$

because **B** is tangent to the walls of the flux tube, and Gauss' Theorem can be applied to the region bounded by those walls, A_1 and A_2. Since $\nabla \cdot \mathbf{B} = 0$ in this region, and $\hat{n}_A \cdot \mathbf{B} = 0$, one has (6.2.79). The flux (6.2.79) is called the flux through the tube, or the strength of the tube.

If a flux tube lies in a perfectly conducting material at time $t = 0$, the particles that make up the tube at $t = 0$ will be a flux tube at every later time, because of (6.2.62). Furthermore, the strength of the flux tube remains constant as it moves with the fluid, because of (6.2.78).

6.3 Some Simple Dynamic Problems

We briefly discuss two problems involving the motion of perfectly conducting fluids. In these applications of the theory we begin to see

how the magnetic field permeating the fluid gives it a kind of internal
pressure and stiffness that nonconducting or nonmagnetic systems do
not possess. These properties arise of course from the forces created
when the magnetic field interacts with the electric current, the Lorentz
force. In order to get a feeling for the Lorentz force, $\mathbf{L} = \mathbf{J} \times \mathbf{B}$, which
acts on a conducting fluid, it is helpful to think of it as the divergence
of a stress tensor, the Maxwell stress tensor. The details are as follows.

6.3.1 The Maxwell Stress Tensor

The tensor $\mathbf{M}$, whose Cartesian components are

$$M_{ij} = \frac{1}{\mu_0}[B_i B_j - \tfrac{1}{2} B^2 \delta_{ij}], \qquad (6.3.1)$$

is called the *Maxwell stress tensor*. Actually, $\mathbf{M}$ contains an electric
part too, but (6.2.14) justifies neglecting it. The electric force $\rho\mathbf{E}$ was
omitted from (6.2.1) for the same reason. The Lorentz force density
$\mathbf{L}$, the force per unit volume on a small packet of fluid, is given by

$$L_j = \partial_i M_{ij}. \qquad (6.3.2)$$

To see this, we note from the pre-Maxwell equations that

$$\mu_0 \mathbf{L} = \mu_0 \mathbf{J} \times \mathbf{B} = (\nabla \times \mathbf{B}) \times \mathbf{B}.$$

In Cartesian coordinates

$$\mu_0 \mathbf{J} \times \mathbf{B}_i = \epsilon_{ijk}(\nabla \times \mathbf{B})_j B_k = \epsilon_{ijk}\epsilon_{j\ell m}(\partial_\ell B_m) B_k$$
$$= (\delta_{im}\delta_{k\ell} - \delta_{i\ell}\delta_{km})(\partial_\ell B_m) B_k$$
$$= (\partial_k B_i)B_k - (\partial_i B_k)B_k.$$

But

$$(\partial_i B_k)B_k = \tfrac{1}{2}\partial_i(B_k B_k) = \tfrac{1}{2}\partial_i B^2 \text{ and } \partial_k B_k = 0,$$

and so

$$\partial_k(B_i B_k) = (\partial_k B_i)B_k.$$

And therefore, finally,

$$\mu_0(\mathbf{J} \times \mathbf{B})_i = \partial_k[B_i B_k - \tfrac{1}{2} B^2 \delta_{ik}],$$

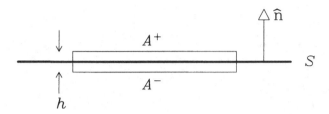

Figure 6.3.1: Section through a volume in the shape of a flat pillbox that straddles the surface S.

which is the desired result.

If V is any region, with boundary ∂V and outward unit normal $\hat{\mathbf{n}}$, the total Lorentz force $\mathcal{L} = \int_V d^3\mathbf{r}\, \mathbf{L}$ on V can be written

$$\mathcal{L}_j = \int_V d^3\mathbf{r}\, L_j = \int_V d^3\mathbf{r}\, \partial_i M_{ij} = \int_{\partial V} d^2\mathbf{r}\, n_i M_{ij}. \qquad (6.3.3)$$

The pressure force exerted on V, by the second form of Gauss' Theorem (7.3.6), is

$$\mathcal{P} = -\int_{\partial V} d^2\mathbf{r}\, \hat{\mathbf{n}} p = -\int_V d^3\mathbf{r}\, (\boldsymbol{\nabla} p),$$

so the total $\mathcal{P} + \mathcal{L}$ is

$$\mathcal{P}_j + \mathcal{L}_j = \int_{\partial V} d^2\mathbf{r}\, [-n_j p + n_i M_{ij}]. \qquad (6.3.4)$$

Now consider a surface S across which $\mathbf{B}$ is discontinuous. In the real world, $\mathbf{B}$ would change very rapidly but continuously in the direction $\hat{\mathbf{n}}$ normal to S because of concentrations of current flowing parallel to the surface; we idealize this situation by allowing discontinuous $\mathbf{B}$. Consider a pillbox (Figure 6.3.1) whose top, A^+, and bottom, A^-, are very close to S, and whose thickness h is very small. As $h \to 0$, the total pressure plus Lorentz force on the pillbox will approach

$$\Delta(\mathcal{P}_j + \mathcal{L}_j) = \int_A d^2\mathbf{r}\, [-n_j p + n_i M_{ij}]_-^+.$$

The integrals over the pillbox of all other terms in (6.2.1) will tend to zero with h, so in the limit we must have

$$\int_A d^2\mathbf{r}\, [-n_j p + n_i M_{ij}]_-^+ = 0$$

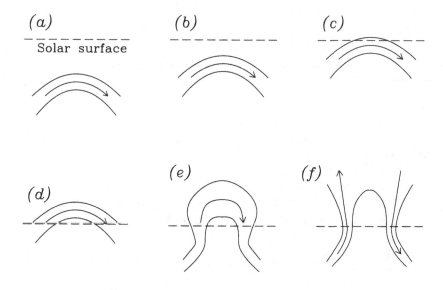

Figure 6.3.2: The birth of a sunspot pair. Snapshots of a flux tube floating up through the solar photosphere; it is lighter and cooler than the surrounding plasma because of the higher magnetic field inside.

for every patch A on S. This means

$$-n_j[p]_-^+ + n_i\,[M_j]_-^+ = 0. \tag{6.3.5}$$

If S is the surface of a flux tube, (6.3.1) shows that

$$n_i M_{ij} = -\frac{1}{2\mu_0} B^2 n_j,$$

so (6.3.5) is

$$-n_j \left[p + \frac{1}{2\mu_0} B^2 \right]_-^+ = 0. \tag{6.3.6}$$

The quantity $B^2/2\mu_0$ can thus be thought of as a magnetic pressure. At a jump discontinuity in p and $\mathbf{B}$, the total pressure,

$$p + \frac{1}{2\mu_o} B^2 = p_{\text{total}} \tag{6.3.7}$$

must be continuous.

6.3.2 Sunspots

The following is a simple physical model for the generation of sunspots. We consider a flux tube in which $\mathbf{B}$ is larger than in the surrounding medium; the tube and the medium are perfectly conducting, a good approximation in the solar photosphere, which is a plasma. Since p_{total} is continuous across the wall of the tube, the p in (6.3.7) will be less in the tube than outside. In the sun the pressure includes radiation pressure as well as gas pressure, but radiation pressure is proportional to the fourth power of absolute temperature and so is negligible in the outer part of the sun. Assuming the temperature is roughly constant across the tube, the gas density in the tube will be less than outside. The tube will be buoyant, and will float to the surface. If the ends of the tube are pinned or delayed, subsequent snapshots will be as shown in Figure 6.3.1. As the tube tries to rise above the surface, fluid can drain out of it along the lines of force of $\mathbf{B}$. This does not violate (6.2.62), the condition that in a perfect conductor, the magnetic lines are permanently attached to the fluid particles. Eventually it becomes a sunspot pair. The sunspots are visible because they are dark, and they are dark because they are cool. The strong $\mathbf{B}$ in them (~ 1 T) inhibits convection and heat transfer from the hot region below, thus making them cool. For more details see Parker (1979).

6.3.3 Alfvén Waves

Consider an incompressible, perfectly conducting, nonviscous fluid filling all of space, initially at rest, with no body forces acting on it. It contains a uniform magnetic field $B\hat{\mathbf{x}}_3$, $B > 0$. Suppose that somehow it is given an impulse, so as to acquire everywhere a small initial velocity $\mathbf{u}(\mathbf{r}, 0)$. What will it do?

We linearize the problem. The magnetic field will be

$$\mathbf{B} = B(\hat{\mathbf{x}}_3 + \mathbf{b}) \tag{6.3.8}$$

where $\mathbf{b}$ is dimensionless and we assume $|\mathbf{b}| \ll 1$. We will neglect $|\mathbf{u}|^2$, $|\mathbf{u}||\mathbf{b}|$, and $|\mathbf{b}|^2$. Then inserting (6.3.8) into the dynamo equation (6.2.26) and dropping the neglected terms, we find

$$\partial_t B\mathbf{b} = \boldsymbol{\nabla} \times [\mathbf{u} \times (B\hat{\mathbf{x}}_3)]$$

or

$$\partial_t \mathbf{b} = \boldsymbol{\nabla} \times (\mathbf{u} \times \hat{\mathbf{x}}_3)$$

$$= (\boldsymbol{\nabla} \cdot \hat{\mathbf{x}}_3 + \hat{\mathbf{x}}_3 \cdot \boldsymbol{\nabla})\mathbf{u} - (\boldsymbol{\nabla} \cdot \mathbf{u} + \mathbf{u} \cdot \boldsymbol{\nabla})\hat{\mathbf{x}}_3$$

where we have applied a vector operator identity derived from (7.2.68). From the fact that $\hat{\mathbf{x}}_3$ is constant and the fluid is assumed to be incompressible, so that $\boldsymbol{\nabla} \cdot \mathbf{u} = 0$, this reduces to

$$\partial_t \mathbf{b} = \partial_3 \mathbf{u}. \tag{6.3.9}$$

The exact momentum equation without viscous terms is

$$\breve{\rho}[\partial_t \mathbf{u} + (\mathbf{u} \cdot \boldsymbol{\nabla})\mathbf{u}] = -\boldsymbol{\nabla} p + \frac{1}{\mu_0}(\boldsymbol{\nabla} \times \mathbf{B}) \times \mathbf{B}.$$

We assume $\breve{\rho}$, the mass density of the fluid, is constant. Linearizing the momentum equation gives

$$\breve{\rho}\partial_t \mathbf{u} = -\boldsymbol{\nabla} p + \frac{B^2}{\mu_0}(\boldsymbol{\nabla} \times \mathbf{b}) \times \hat{\mathbf{x}}_3.$$

Now $(\boldsymbol{\nabla} \times \mathbf{b}) \times \hat{\mathbf{x}}_3 = -\boldsymbol{\nabla} b_3 + \partial_3 \mathbf{b}$, so the above becomes

$$\breve{\rho}\partial_t \mathbf{u} = -\boldsymbol{\nabla}\left(p + \frac{B^2}{\mu_0}b_3\right) + \frac{B^2}{\mu_0}\partial_3 \mathbf{b}. \tag{6.3.10}$$

Take the divergence of this equation; the fluid is incompressible and $\boldsymbol{\nabla} \cdot \mathbf{b} = 0$, so (6.3.10) implies

$$\nabla^2\left(p + \frac{B^2}{\mu_0}b_3\right) = 0. \tag{6.3.11}$$

We assume that the disturbance does not extend to infinity. Then

$$p + \frac{B^2}{\mu_0}b_3 \to p_\infty \quad \text{a constant}$$

as $|\mathbf{r}| \to \infty$. If we define $U = p + B^2 b_3/\mu_0$ then $\boldsymbol{\nabla} U$ is a vector field whose divergence and curl both vanish everywhere, and $\boldsymbol{\nabla} U \to 0$

at infinity; hence from the discussion in section 7.3.3, ∇U vanishes everywhere. This implies

$$p + \frac{B^2}{\mu_0} b_3 = p_\infty \quad \text{everywhere.} \qquad (6.3.12)$$

Equation (6.3.12) has a simple physical interpretation. The total pressure, hydrostatic plus magnetic, is

$$p_{\text{total}} = p + \frac{1}{2\mu_0} |\mathbf{B}|^2 = p + \frac{1}{2\mu_0} B^2 + \frac{1}{\mu_0} B^2 b_3 \frac{B^2}{2\mu_0} |\mathbf{b}|^2.$$

Thus neglecting the term in $|\mathbf{b}|^2$, we see that (6.3.12) implies that p_{total} is constant everywhere, despite the disturbance caused by $\mathbf{b}$ and $\mathbf{u}$.

Because of (6.3.12) we can write (6.3.10) as

$$\partial_t \mathbf{u} = \frac{B^2}{\mu_0 \check{\rho}} \partial_3 \mathbf{b}. \qquad (6.3.13)$$

Differentiating (6.3.13) in time and inserting (6.3.9) gives

$$\partial_t^2 \mathbf{u} = \alpha^2 \partial_3 \mathbf{u} \qquad (6.3.14)$$

where α has the units of velocity and is called the Alfvén speed:

$$\alpha = B / \sqrt{\mu_0 \check{\rho}}. \qquad (6.3.15)$$

Differentiating (6.3.9) and proceeding in a similar manner, we see that $\mathbf{b}$ obeys the same equation as $\mathbf{u}$:

$$\partial_t^2 \mathbf{b} = \alpha^2 \partial_3 \mathbf{b}. \qquad (6.3.16)$$

Equations (6.3.13, 6.3.16) describe the nondispersive propagation of a simple wave. Any initial disturbance that makes $\mathbf{u} \neq \mathbf{0}$ will propagate along the lines of force, in the $\pm \hat{\mathbf{x}}_3$ directions with the Alfvén speed. There will be no interaction between values of the disturbance on different lines of force, which act like plucked strings under tension.

In this analogy the field line is under tension of B^2/μ_0, and it has a mass $\breve{\rho}$ per unit length.

In the earth's core, incompressibility is probably not a bad approximation for small disturbances other than seismic waves and large-scale vertical motions. However, the Coriolis force is not necessarily negligible. The linearized momentum equation is (6.3.10) with, on the left, the added term $2\breve{\rho}\mathbf{\Omega} \times \mathbf{u}$, where $\mathbf{\Omega}$ is the angular velocity vector of the earth's rotation. The effects of rotation on Alfvén wave propagation are negligible if

$$\frac{2\breve{\rho}\Omega u}{B^2 b/\mu_0 L} \ll 1 \quad \text{or} \quad \frac{2\Omega u L}{\alpha^2 b} \ll 1.$$

If rotation does not much affect the Alfvén wave, then

$$\partial_t \mathbf{u} = \alpha^2 \partial_3 \mathbf{b}.$$

But in a wave with speed α, the length and time scales are related by $L = \alpha T$, so

$$u \approx \alpha b.$$

Then the condition that the Coriolis force have negligible effect on Alfvén waves is

$$\frac{2\Omega L}{\alpha} \ll 1 \tag{6.3.17}$$

or

$$L \ll \frac{\alpha}{2\Omega} = \frac{B}{2\Omega\sqrt{\mu_0\breve{\rho}}}. \tag{6.3.18}$$

In the core, if $B = 5 \times 10^{-4}$ T, the dipole strength at the core surface, then (6.3.17) gives $L \ll 30$ m. Even if B is 100 times larger, still $L \ll 3$ km if Ω is to be negligible. In short, there are probably no pure Alfvén waves in the core at wavelengths whose effects are observable at the surface. The waves in the core are dominated by rotation. More detailed analysis shows that there is a narrow band of propagator directions normal to the angular velocity vector, $\mathbf{\Omega}$, in which Alfvén waves are possible with little perturbation by rotational effects. For more on the effect of rotation and buoyancy on waves in the core see Chapter 10 of Moffatt (1978).

6.4 Application of Perfect Conductor Theory to the Core

The main geophysical application of the theory we have been developing is in the attempt to estimate the fluid motion at the top of the core. We show that it is very plausible to assume the core may be regarded as perfectly conducting fluid on the time scales for which detailed magnetic fields have been observed at the earth's surface. Then, as might be expected, the behavior of the magnetic field at the surface of the core is essentially governed by the velocity at the surface, although one must be careful about what one means by this velocity. The magnetic field can be estimated at the core, so that it is natural to ask whether knowledge of the field at the surface of the core (and its time derivative) is enough to determine the fluid velocity. The answer is surprisingly complicated: we will see that we can estimate the component of the velocity perpendicular to certain curves, the lines where the magnetic field normal to the core vanishes. There are certain other points on the core where velocity information may be obtained, but the general velocity field cannot be derived from the magnetic data.

6.4.1 The Hypothesis of Roberts and Scott

As we saw in Chapter 4, that part of the geomagnetic field $\mathbf{B}$ described by spherical harmonics of low degree ℓ (at present probably $1 \leq \ell \leq 10$) can be extrapolated down from ∂V, the surface of the earth, to ∂K, the core–mantle boundary, if the conductivity of the mantle is negligible. The same is true of $\partial_t \mathbf{B}$, the secular variation. We have seen (section 5.4) that the electrical conductivity of the mantle can attenuate the time-varying magnetic fields produced by the core. Backus (1983) showed that the major effect of mantle conductivity on spherical harmonics of degree ℓ in $\mathbf{B}$ and $\partial_t \mathbf{B}$ is that the output on ∂V is observed at time $\tau_1(\ell)$ later than the input on ∂K. A secondary effect is to smooth out the input periods shorter than a "smoothing time" $\sigma_2(\ell)$. Data from the 1969 magnetic jerk (see Chapter 1) suggest that $\tau_1(\ell) = \tau_1$, independent of ℓ, for low ℓ, but that the $\sigma_2(\ell)$ are significantly different. That jerk also suggests $\tau_1 \leq 13$ years, and possibly $\tau_1 \leq 7$ years. It is a firm result (see Backus, 1983) that $\sigma_2(\ell) \leq \tau_1(\ell)\sqrt{2/3}$, so $\sigma_2(\ell) \leq 11$ years and quite possibly $\sigma_2(\ell) \lesssim 4$ years.

Therefore, if we do ignore mantle conductivity in extrapolating $\mathbf{B}$ and $\partial_t \mathbf{B}$ down from ∂V to ∂K, we will be seeing ∂K as it was τ_1

years ago, and with periods greater than $\sigma_2(\ell)$ attenuated. The delay does no harm (we see the nearest nonsolar star with a delay of four years), and neither does the attenuation if we look only at periods $T \geq 5$ or 10 years.

There is a possibility (Backus, 1982) that if the mantle conductivity satisfies

$$-r\partial_r \ln \sigma \gg 1 \qquad (6.4.1)$$

then the electric field $\mathbf{E}$ can be extrapolated down from ∂V to ∂K without any information about σ other than (6.4.1). However, the data for $\mathbf{E}$ on ∂V are not available now, and the total $\mathbf{E}$ that would be measured on ∂V includes a term from long-period, large-scale oceanic fluid flow that is as large as the $\mathbf{E}$ produced by the core. This oceanic dynamo field must be accurately calculated and removed before the core-generated $\mathbf{E}$ can be obtained on ∂V.

If we can observe $\mathbf{B}$ and $\partial_t\mathbf{B}$, and perhaps $\mathbf{E}$, in the mantle just above ∂K, what can we learn about the core? In 1965 Roberts and Scott proposed a crucial simplification for understanding the secular variation $\partial_t\mathbf{B}$ at periods so far observable ($T \lesssim 100$ years). They pointed out that if we study $\mathbf{B}$ in the core at length scales L and time scales T, in the dynamo equation in the core we have $|\nabla \times \eta\nabla \times \mathbf{B}| \ll |\partial_t\mathbf{B}|$ if

$$T \ll \mu_0\sigma L^2 = L^2/\eta \qquad (6.4.2)$$

where σ is conductivity of the core. Under these circumstances we can treat the core as a perfect conductor, satisfying

$$\mathbf{E} = \mathbf{u} \times \mathbf{B} \qquad (6.4.3)$$

$$\partial_t\mathbf{B} = \nabla \times (\mathbf{u} \times \mathbf{B}) = -\nabla \times \mathbf{E}. \qquad (6.4.4)$$

If $\sigma = 3 \times 10^5$ S/m so that $\eta \approx 2.5$ m^2/s, and if $L = 3.48 \times 10^6$ m, the core radius, then (6.4.2) gives $T \ll 150,000$ years. We must treat this very large estimate with considerable caution, however. Recall from subsection 5.4.4 that the slowest decaying mode for the core has a mean life of only 15,000 years. The analysis of a spatially periodic field given in subsection 6.2.3 suggests that a better estimate for the characteristic diffusion time would be a factor of π^2 smaller than the one in (6.4.2). Let us accept this correction factor. Recalling that the wavelength of a spherical harmonic of degree ℓ on the core's surface is $2\pi a/(\ell + \frac{1}{2})$, where a is the core radius, and taking the scale L to be half a wavelength, we find for diffusion times longer than 100 years, $\ell < 38$. An analysis based on (5.4.45) gives essentially the same answer. We saw in Chapter 4 that our knowledge of the core field does not extend beyond $\ell = 13$, so the perfect conductor flux approximation appears safe for all observable fields. The vertical scales of $\mathbf{B}$ in the core are quite unknown. If they also satisfy $L \geq c/10$, then for $T < 100$ years it is acceptable to treat the core as a perfect conductor. The length scale L used in these rough calculations must be the *smallest* scale in which we are interested, because we need an *upper bound* on $|\nabla \times \eta \nabla \times B| \approx \pi^2 \eta B/L^2$ in order to justify neglecting it.

If the vertical scale of $\mathbf{B}$ in the core does satisfy $L \geq c/10$, then for $1 \leq \ell \leq 10$ we can calculate $\partial_t \mathbf{B}$ from $\mathbf{B}$ as if the core were a perfect conductor, as long as we restrict attention to periods $T < 100$ years. If we also want to treat the mantle as an insulator, we also need $T > 10$ years. In this range, $10 < T < 100$ years, the Roberts and Scott hypothesis, hereafter called RSH, is valid. The hypothesis is of course the same as saying the frozen-in-field approximation is valid, and is often referred to as the frozen-flux hypothesis. It makes predictions about the relationships of $\mathbf{B}$ to $\partial_t \mathbf{B}$, and some of these can be used as a test of the hypothesis, while others give information about $\mathbf{u}$ in the core just below the surface of the core, ∂K.

If we adopt RSH, then the magnetic diffusivity $\eta = 0$ in the core, so surface currents are possible on the core's surface. Therefore $\mathbf{B}_S$ in the mantle just above ∂K tells us nothing about $\mathbf{B}_S$ in the core just below ∂K, nor does $\partial_t \mathbf{B}_S^+$ give $\partial_t \mathbf{B}_S$. However, the radial components, B_r and $\partial_t B_r$, are continuous across ∂K. If we can observe them at the bottom of the mantle, we can see them at the top of the core. The large jump in electrical conductivity across ∂K will permit ∂K to accumulate surface charge, so $[E_r]_-^+ \neq 0$ on ∂K. However $[\mathbf{E}_S]_-^+ = 0$, so if we can see $\mathbf{E}_S$ at the bottom of the mantle, we can see it at the top of the core.

As we noted in section 6.2, estimates of the viscosity differ widely. The kinematic viscosity, ν (viscosity/density), could be as low as 10^{-7} m^2/s. It seems almost certain that $\nu \ll \eta$, so if we take $\eta = 0$, we may also take $\nu = 0$. The significance of this fact arises from the observation that ν is the diffusivity of the vorticity $\boldsymbol{\omega} = \nabla \times \mathbf{u}$. In an incompressible viscous fluid of constant density, the vorticity satisfies the dynamo equation $\partial_t \boldsymbol{\omega} = \boldsymbol{\nabla} \times (\mathbf{u} \times \boldsymbol{\omega} - \nu \boldsymbol{\nabla} \times \boldsymbol{\omega})$; our solution of that equation was first given by Helmholtz in the nineteenth century as the solution of his vorticity equation. With $\nu = 0$ the fluid in the core need not have $\mathbf{u} = \mathbf{0}$ at the solid boundary ∂K. The actual fluid does have $\mathbf{u} = \mathbf{0}$ there, and since $\eta > 0$, it carries no surface current. However, η and ν are so small that there are very thin layers of concentrated vorticity $\boldsymbol{\nabla} \times \mathbf{u}$ and concentrated current $\boldsymbol{\nabla} \times \mathbf{B}/\mu_0$, which are called *boundary layers*. Across the magnetic boundary layer, $\mathbf{B}_S$ changes very rapidly, and across the viscous boundary layer, $\mathbf{u}_S$ drops rapidly from its "free stream" value just below the boundary layer to $\mathbf{0}$ on ∂K. Throughout the viscous boundary layer $u_n = 0$ to lowest order in ν. For details see Backus (1968). We treat both boundary layers as of infinitesimal thickness, so $[\mathbf{B}_S]_-^+$ and $\mathbf{u}_S \neq 0$ on ∂K.

Much of what we say depends only on ∂K's being a spheroid, not a sphere, so we will carry out the discussion for a spheroidal ∂K, with outward unit normal $\hat{\mathbf{n}}$, and $B_n = \hat{\mathbf{n}} \cdot \mathbf{B}$, $\mathbf{B}_S = \mathbf{B} - \hat{\mathbf{n}} B_n$. If we can measure B_n^- and $\mathbf{E}_S^-$ on ∂K (i.e., B_n and $\mathbf{E}_S$ in the core just below ∂K) then we can find $\mathbf{u}$ just below ∂K very easily. First, of course, $u_n = \hat{\mathbf{n}} \cdot \mathbf{u} = 0$, so

$$\mathbf{u} = \mathbf{u}_S \text{ in } K \text{ just below } \partial K. \qquad (6.4.5)$$

Second, $\mathbf{E} = -\mathbf{u} \times \mathbf{B}$ and thus

$$\hat{\mathbf{n}} \times \mathbf{E} = +\hat{\mathbf{n}} \times (\mathbf{B} \times \mathbf{u}) = \mathbf{B}(\hat{\mathbf{n}} \cdot \mathbf{u}) - \mathbf{u}(\hat{\mathbf{n}} \cdot \mathbf{B}) = -B_n \mathbf{u}_S.$$

And third, $\hat{\mathbf{n}} \times \mathbf{E} = \hat{\mathbf{n}} \times \mathbf{E}_S$, so that

$$\mathbf{u}_S = -(\hat{\mathbf{n}} \times \mathbf{E}_S)/B_n \text{ in } K \text{ just below } \partial K. \qquad (6.4.6)$$

If we do not know $\mathbf{E}_S$ on ∂K, then all we have there is $\partial_t B_n$ and B_n. It is a remarkable fact that the dynamo equation (6.4.4) connects these two measured quantities and $\mathbf{u}_S$ on ∂K, without requiring us (or enabling us) to look deeper in the core. This is the great simplification produced by RSH. To obtain this result, dot $\hat{\mathbf{n}}$ into (6.4.4). Then on ∂K

$$\partial_t B_n = \hat{\mathbf{n}} \cdot \boldsymbol{\nabla} \times (\mathbf{u} \times \mathbf{B}). \qquad (6.4.7)$$

We would like to use (6.4.7) in K below ∂K. To do so we must define $\hat{\mathbf{n}}$ in such a layer, i.e., off the surface ∂K. A particularly simple way

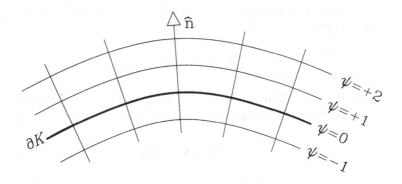

Figure 6.4.1: Construction of a vector field matching the normal to ∂K on that surface by imitating spheres.

to do so is to imitate spheres; see Figure 6.4.1. Consider the straight lines that are perpendicular to ∂K where they intersect ∂K. On each such straight line, let ψ be the distance from the point of intersection of that line with ∂K. Then $\psi(\mathbf{r})$ is defined in a layer of finite thickness near ∂K, and the straight lines normal to ∂K are normal to all the level surfaces of ψ. If we define

$$\hat{\mathbf{n}} = \nabla\psi$$

then this agrees with the ordinary $\hat{\mathbf{n}}$ on ∂K, and it also satisfies

$$\nabla \times \hat{\mathbf{n}} = \mathbf{0} \quad \text{and} \quad \partial_n \hat{\mathbf{n}} = 0$$

where $\partial_n = \hat{\mathbf{n}} \cdot \nabla$ is the derivative in direction $\hat{\mathbf{n}}$. Then (6.4.7) holds in a layer below ∂K. For any vector field $\mathbf{v}$ in that layer

$$\nabla \cdot (\hat{\mathbf{n}} \times \mathbf{v}) = \mathbf{v} \cdot (\nabla \times \hat{\mathbf{n}}) - \hat{\mathbf{n}} \cdot \nabla \times \mathbf{v} = -\hat{\mathbf{n}} \cdot (\nabla \times \mathbf{v}),$$

so in that layer (6.4.7) is

$$\partial_t B_n = -\nabla \cdot [\hat{\mathbf{n}} \times (\mathbf{u} \times \mathbf{B})]$$

or

$$\partial_t B_n = -\nabla \times [B_n \mathbf{u} - u_n \mathbf{B}]. \tag{6.4.8}$$

It is shown in the mathematical appendix (subsections 7.2.8 and 7.2.9) that, for any vector field $\mathbf{v}$,

$$(\hat{\mathbf{n}}\partial_n) \cdot \mathbf{v} = \hat{\mathbf{n}} \cdot \partial_n \mathbf{v}.$$

But we also have $\partial_n \hat{\mathbf{n}} = 0$, so $\hat{\mathbf{n}} \cdot \partial_n \mathbf{v} = \partial_n (\hat{\mathbf{n}} \cdot \mathbf{v}) = \partial_n v_n$. Therefore, if we define the surface gradient in the usual way (see subsection 7.4.3)

$$\boldsymbol{\nabla}_S = \boldsymbol{\nabla} - \hat{\mathbf{n}}\partial_n$$

then

$$\boldsymbol{\nabla} \cdot \mathbf{v} = (\hat{\mathbf{n}}\partial_n + \boldsymbol{\nabla}_S) \cdot \mathbf{v}(\hat{\mathbf{n}}\partial_n) \cdot \mathbf{v} + \boldsymbol{\nabla}_S \cdot \mathbf{v}$$

$$= \partial_n v_n + \boldsymbol{\nabla}_S \cdot \mathbf{v}.$$

But if $\mathbf{v} = B_n \mathbf{u} - u_n \mathbf{B}$, $v_n = 0$ in the whole layer below ∂K, so $\partial_n v_n = 0$ there, and $\boldsymbol{\nabla} \cdot \mathbf{v} = \boldsymbol{\nabla}_S \cdot \mathbf{v}$. Moreover, $\mathbf{v} = \mathbf{v}_S$ in this case, so (6.4.8) can be written

$$\partial_t B_n = -\boldsymbol{\nabla}_S \cdot [B_n \mathbf{u}_S - u_n \mathbf{B}_S]. \qquad (6.4.9)$$

This is true in the whole layer below ∂K. But $\boldsymbol{\nabla}_S \cdot \mathbf{v}$ on ∂K depends only on the values of $\mathbf{v}$ on ∂K, and not on values of $\mathbf{v}$ elsewhere. And on ∂K, $u_n = 0$. Therefore,

$$\partial_t B_n + \boldsymbol{\nabla}_S \cdot (B_n \mathbf{u}_S) = 0 \quad \text{on} \quad \partial K. \qquad (6.4.10)$$

the equation derived by Roberts and Scott for spheres. It has a simple physical interpretation: Consider a patch P of fluid particles moving on ∂K. The magnetic flux $\int_P d^2\mathbf{r} \, B_n$ through this patch is conserved as the patch moves, and is calculated from the flux density B_n per unit area. We can think of $\int_P d^2\mathbf{r} \, B_n$ as the "mass" of the two-dimensional patch, since this "mass" is conserved as the patch moves. Then (6.4.10) is simply the continuity equation for conservation of this "mass" in the two-dimensional motion on the surface ∂K.

Now we regard B_n at instant t as given by our measurements. If we know $\mathbf{u}_S$ on ∂K we could calculate $\partial_t B_n$ from (6.4.10). The actual situation is that we know $\partial_t B_n$ and would like to find $\mathbf{u}_S$. This is an inverse problem, the direct problem being to find $\partial_t B_n$ from $\mathbf{u}_S$. It is clearly linear; $\partial_t B_n$ depends linearly on $\mathbf{u}_S$.

Kahle, Ball, and Vestine (1967) made the following suggestion for solving the inverse problem to find $\mathbf{u}_S$ when B_n and $\partial_t B_n$ are given and ∂K is a sphere. Madden and Le Mouël (1982), among others (see Bloxham and Jackson, 1991), repeated the suggestion with more recent satellite data for $\mathbf{B}$ and $\partial_t \mathbf{B}$. Since $\mathbf{u}_S$ is a tangent vector field on ∂K, there are by the surface form of Helmholtz's representation (subsection 7.4.7) unique scalar fields g and h on ∂K such that

$$\mathbf{u}_S = \boldsymbol{\nabla}_S g + \boldsymbol{\Lambda}_S h \quad \text{and} \quad \langle g \rangle_{\partial K} = \langle h \rangle_{\partial K} = 0. \qquad (6.4.11)$$

Here $\boldsymbol{\Lambda}_S = \hat{\mathbf{n}} \times \boldsymbol{\nabla}_S$; note that $\boldsymbol{\Lambda}_S = \frac{1}{r}\boldsymbol{\Lambda}$ if $\partial K = S(r)$. Kahle, Ball, and Vestine argue that, since we know $\partial_t B_n$ and B_n only at low spherical harmonic degrees ℓ, why not approximate the spherical harmonic expansions of g and h by truncating them too? That is, write

$$g = \sum_{\ell=1}^{L} \sum_{m=-\ell}^{\ell} g_\ell^m \beta_\ell^m(\hat{\mathbf{r}})$$

$$\qquad (6.4.12)$$

$$h = \sum_{\ell=1}^{L} \sum_{m=-\ell}^{\ell} h_\ell^m \beta_\ell^m(\hat{\mathbf{r}})$$

where L is, say, 10. Then instead of trying to make $\mathbf{u}_S$ satisfy (6.4.10) exactly, one can choose the g_ℓ^m and h_ℓ^m in (6.4.12) to minimize

$$\mathcal{E}^2 = \langle [\partial_t B_n + \boldsymbol{\nabla}_S \cdot (B_n \mathbf{u}_S)]^2 \rangle_{\partial K}. \qquad (6.4.13)$$

There is a rather subtle flaw in this scheme (Backus, 1968). The representation theorem (6.4.11) can be applied not only to $\mathbf{u}_S$ but also to $B_n \mathbf{u}_S$. There are unique scalars G and H on ∂K such that

$$B_n \mathbf{u}_S = \boldsymbol{\nabla}_S G + \boldsymbol{\Lambda}_S H \quad \text{and} \quad \langle G \rangle_{\partial K} = \langle H \rangle_{\partial K} = 0. \qquad (6.4.14)$$

Since we know B_n, finding G and H from $\partial_t B_n$ is as good a way to get $\mathbf{u}_S$ as is finding g and h in (6.4.11). And finding G and H from $\partial_t B_n$ is almost trivial: if we substitute (6.4.14) in (6.4.10) we get

$$\partial_t B_n + \nabla_S^2 G = 0. \qquad (6.4.15)$$

Since $\langle G \rangle_{\partial K} = 0$, and of course $\langle \partial_t B_n \rangle = 0$, (6.4.15) determines G uniquely by (5.1.11). And if we use this G in (6.4.14), we will get a

$\mathbf{u}_S$ that satisfies (6.4.10), *no matter how we choose H*. Therefore we cannot determine H in (6.4.14) from B_n and $\partial_t B_n$ on ∂K and a part of $\mathbf{u}_S$ is completely indeterminate. It turns out, as we will see next, that while the surface velocity cannot be found from the magnetic field and its time derivative at an arbitrary point, there are certain special points on the core where something definite can be said.

6.4.2 Null-Flux Curves

We can learn a little about H. Since $\langle B_n \rangle_{\partial K} = 0$, there must be curves on ∂K where $B_n = 0$. We call such curves *null-flux curves*. On any null-flux curve, (6.4.14) gives $\boldsymbol{\nabla}_S G + \boldsymbol{\Lambda}_S H = \mathbf{0}$, so

$$\hat{\mathbf{n}} \times \boldsymbol{\nabla}_S G + \hat{\mathbf{n}} \times \boldsymbol{\Lambda}_S H = 0.$$

From the definition of $\boldsymbol{\Lambda}_S$ we have $\hat{\mathbf{n}} \times \boldsymbol{\Lambda}_S = \hat{\mathbf{n}} \times (\hat{\mathbf{n}} \times \boldsymbol{\nabla}_S) = -\boldsymbol{\nabla}_S$, so

$$\boldsymbol{\nabla}_S H = \boldsymbol{\Lambda}_S G \qquad \text{on every null-flux curve.} \qquad (6.4.16)$$

We know G and therefore we can integrate (6.4.16) around each null-flux curve. Thus H is determined up to a different additive constant on each null-flux curve. And the directional derivative of H normal to every null-flux curve is also determined by (6.4.16). But, once H and its normal derivative on each null-flux curve are calculated, H is arbitrary and unknown off the null-flux curves. At best we can expect that the Kahle, Ball, and Vestine technique will give one of a gigantic collection of $\mathbf{u}_S$, all of which are exactly in accord with $\partial_t B_n$ and B_n on ∂K.

Despite the almost complete failure of $\partial_t B_n$ and B_n to give information about H on ∂K, they do give some information about $\mathbf{u}_S$. Equation (6.4.10) can be written

$$\partial_t B_n + (\mathbf{u}_S \cdot \boldsymbol{\nabla}_S) B_n + B_n (\boldsymbol{\nabla}_S \cdot \mathbf{u}_S) = 0, \qquad (6.4.17)$$

and this is

$$D_t B_n + B_n (\boldsymbol{\nabla}_S \cdot \mathbf{u}_S) = 0. \qquad (6.4.18)$$

If we follow a particular particle, labeled $\mathbf{x}$, as it moves on ∂K, and if we define $(\boldsymbol{\nabla}_S \cdot \mathbf{u}_S)^L(\mathbf{x}, t) = f(\mathbf{x}, t)$, then (6.4.18) can be integrated immediately as

$$B_n^L(\mathbf{x}, t) = B_n^L(\mathbf{x}, 0) e^{-\int_0^t d\tau\, f(\mathbf{X}, \tau)}. \qquad (6.4.19)$$

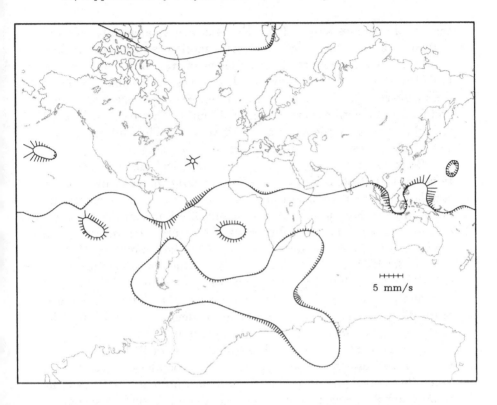

Figure 6.4.2: The surface velocity components normal to the null-flux curves for the field model IGRF1980. The map is drawn in Mercator projection, which preserves angles locally but not scale. The lines normal to the null-flux curves show the magnitude of the local velocity component. The sizes of the normal velocity vectors are drawn at a constant scale except on the imploding curve under the western Pacific, where the lines have been reduced in size by a factor of 4. Note that the velocity components parallel to the null-flux curves are unknown.

If $B_n = 0$ at particle **x** at time 0, then $B_n = 0$ at particle **x** at all later times. This can also be seen from (6.2.62). If $B_n = 0$ at $t = 0$, then the vector $\epsilon \mathbf{B}/\breve{\rho}$, where ϵ is a small quantity, at $t = 0$ connects two nearby particles on ∂K. They remain on ∂K as they move, and $\epsilon \mathbf{B}/\breve{\rho}$ continues to connect them, so B_n continues to be 0 at these particles.

It follows that if a curve $C(t)$ moves on ∂K so as always to consist of the same fluid particles, and if $C(0)$ is a null-flux curve, then $C(t)$ is a null-flux curve for all t. The null-flux curves move with the fluid on ∂K, even though they are not lines of force of **B**. If we measure B_n at different times, or B_n and $\partial_t B_n$ at one time, we can watch the null-flux curves move. If $\hat{\boldsymbol{v}}$ is the unit vector tangent to ∂K and normal to a null-flux curve $C(t)$ whose motion is observed, we can see $\hat{\boldsymbol{v}} \cdot \mathbf{u}_S$. We cannot see the component of $\mathbf{u}_S$ along $C(t)$. Booker (1969) was able to detect four null-flux curves, three small ones and the magnetic equator. Along these curves, $|\hat{\boldsymbol{v}} \cdot \mathbf{u}_S|$ ranged between 0 and 0.8 mm/s, with a mean around 0.3 mm/s. The reader will recall that this is roughly the westward drift velocity. The null-flux curves were drifting west, but were also changing their shapes.

Booker's investigation relies on the spherical harmonic description GSFC (12/66 field modelGSFC (12/66), which is by modern standards a primitive spherical harmonic model of the geomagnetic field. Modern field models based on MAGSAT measurements show eight or sometimes nine null-flux curves. In Figure 6.4.2 we illustrate a repetition of Booker's analysis based upon the field model IGRF1980, field model IGRF1980 (see Langel, 1987), which is representative of several modern solutions; it is an expansion to degree 10 for the main field and to order 8 for $\partial_t \mathbf{B}$. We see eight null-flux curves. The mean of $|\hat{\boldsymbol{v}} \cdot \mathbf{u}_S|$ is 0.41 mm/s; there is a small tendency to a westward drift of the curves but only at an average speed of 0.05 mm/s.

At this point we find an observational test of RSH. Let P be a patch of area on ∂K, and suppose ∂P, the boundary curve of P, is a null-flux curve. Let $\hat{\boldsymbol{v}}$ be the unit outward normal to ∂P, tangent to ∂K. For any tangent vector field $\mathbf{v}_S$, there is a surface version of Gauss' Theorem (see Theorem 2, subsection 7.4.6, and Figure 7.4.3):

$$\int_P d^2\mathbf{r} \, \boldsymbol{\nabla}_S \cdot \mathbf{v}_S = \int_{\partial P} d\ell \, \hat{\boldsymbol{v}} \cdot \mathbf{v}_S$$

where $d\ell$ is an element of length. If we apply this to (6.4.10) we get

$$\int_P d^2\mathbf{r} \, \partial_t B_n + \int_{\partial P} d\ell \, \hat{\boldsymbol{v}} \cdot B_n \mathbf{u}_S = 0.$$

But if ∂P is a null-flux curve, $B_n = 0$ there, and

$$\int_P d^2\mathbf{r} \, \partial_t B_n = 0 \qquad\qquad (6.4.20)$$

if ∂P is a null-flux curve. If RSH is correct, (6.4.20) follows. And the null-flux curves can be found by observing B_n, so (6.4.20) can be tested with the observed $\partial_t B_n$. Booker (1969) found that within his rather large estimated error limits of the 1965 field model, equation (6.4.20) was verified, although the errors were so large that the check was not spectacularly reassuring. A serious problem is in obtaining credible error estimates for the coefficients of a spherical harmonic expansion. Given this difficulty and the lack of agreement among different studies in the values for the coefficients of $\partial_t \mathbf{B}$, a somewhat different approach is now preferred for testing the RSH.

An equivalent check has been performed on recent satellite data and seems to work well. To be specific, we note the following consequence of (6.4.20). Let ∂K be divided into patches $P_1^+, \ldots, P_M^+, P_1^-, \ldots, P_N^-$, each having a null-flux curve for boundary. In any P_i^+, $B_n > 0$, while in any P_j^-, $B_n < 0$. Then from (6.4.20),

$$\sum_{i=1}^{M} \int_{P_i^+} d^2\mathbf{r}\, \partial_t B_n - \sum_{j=1}^{N} \int_{P_j^-} d^2\mathbf{r}\, \partial_t B_n = 0. \qquad (6.4.21)$$

If $\operatorname{sgn} x = +1$ when $x > 0$, $\operatorname{sgn} x = -1$ when $x < 0$, and $\operatorname{sgn} 0 = 0$, then we can write the foregoing equation as

$$0 = \sum_{i=1}^{M} \int_{P_i^+} d^2\mathbf{r}\, (\partial_t B_n)(\operatorname{sgn} B_n) + \sum_{j=1}^{N} \int_{P_j^-} d^2\mathbf{r}\, (\partial_t B_n)(\operatorname{sgn} B_n).$$

$$(6.4.22)$$

But $(\partial_t B_n)(\operatorname{sgn} B_n) = \partial_t |B_n|$, so since $P_1^+, \ldots, P_M^+, P_1^-, \ldots, P_N^-$ cover ∂K, we finally get

$$\int_{\partial K} d^2\mathbf{r}\, \partial_t |B_n| = 0. \qquad (6.4.23)$$

In 1978 Raymond Hide suggested that if we didn't know the core radius c, we could find it by extrapolating B_n and $\partial_t B_n$ down from the earth's surface $S(a)$ to various spheres $S(c)$ until we found the one on which (6.4.23) is true. Hide and Malin (1981) found c this way to 10 percent by minimizing the square of the left side of (6.4.21); Voorhies and Benton (1982) reduced this to 2 percent. That the method works so well is an indirect check on the RSH prediction (6.4.20). Hide (1978) suggested using this method to find the radii of the conducting cores of Jupiter and Saturn. Its advantage over the convergence test for $\langle \mathbf{B}^2 \rangle_{S(c)}$ described in Chapter 4 is that it gives c rather than a lower bound for c. The reader will find a thorough discussion of the tests of RSH in Bloxham et al. (1989).

Whaler (1980) suggested another way to use (6.4.10) to extract information about $\mathbf{u}_S$ from measurements of $\partial_t B_n$ and B_n. She pointed out that at each maximum or minimum of B_n, $\boldsymbol{\nabla}_S B_n = 0$. At these extrema, (6.4.17) gives

$$\boldsymbol{\nabla}_S \cdot \mathbf{u}_S = -\partial_t B_n / B_n, \qquad (6.4.24)$$

a measurable quantity. Thus $\boldsymbol{\nabla}_S \cdot \mathbf{u}_S$ can be measured from the data at all extremal points of B_n. Whaler found 19 extrema of B_n, with an average value of $|\boldsymbol{\nabla}_S \cdot \mathbf{u}_S|$ equal to 4.5×10^{-3} year^{-1}.

Following the suggestion of an interested bystander, Whaler extracted even more information from (6.4.10). Let P now be a patch of area on ∂K whose boundary curve ∂P is *any* level line of B_n (a line of constant B_n), not just a null-flux curve. Integrating (6.4.10) over P and using the surface version of Gauss' Theorem (7.4.19) twice, we get

$$0 = \int_P d^2\mathbf{r}\, \partial_t B_n + \int_{\partial P} d\ell\, \hat{\boldsymbol{v}} \cdot (B_n \mathbf{u}_S)$$

$$= \int_P d^2\mathbf{r}\, \partial_t B_n + B_n \int_{\partial P} d\ell\, \hat{\boldsymbol{v}} \cdot \mathbf{u}_S,$$

so

$$\int_P d^2\mathbf{r}\, \partial_t B_n + B_n \int_P d^2\mathbf{r}\, (\boldsymbol{\nabla}_S \cdot \mathbf{u}_S) = 0 \qquad (6.4.25)$$

where in (6.4.25) B_n is the constant value of B_n on the level line ∂P. From (6.4.25) it is possible to get average values of $\boldsymbol{\nabla}_S \cdot \mathbf{u}_S$ between pairs of level curves on ∂K. She found $\langle |\boldsymbol{\nabla}_S \cdot \mathbf{u}_S| \rangle_{\partial K} \approx 3 \times 10^{-3}$ year^{-1}, in tolerable agreement with the value, 4.5×10^{-3} year^{-1}, obtained from stationary points. One would not expect exact agreement: two different kinds of average of $\boldsymbol{\nabla}_S \cdot \mathbf{u}_S$ are being computed.

Whaler concludes that $\boldsymbol{\nabla}_S \cdot \mathbf{u}_S$ is so small that it could be 0, in which case $\partial_r u_r = 0$ under ∂K, so $u_r = 0$ there, and the region just under ∂K may be stably stratified. We would prefer to be a little more cautious. The details of the argument are these: recall $\breve{\rho}$ is the mass density in the core. Then mass conservation requires

$$\partial_t \breve{\rho} + \boldsymbol{\nabla} \cdot (\breve{\rho}\mathbf{u}) = 0 \qquad \text{in the core.}$$

In the core, $\breve{\rho}$ is almost entirely determined by depth, which determines temperature and pressure. Therefore the Eulerian quantity

$\check{\rho}^E(\mathbf{r}, t)$ depends on r but not $\hat{\mathbf{r}}$ or t. Thus

$$\partial_t \check{\rho} = 0 \qquad \text{in the core.}$$

From the mathematical preface, for any vector field $\mathbf{v}$ on any $S(r)$,

$$\boldsymbol{\nabla} \cdot \mathbf{v} = \partial_r v_r + \frac{2}{r} v_r + \boldsymbol{\nabla}_S \cdot \mathbf{v}_S. \qquad (6.4.26)$$

Set $\mathbf{v} = \check{\rho} \mathbf{u}$. Then $\boldsymbol{\nabla} \cdot \mathbf{v} = 0$ so $\check{\rho} \boldsymbol{\nabla} \cdot \mathbf{u} + \mathbf{u} \cdot \boldsymbol{\nabla} \check{\rho} = 0$. On ∂K, $u_r = 0$, so $\boldsymbol{\nabla} \cdot \mathbf{u} = 0$. Then from (6.4.26),

$$\partial_r u_r = -\boldsymbol{\nabla}_S \cdot \mathbf{u}_S \qquad \text{on } \partial K. \qquad (6.4.27)$$

We would prefer to modify Whaler's conclusion as follows. Since $\boldsymbol{\nabla} \cdot (\check{\rho} \mathbf{u}) = 0$ in the core and $\langle \check{\rho} u_r \rangle_{\partial K} = 0$ because $u_r = 0$ there, we can assert that $\check{\rho} \mathbf{u}$ is solenoidal (see subsection 5.2.1). Therefore there are scalar fields p, q for a separation into poloidal and toroidal parts:

$$\begin{aligned}
\check{\rho} \mathbf{u} &= \boldsymbol{\nabla} \times \boldsymbol{\Lambda} p + \boldsymbol{\Lambda} q \\
&= \hat{\mathbf{r}} \frac{1}{r} \nabla_1^2 p - \boldsymbol{\nabla}_1 \frac{1}{r} \partial_r r p + \boldsymbol{\Lambda} q \qquad (6.4.28) \\
&= \hat{\mathbf{n}} r \nabla_S^2 p - \boldsymbol{\nabla}_S \partial_r r p + \boldsymbol{\Lambda}_S q r.
\end{aligned}$$

If there are vertical motions just below ∂K, they are described by p, the poloidal scalar, since the toroidal velocities are everywhere horizontal; on ∂K the poloidal velocity is given by

$$\boldsymbol{\nabla}_S \left[-\frac{1}{\check{\rho}} \partial_r r p \right],$$

while the whole $\mathbf{u}_S$ is

$$\mathbf{u}_S = \boldsymbol{\nabla}_S \left[-\frac{1}{\check{\rho}} \partial_r r p \right] + \boldsymbol{\Lambda}_S \left(\frac{q r}{\check{\rho}} \right) = \boldsymbol{\nabla}_S g + \boldsymbol{\Lambda}_S h.$$

Thus g describes that part of $\mathbf{u}_S$ that comes from vertical motion, and h describes the part that involves no radial component. Whaler has estimated $\boldsymbol{\nabla}_S \cdot \mathbf{u}_S$, and clearly $\boldsymbol{\nabla}_S \cdot \mathbf{u}_S = \nabla_S^2 g$. If the dominant spherical harmonic degree in g is ℓ, then

$$|\nabla_S^2 g| \approx \frac{\ell(\ell+1)}{c^2}|g|,$$

while

$$|\boldsymbol{\nabla}_S g| \approx \frac{(\ell+\frac{1}{2})}{c}|g|.$$

Thus

$$|\boldsymbol{\nabla}_S g| \approx \frac{c|\nabla_S^2 g|}{(\ell+\frac{1}{2})} \approx \frac{c\boldsymbol{\nabla}_S \cdot \mathbf{u}_S}{(\ell+\frac{1}{2})}.$$

Then the fraction of $\mathbf{u}_g$ that is produced by radial motion is

$$\frac{|\boldsymbol{\nabla}_S g|}{|\mathbf{u}_S|} \approx \frac{(\boldsymbol{\nabla}_S \cdot \mathbf{u}_S)c}{|\mathbf{u}_S|(\ell+\frac{1}{2})}.$$

Taking $\boldsymbol{\nabla}_S \cdot \mathbf{u}_S = 3 \times 10^{-3}$ year^{-1} and $|\mathbf{u}_S| = 0.3$ mm/s $= 10$ km/year (the westward drift value and the one normal velocity to the null-flux curves), we find

$$\frac{|\boldsymbol{\nabla}_S g|}{|\mathbf{u}_S|} \approx \frac{1.1}{\ell+\frac{1}{2}}. \qquad (6.4.29)$$

Only for convection cells with $\ell = 1$ or 2 can most of $\mathbf{u}_S$ on ∂K be due to vertical motion. For $\ell \geq 3$, most of $\mathbf{u}_S$ is purely toroidal. Booker's and the modern "map" in Figure 6.4.2 of $\hat{\boldsymbol{v}} \cdot \mathbf{u}_S$ on null-flux curves suggests that ℓ is not small for $\mathbf{u}_S$, and in this sense Whaler is correct. Probably only a small part of $\mathbf{u}_S$ is not toroidal on ∂K.

If one is willing to make assumptions about the nature of the fluid flow, one can learn correspondingly more. For example, if $\mathbf{u}$ is constrained to be toroidal near the top of the outer core, the surface flow can be obtained from the magnetic field data, except at those points where $\boldsymbol{\nabla}_S \cdot B_n = 0$. For a review of the various assumptions and the results, the reader is referred to the article by Bloxham and Jackson (1991).

6.5 Kinematic Dynamos

As we have seen, when the magnetic diffusivity $\eta = 0$ in the dynamo equation,

$$\partial_t \mathbf{B} = \boldsymbol{\nabla} \times (\mathbf{u} \times \mathbf{B}) - \boldsymbol{\nabla} \times (\eta \boldsymbol{\nabla} \times \mathbf{B}) \qquad (6.5.1)$$

$|\mathbf{B}|$ usually grows. And when $\mathbf{u} = \mathbf{0}$, $\mathbf{B}$ dies away exponentially. The obvious question is whether when $\eta > 0$ and $\mathbf{u} \neq \mathbf{0}$ these two effects can balance, to produce a more or less steady field. In physical terms, when the conducting fluid flows across the magnetic lines of force, the resulting electromotive force $\mathbf{u} \times \mathbf{B}$ will drive a current against ohmic losses. Is it possible for this current to be exactly what is needed to sustain the original magnetic field?

To some people, the existence of such a possibility may seem trivial. To others it may look very implausible, smacking of violation of some sort of conservation law. At any rate, there is no violation of the law of conservation of energy. Whoever tries to make a field by such a "self-sustaining dynamo" mechanism will have to supply energy as he or she pushes the fluid across the lines of force.

Larmor (1919) originally suggested such a mechanism for producing sunspot magnetic fields. The suggestion received a serious setback in 1934 when Cowling proved that axisymmetric magnetic fields cannot be maintained by a self-sustaining dynamo. There is no velocity field $\mathbf{u}$ whatever, however maintained, which can produce an axisymmetric magnetic field, if the conductor has the same axis of symmetry. Cowling's is one of a number of theorems that prohibit dynamo activity when certain symmetry or other special conditions apply. We discuss another of these due to Bullard and Gellman(1954). Others can be found in Moffatt's (1978) book. The essence of the antidynamo theorems is that they show that many seemingly natural simplifications are in fact fatal to dynamo action; motions supporting the dynamo are evidently quite complex, and this is what makes the kinematic theory difficult.

The disturbing possibility arose that there might be a quite general antidynamo theorem, and indeed many mistakenly believed Cowling's result was that result, but in 1958, mathematically rigorous demonstrations were finally provided of self-sustaining dynamos. Thereafter the issue became that of constructing dynamo systems with physically reasonable properties and behavior, since the first rigorous models could not be considered plausible candidates for the earth's core or for magnetic field generation in interstellar space, which was the focus of much important work.

Because the subject of kinematic dynamos is very large, we can give only a brief treatment, touching on some of the essential topics. Readers will find more thorough treatments in Moffatt (1978), Gubbins and Roberts (1987), and Roberts and Gubbins

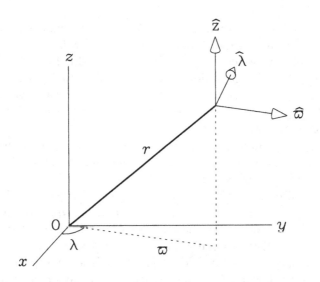

Figure 6.5.1: Cylindrical coordinate system used in the proof of Cowling's Theorem.

(1987). We first prove Cowling's Theorem in a modern version. Then we give a brief history of the early ideas and on the way prove another antidynamo theorem, which shows that velocity fields lacking a vertical component must be excluded. We give only a short description of the mathematically rigorous dynamos, because a proper development would occupy too much space. Some results of numerical studies are mentioned, and the section concludes with a somewhat more detailed description of mean field dynamos.

6.5.1 Cowling's Theorem

Now we prove the historically most important antidynamo theorem, the one stated by Cowling (1934). Our proof is completely different from the original, which contained an error. Furthermore, Cowling assumed axisymmetric motions, but this turns out to be unnecessarily restrictive. We study a bounded, axisymmetric conducting body set in an infinite insulator. Inside the conductor, fluid motions of any kind are allowed, but the magnetic field, which tends to zero at large distances, remains axisymmetric everywhere. We will show that the axisymmetric magnetic field cannot grow or even remain constant in time – it must decay away and eventually vanish.

Referring to Figure 6.5.1, let $\hat{x}, \hat{y}, \hat{z}$ be an orthonormal basis for R^3. Let ϖ, λ, z be cylindrical coordinates relative to this basis, so

$$\mathbf{r}(\varpi, \lambda, z) = \hat{x}\varpi\cos\lambda + \hat{y}\varpi\sin\lambda + \hat{z}z.$$

Let $\hat{\boldsymbol{\varpi}}, \hat{\boldsymbol{\lambda}}, \hat{\mathbf{z}}$ be unit vectors in the directions of increasing ϖ, λ, z respectively. A vector field $\mathbf{v}(\mathbf{r})$ is *axisymmetric about the $\hat{z}$ axis* if

$$\mathbf{v} = \hat{\boldsymbol{\varpi}}v_\varpi(\varpi, z) + \hat{\boldsymbol{\lambda}}v_\lambda(\varpi, z) + \hat{\mathbf{z}}v_z(\varpi, z). \qquad (6.5.2)$$

That is, $\partial_\lambda v_\varpi = \partial_\lambda v_\lambda = \partial_\lambda v_z = 0$. Note: this is not the same as $\partial_\lambda \mathbf{v} = \mathbf{0}$.

The usual formulas for $\nabla \cdot \mathbf{v}$ and $\nabla \times \mathbf{v}$ in cylindrical coordinates become slightly simpler when $\mathbf{v}$ is axisymmetric. Then they are

$$\nabla \cdot \mathbf{v} = \frac{1}{\varpi}\partial_\varpi(\varpi v_\varpi) + \partial_z v_z \qquad (6.5.3)$$

$$\nabla \times \mathbf{v} = -\hat{\boldsymbol{\varpi}}\partial_z v_\lambda + \hat{\mathbf{z}}\frac{1}{\varpi}\partial_\varpi(\varpi v_\lambda) + \hat{\boldsymbol{\lambda}}(\partial_z v_\varpi - \partial_\varpi v_z). \qquad (6.5.4)$$

Since $\mathbf{B}$ is axisymmetric and solenoidal, from (6.5.3)

$$\partial_\varpi(\varpi B_\varpi) + \partial_z(\varpi B_z) = 0.$$

Thus there is a scalar p such that

$$\varpi B_\varpi = -\partial_z p \quad \text{and} \quad \varpi B_z = \partial_\varpi p. \qquad (6.5.5)$$

This scalar is determined up to an additive constant, which we choose so that $p(0,0) = 0$. Then $\partial_z p = -\varpi B_\varpi$ implies $p(0, z) = 0$, and $\partial_\varpi p = \varpi B_z$ then implies that as $\varpi \to 0$, p behaves like $B_z\varpi^2/2$. Thus there is a unique p satisfying (6.5.5) and also

$$\varpi^{-2}p(\varpi, z) \text{ and } \varpi^{-1}\partial_\varpi p(\varpi, z) \text{ are bounded as } \varpi \to 0. \qquad (6.5.6)$$

We can write (6.5.5) as

$$\hat{\boldsymbol{\varpi}}B_\varpi + \hat{\mathbf{z}}B_z = \nabla \times \left(p\frac{\hat{\boldsymbol{\lambda}}}{\varpi}\right) = \nabla p \times \frac{\hat{\boldsymbol{\lambda}}}{\varpi}.$$

If we define $q = \varpi B_\lambda$, we conclude that every solenoidal axisymmetric **B** can be written

$$\mathbf{B} = \nabla p \times \frac{\hat{\lambda}}{\varpi} + q\frac{\hat{\lambda}}{\varpi} \qquad (6.5.7)$$

where p and q behave like ϖ^2 as $\varpi \to 0$ for fixed z, and p and q are uniquely determined by these requirements.

The fields $\nabla p \times \hat{\lambda}/\varpi$ and $q\hat{\lambda}/\varpi$ are poloidal and toroidal in the sense defined in Chapter 5, but p and q are not their poloidal and toroidal scalars. Obviously the curl of $q\hat{\lambda}/\varpi$ is $\nabla q \times \hat{\lambda}/\varpi$, poloidal. The curl of $\nabla p \times \hat{\lambda}/\varpi$ can be computed from (6.5.4). It is

$$\nabla \times \left(\nabla p \frac{\hat{\lambda}}{\varpi} \right) = -\Delta_1 p \frac{\hat{\lambda}}{\varpi}$$

where

$$\Delta_1 p = \partial_\varpi^2 p - \frac{1}{\varpi}\partial_\varpi p + \partial_z^2 p$$

$$= \nabla^2 p - \frac{2}{\varpi}\partial_\varpi p \qquad (6.5.8)$$

$$= \nabla \cdot \left(\nabla p - \frac{2\,\hat{\varpi}p}{\varpi} \right). \qquad (6.5.9)$$

The operator Δ_1 behaves in many respects like the cylindrical version of ∇^2.

If **B** is axisymmetric, so is $\mathbf{J} = \nabla \times \mathbf{B}/\mu_0$. Hence so is the vector potential **A**, computed from **J** via the Coulomb integral. Thus there are scalars $\tilde{p}$ and $\tilde{q}$ such that

$$\mathbf{A} = \nabla \tilde{p} \times \frac{\hat{\lambda}}{\varpi} + \tilde{q}\frac{\hat{\lambda}}{\varpi}.$$

Then

$$\mathbf{B} = \nabla \times \mathbf{A} = \nabla \tilde{q} \times \frac{\hat{\lambda}}{\omega} - \frac{\hat{\lambda}}{\varpi}\Delta_1\tilde{p}.$$

Thus $\tilde{q} = p$ and $\Delta_1\tilde{p} = -q$, and

$$\mathbf{A} = \nabla \tilde{p} \times \frac{\hat{\lambda}}{\varpi} + p\frac{\hat{\lambda}}{\varpi}. \qquad (6.5.10)$$

Now the electric field $\mathbf{E}$ is given both by $\mathbf{E} = \eta \nabla \times \mathbf{B} - \mathbf{u} \times \mathbf{B}$ and by $\mathbf{E} = -\partial_t \mathbf{A} - \nabla \phi$, where ϕ is the electrostatic potential. Hence

$$\partial_t \mathbf{A} + \nabla \phi = \mathbf{u} \times \mathbf{B} - \eta \nabla \times \mathbf{B}.$$

If we dot $\hat{\boldsymbol{\lambda}}$ into this equation and use (6.5.10) and

$$\nabla \times \mathbf{B} = \nabla q \times \frac{\hat{\boldsymbol{\lambda}}}{\varpi} - \frac{\hat{\boldsymbol{\lambda}}}{\varpi} \Delta_1 p \qquad (6.5.11)$$

we obtain

$$\frac{1}{\varpi} \partial_t p + \partial_\lambda \phi = u_z B_\varpi - u_\varpi B_z + \frac{\eta}{\varpi} \Delta_1 p. \qquad (6.5.12)$$

For any fixed ϖ, z, we integrate (6.5.12) with respect to λ from 0 to 2π and divide by 2π. The result is

$$\frac{1}{\varpi} \partial_t p = \langle u_z \rangle B_\varpi - \langle u_\varpi \rangle B_z + \frac{\langle \eta \rangle}{\varpi} \Delta_1 p \qquad (6.5.13)$$

where for any $f(\varpi, \lambda, z)$,

$$\langle f \rangle (\varpi, z) = \frac{1}{2\pi} \int_0^{2\pi} d\lambda \, f(\varpi, \lambda, z). \qquad (6.5.14)$$

Using (6.5.5), and multiplying (6.5.13) by ϖ, we obtain

$$\langle \eta \rangle \Delta_1 p = \partial_t p + \langle u_\varpi \rangle \partial_\varpi p + \langle u_z \rangle \partial_z p$$
$$= \partial_t p + \mathbf{U} \cdot \nabla p \qquad (6.5.15)$$

where

$$\mathbf{U} = \hat{\boldsymbol{\varpi}} \langle u_\varpi \rangle + \hat{\mathbf{z}} \langle u_z \rangle.$$

Equation (6.5.15) holds in the conductor (e.g., the earth's core). Outside the conductor, $\nabla \times \mathbf{B} = \mathbf{0}$ so, from (6.5.11),

$$\Delta_1 p = 0 \quad \text{and} \quad p \to 0 \text{ as } |\mathbf{r}| \to \infty. \qquad (6.5.16)$$

In fact, because the magnetic field in the insulator dies away at least as fast as a dipole field, it is easily shown that p declines like $1/r$ or faster. If Δ_1 were ∇^2 (see (6.5.8)), since the right side of (6.5.15) is the material time derivative $D_t p$ for advection by the velocity $\mathbf{U}$, equations (6.5.15) and (6.5.16) would describe the passive diffusion of a quantity, like heat or salt, with concentration p in a medium that is being stirred with velocity $\mathbf{U}$. Equation (6.5.16) would be a leak to infinity, so p should decay exponentially. Cowling's original proof of this was incorrect but repairable. For one repair device, see Backus (1957); the following is a simplified version.

To complete the argument we simplify things and assume that within the core K the diffusivity is uniform and furthermore that the flow is incompressible, so that $\nabla \cdot \mathbf{U} = 0$. Recall that by definition $1/\eta = \mu_0 \sigma$, where now the conductivity is constant. We may write (6.5.15)

$$\Delta_1 p = \mu_0 \sigma (\partial_t p + \mathbf{U} \cdot \nabla p), \quad \text{in } K$$
$$= 0, \qquad \text{outside } K. \tag{6.5.17}$$

Next we integrate $p \, \Delta_1 p$ over all space: in view of (6.5.17) we have

$$\int_{R^3} d^3\mathbf{r} \; p\Delta_1 p = \mu_0\sigma \int_K d^3\mathbf{r} \; (p\partial_t p + p\mathbf{U} \cdot \nabla p).$$

On the right we apply the familiar identity $\nabla \cdot s\mathbf{A} = s\nabla \cdot \mathbf{A} + \nabla s \cdot \mathbf{A}$ which together with incompressibility gives

$$p\mathbf{U} \cdot \nabla p = \nabla \cdot (\mathbf{U} p^2 / 2).$$

Thus

$$\int_{R^3} d^3\mathbf{r} \; p\Delta_1 p = \tfrac{1}{2}\mu_0\sigma \int_K d^3\mathbf{r} \; [\partial_t p^2 + \nabla \cdot (\mathbf{U} p^2)].$$

But from Gauss' Theorem

$$\int_K d^3\mathbf{r} \; \nabla \cdot (\mathbf{U} p^2) = \int_{\partial K} d^2\mathbf{r} \; \hat{\mathbf{n}} \cdot \mathbf{U} p^2 = 0$$

because the normal velocity at the core's surface vanishes. Thus we arrive at the relationship

$$\tfrac{1}{2}\mu_0\sigma\partial_t \int_K d^3\mathbf{r} \; p^2 = \int_{R^3} d^3\mathbf{r} \; p\Delta_1 p. \tag{6.5.18}$$

The remarkable feature of this equation is that it does not contain the velocity field. What we need to show next is that the right side of the equation is always negative, which would show that p decays away in time.

To prove that the integral on the right of (6.5.18) is negative we exploit the behavior of p for large r and use the expression for Δ_1 in (6.5.9): consider

$$\boldsymbol{\nabla} \cdot \left(p\boldsymbol{\nabla}p - 2p^2 \frac{\hat{\boldsymbol{\varpi}}}{\varpi}\right) = p\Delta_1 p + \boldsymbol{\nabla}p \cdot \boldsymbol{\nabla}p - \frac{2p\,\hat{\boldsymbol{\varpi}}}{\varpi} \cdot \boldsymbol{\nabla}p$$

$$= p\Delta_1 p + |\boldsymbol{\nabla}p|^2 - \boldsymbol{\nabla} \cdot \left(\frac{\hat{\boldsymbol{\varpi}}}{\varpi}p^2\right) + p^2\boldsymbol{\nabla} \cdot \left(\frac{\hat{\boldsymbol{\varpi}}}{\varpi}\right).$$

The last term vanishes identically because of (6.5.3). Thus we conclude that

$$p\Delta_1 p = -|\boldsymbol{\nabla}p|^2 + \boldsymbol{\nabla} \cdot \left(p\boldsymbol{\nabla}p - p^2 \frac{\hat{\boldsymbol{\varpi}}}{\varpi}\right).$$

When this expression is integrated over all space, the second term can be converted by Gauss' Theorem to a surface integral over a large sphere, whose radius grows without bound. It is easily confirmed, from the previously mentioned property that $p \to C/r$ on this sphere, that the contribution from the surface integral vanishes. Hence

$$\tfrac{1}{2}\mu_0\sigma\partial_t \int_K d^3\mathbf{r}\, p^2 = -\int_{R^3} d^3\mathbf{r}\, |\boldsymbol{\nabla}p|^2.$$

Because the integral of p^2 can never be negative, this equation shows that it must tend toward zero or a positive constant from above. Unless p is constant throughout space at this final state, the relation shows the integral must continue to decrease, which contradicts its being the limiting value. But if p is a constant everywhere (and this constant must be zero from (6.5.6) or (6.5.16)), the limiting state of the magnetic field is such that it too must vanish. A quantitative, and perhaps more satisfying, version of this final phase of the proof will be given in the next section.

6.5.2 Elsasser; Blackett and Runcorn; Bullard and Gellman

In 1947, Patrick Blackett, discouraged by Cowling's Theorem, suggested modifying Maxwell's equations to include a term that would be a source for **B** and which was proportional to angular momentum. The first magnetic star had just been discovered, and its ratio of magnetic moment to angular momentum seemed to support Blackett's relationship.   Blackett's equations predicted that as one moved closer to the center of the earth, the observed magnetic field ought to decrease.   But Keith Runcorn showed experimentally that, on descending into a deep English coal mine, **B** increased instead, just as would be expected for a core source.

Independently, Walter Elsasser (1946) and Edward Bullard (1949) suggested that maybe the earth's core evaded Cowling's Theorem by making essential use of the observed nonaxisymmetry of the geomagnetic field. Elsasser claimed to have proved the existence of kinematic dynamos. Bullard was more cautious. Both showed that there were feedback loops in which toroidal fluid shear could generate a toroidal from a poloidal **B** and radial motion could generate poloidal from toroidal **B**. The loops had serious leaks into high wavenumber components, so their existence was not enough to establish that **B** could be sustained. In addition, Elsasser used the wrong equation. He worked with the magnetic vector potential introduced by Hertz. If **A** is our usual vector potential, $\mathbf{B} = \nabla \times \mathbf{A}$, $\nabla \cdot \mathbf{A} = 0$, and $\mathbf{E} = -\partial_t \mathbf{A} - \nabla \phi$, so the dynamo equation in terms of Ohm's law is

$$\partial_t \mathbf{A} + \nabla \phi = \mathbf{u} \times \nabla \times \mathbf{A} - \eta \nabla \times \nabla \times \mathbf{A}.$$

Elsasser defined
$$\tilde{\mathbf{A}} = \mathbf{A} + \nabla \int dt \, \phi,$$

so $\mathbf{E} = -\partial_t \tilde{\mathbf{A}}$ and the dynamo equation is

$$\partial_t \tilde{\mathbf{A}} = \mathbf{u} \times (\nabla \times \tilde{\mathbf{A}}) - \eta \nabla \times \nabla \times \tilde{\mathbf{A}}.$$

But Elsasser continued to use $\nabla \cdot \tilde{\mathbf{A}} = 0$ when in fact

$$\nabla \cdot \tilde{\mathbf{A}} = - \int dt \, \nabla \cdot \mathbf{E} \neq 0.$$

This was noted by Bullard and Gellman (1954).

Bullard and Gellman (1954) carried out numerical computations aimed at exhibiting that certain particular velocity fields **u** could sustain a magnetic field indefinitely in a conducting fluid sphere in vacuo. Improvement in computers have made it possible to do the Bullard and Gellman calculations more accurately, and it is now known that the velocity field **u** they considered will not sustain a magnetic field, although at the time they believed they had found a successful kinematic dynamo.

Nevertheless, the Bullard and Gellman formalism has turned out to be very useful, both for theory and for computation. The generality of the expansion of a solenoidal field into poloidal and toroidal parts had not been proved in 1954, so Bullard and Gellman assumed without proof that

$$\mathbf{B} = \boldsymbol{\nabla} \times \boldsymbol{\Lambda} p + \boldsymbol{\Lambda} q \qquad (6.5.19)$$

and wrote a spherical harmonic expansion for the scalars

$$\begin{pmatrix} p \\ q \end{pmatrix} = \sum_{\ell=1}^{\infty} \sum_{m=-\ell}^{\ell} \begin{pmatrix} p_\ell^m(r,t) \\ q_\ell^m(r,t) \end{pmatrix} \beta_\ell^m(\hat{\mathbf{r}}). \qquad (6.5.20)$$

They substituted (6.5.19) and (6.5.20) in the dynamo equation, along with a similar expression for **u**, since they also assumed $\boldsymbol{\nabla} \cdot \mathbf{u} = 0$. They truncated all spherical harmonic expansions so as to use only $1 \le \ell \le L$, taking $L = 2, 3, 4$; it turned out later (e.g. Gibson and Roberts, 1969) that this was not enough. They obtained equations for $p_\ell^m(r,t)$ and $q_\ell^m(r,t)$, which they solved numerically. Their approach was to seek a steady dynamo, so they asserted $\partial_t p = \partial_t q = 0$ and then had to solve an eigenvalue problem for the critical magnetic diffusivity. They assumed that at this critical value the magnetic field would be steady. In fact it is oscillatory.

With their formalism, Bullard and Gellman were able to prove another antidynamo theorem: if $\boldsymbol{\nabla} \cdot \mathbf{u} = 0$, $u_r = 0$ and η is constant, then B_r dies away with t. In other words, the fluid motions in an incompressible fluid of uniform conductivity without radial components cannot sustain a dynamo – motions confined within spherical shells will not work.

A simplified version of their proof follows. We begin by dotting the dynamo equation with **r**:

$$\partial_t \mathbf{r} \cdot \mathbf{B} = \mathbf{r} \cdot \boldsymbol{\nabla} \times (\mathbf{u} \times \mathbf{B}) - \mathbf{r} \cdot \boldsymbol{\nabla} \times (\eta \boldsymbol{\nabla} \times \mathbf{B}). \qquad (6.5.21)$$

If $\eta(\mathbf{r}) = \eta(r)$, so that $\nabla\eta = \hat{\mathbf{r}}\partial_r\eta$, it follows from the identity $\nabla \times (s\mathbf{A}) = s\nabla \times \mathbf{A} - \mathbf{A} \times \nabla s$ that

$$\mathbf{r} \cdot \nabla \times \eta\nabla \times \mathbf{B} = \eta\mathbf{r} \cdot \nabla \times \nabla \times \mathbf{B}.$$

Now we substitute the expansion (6.5.19) for $\mathbf{B}$ on the right and appeal to a standard result of the Mie representation (5.2.25):

$$\mathbf{r} \cdot \nabla \times \eta\nabla \times \mathbf{B} = -\eta\nabla_1^2\nabla^2 p.$$

But $\nabla_1^2\nabla^2 = \nabla^2\nabla_1^2$ and by (5.2.26) $\nabla_1^2 p = rB_r$. Thus

$$\mathbf{r} \cdot \nabla \times \eta\nabla \times \mathbf{B} = -\eta\nabla^2 rB_r.$$

We introduce the variable $w = rB_r$ and then (6.5.21) becomes

$$\partial_t w = \mathbf{r} \cdot \nabla \times (\mathbf{u} \times \mathbf{B}) + \eta\nabla^2 w.$$

To analyze the first term on the right, note that for any $\mathbf{v}$,

$$\nabla \cdot (\mathbf{r} \times \mathbf{v}) = \mathbf{v} \cdot \nabla \times \mathbf{r} - \mathbf{r} \cdot \nabla \times \mathbf{v} = -\mathbf{r} \cdot \nabla \times \mathbf{v}$$

and

$$\mathbf{r} \times (\mathbf{u} \times \mathbf{B}) = \mathbf{u}rB_r - \mathbf{B}ru_r.$$

Setting $\mathbf{v} = \mathbf{u} \times \mathbf{B}$ and then inserting w for rB_r, we get

$$\partial_t w = \nabla \cdot (\mathbf{u}w - \mathbf{B}ru_r) + \eta\nabla^2 w$$

or, since $\nabla \cdot \mathbf{u} = 0$, and $\nabla \cdot \mathbf{B} = 0$,

$$\partial_t w + \mathbf{u} \cdot \nabla w = (\mathbf{B} \cdot \nabla)(ru_r) + \eta\nabla^2 w \quad \text{in} \quad K. \qquad (6.5.22)$$

This equation gives the material derivative for radial fields in terms of sources in the conductor. Also

$$\nabla^2 w = 0 \quad \text{and} \quad w \to 0 \text{ as } r \to \infty \quad \text{outside} \quad K, \qquad (6.5.23)$$

and w and ∇w are continuous everywhere because $w = \nabla_1^2 p$.

So far we have assumed only $\nabla \cdot \mathbf{u} = 0$. If we now introduce the further assumptions that $u_r = 0$, and $\eta = $ constant, we can multiply (6.5.22) by w and write it

$$\tfrac{1}{2}\partial_t w^2 + \tfrac{1}{2}\mathbf{u} \cdot \nabla w^2 = \eta \nabla \cdot (w\nabla w) - \eta |\nabla w|^2 \quad \text{in} \quad K. \qquad (6.5.24)$$

Multiplying (6.5.23) by ηw gives after some familiar rearrangement

$$0 = \eta \nabla \cdot (w\nabla w) - \eta |\nabla w|^2 \quad \text{outside} \quad K. \qquad (6.5.25)$$

If we integrate (6.5.24) over the core K, (6.5.25) over the region outside K that we call V, where $c \leq r < \infty$, and add the two equations, we obtain, since $\nabla \cdot \mathbf{u} = 0$

$$\tfrac{1}{2}\int_K d^3\mathbf{r} \, \partial_t w^2 + \tfrac{1}{2}\int_K d^3\mathbf{r} \, \nabla \cdot (\mathbf{u}w^2) = \eta \int_K d^3\mathbf{r} \, \nabla \cdot (w\nabla w)$$

$$- \eta \int_K d^3\mathbf{r} \, |\nabla w|^2 + \eta \int_V d^3\mathbf{r} \, \nabla \cdot (w\nabla w) - \eta \int_V d^3\mathbf{r} \, |\nabla w|^2.$$

Now

$$\int_K d^3\mathbf{r} \, \nabla \cdot (w\nabla w) + \int_V d^3\mathbf{r} \, \nabla \cdot (w\nabla w) = -\int_{\partial K} d^2\mathbf{r} \, \hat{\mathbf{n}} \cdot [w\nabla w]^+_-$$

by Gauss' Theorem, and $[w\nabla w]^+_- = \mathbf{0}$ on ∂K. Moreover,

$$\int_K d^3\mathbf{r} \, \nabla \cdot (\mathbf{u}w^2) = \int_{\partial K} d^2\mathbf{r} \, \hat{\mathbf{n}} \cdot (\mathbf{u}w^2) = 0$$

because $\hat{\mathbf{n}} \cdot \mathbf{u} = 0$ on ∂K. Therefore, finally,

$$\tfrac{1}{2}\frac{d}{dt}\int_K d^3\mathbf{r} \, w^2 = -\eta \int_{R^3} d^3\mathbf{r} \, |\nabla w|^2, \qquad (6.5.26)$$

which is exactly like (6.5.18). The argument given for that equation shows the decay of the field with time. In fact we can even bound the rate of decay as follows. Bullard and Gellman showed that

$$\int_{R^3} d^3\mathbf{r} \, |\nabla w|^2 \geq \frac{\pi^2}{c^2}\int_K d^3\mathbf{r} \, w^2 \qquad (6.5.27)$$

for any continuously differentiable function w in R^3 that satisfies (6.5.23). It follows that, if we write $W = \int_K d^3\mathbf{r}\ w^2$,

$$\frac{dW}{dt} \leq -\frac{\pi^2\eta}{c^2}W(t)$$

$$\frac{d}{dt}\ln W(t) \leq -\frac{\pi^2\eta}{c^2}$$

$$\ln W(t) \leq \ln W(0) - \frac{\pi^2\eta}{c^2}t,$$

so

$$W(t) \leq W(0)\ \exp\left(-\frac{\pi^2\eta}{c^2}t\right). \qquad (6.5.28)$$

In short, if η is constant, $\nabla \cdot \mathbf{u} = 0$ and $u_r = 0$, then the root-mean-square value of rB_r in K dies away at least as fast as if $\mathbf{u} = \mathbf{0}$, since from subsection 5.4.4 the factor in the exponent $\pi^2\eta/c^2 = \pi^2/\mu_0\sigma c^2$ corresponds to the slowest free decay of any poloidal magnetic field in a rigid sphere of conductivity σ.

The only serious gap in the Bullard and Gellman proof was that they did not prove that *all* solenoidal **B**'s have the form (6.5.19), something we now know to be the case.

In addition to the foregoing formal and mathematical results, Elsasser, Bullard, and Gellman emphasized a very important physical idea. It is troublesome if not actually difficult to manufacture poloidal fields (see (6.5.22)), but if a poloidal field is present, toroidal fields are very easy to manufacture by toroidal shear, i.e., by velocity fields of the form

$$\mathbf{u} = r\sin\theta\omega(r,\theta)\hat{\boldsymbol{\lambda}}.$$

These stretch $\hat{\mathbf{r}}B_r$ into a field with a large $\hat{\boldsymbol{\lambda}}$ component. The ease with which strong toroidal fields can be generated and amplified, compared with the comparative difficulty in creating poloidal fields, has led to speculation that toroidal field strengths in the core may be many (perhaps hundreds of) times larger than the poloidal components; but because the toroidal magnetic field cannot reach the earth's surface, these strong fields would be undetectable. See also subsection 5.4.2.

6.5.3 Rigorous Dynamos

The important question of the existence of dynamos is too complex for us to deal with in any but the most cursory way. Therefore in this section we give a brief narrative of the early efforts to produce a convincing kinematic dynamo. In 1958 it was still an open question whether there were any continuous motions of a simply connected lump of electrically conducting fluid that could maintain a magnetic field. Was the complicated topology of the wires in an industrial dynamo (in Figure 6.1.1 the conductor is not simply connected) really necessary? Bullard and Gellman (1954) observed that the multiple connectedness of Figure 6.1.1 was not essential, but they did not prove this. They thought the chirality of that figure was the issue. Between the working dynamos and the antidynamo theorems of Cowling and Bullard and Gellman was a large terra incognita, and presumably somewhere in that unexplored region was the earth's core.

Several workers claimed to have settled the question positively, but all the published "proofs" had flaws, and the numerical work was unconvincing and, as we have already noted, has since been shown, on larger computers, to be wrong. The situation was rather like that in the theory of turbulence, with the crucial exception that in dynamo theory for fluids there were no experimental data. Indeed, one "proof" of the existence of fluid dynamos was Batchelor's (1950) observation that the Helmholtz vorticity equation for $\omega = \nabla \times \mathbf{u}$ is identical with the dynamo equation for $\mathbf{B}$, and anyone who looks in a bathtub can see the generation of vorticity by the fluid motion. The flaw here is that the boundary conditions for ω and $\mathbf{B}$ are different. Vorticity is generated by slip at the boundary, and $\mathbf{B}$ turns out to be generated within the fluid.

To create a working laboratory model of a homogeneous fluid dynamo is no easy task. One must exceed some critical value of the magnetic Reynolds number, which is a dimensionless number that expresses the ratio of the advective term in the dynamo equation to the diffusive loss term:

$$\mathcal{R}_M \approx \frac{|\nabla \times (\mathbf{u} \times \mathbf{B})|}{|\nabla \times (\eta \nabla \times \mathbf{B})|}.$$

For a sphere of radius c,

$$\mathcal{R}_M = \mu_0 \sigma c u. \tag{6.5.29}$$

Obviously one should choose a liquid with good electrical conductivity; two possible candidates are liquid sodium, with $\sigma \approx 10^7$ S/m, or mercury, with 10^6 S/m. Hence $\mathcal{R}_M = 4\pi cu$. If $c = 1$ m, $u = \mathcal{R}_M/4\pi$. Probably the critical $\mathcal{R}_M$ to initiate dynamo action is more than 100, although this is still not known. If so, velocities on the order of $u \geq 8$ m/s are needed to start a dynamo. A spherical container of liquid sodium 2 m in diameter, stirred at 8 or 10 m/s, would be a formidable object. One was almost built in the basement of the MIT Mathematics Building about 20 years ago, but the administration learned of it.

Herzenberg and Lowes (1957) built an apparatus consisting of two small solid rotors inside a large solid block, with the electrical contact between rotors and block being mercury. With nonferromagnetic material for rotors and block, they could not achieve $\mathcal{R}_M$ large enough for generation of a magnetic field. By using iron they could replace μ_0 in (6.5.29) by $10^4\mu_0$, and were able to generate $\mathbf{B}$, but this was done later (see below), and the possibility that permanent magnetism was involved could not be completely ruled out.

Uriel Frisch (private communication) points out that the liquid metal–cooled breeder reactors in France may have large enough linear dimensions and u to produce a supercritical $\mathcal{R}_M$, in which case the pumps for cooling the reactor will encounter an unexpectedly large resistance from ohmic dissipation in the dynamo they drive, and there will finally be good experimental data on self-regenerative fluid dynamos, unless the experiment works too well.

The problem of the existence of kinematic dynamos was settled mathematically in 1958 by two independent examples of fluid flows that could be proved rigorously to maintain magnetic fields (Herzenberg, 1958; Backus, 1958). Neither was very convincing as a likely motion of the earth's core, but once the existence issue was settled, people felt more comfortable about using approximate methods, often borrowed from turbulence theory, in which existence was assumed.

Herzenberg's dynamo was a large, rigid copper sphere in which were embedded two smaller spheres rotating rigidly about axes oblique to one another. His dynamo was steady. Backus' motion was unsteady but continuous. A large axisymmetric shear produced a large toroidal field from the poloidal field. From this, a poloidal fluid motion produced more poloidal $\mathbf{B}$ from the toroidal $\mathbf{B}$. The problem both workers had to face was how to keep the energy in $\mathbf{B}$ at high wave numbers small enough that the series they used as solutions could be shown to converge. Herzenberg solved this problem by making his two inner spheres very small. Backus solved it by halting the motion occasionally to let the high-wavenumber components of **B** decay ohmically.

In 1964 Braginsky found a more realistic class of motions able to maintain steady dynamos. They, like Backus' motion, were in the tradition of Elsasser and of Bullard and Gellman (fluid axisymmetric shear produces toroidal **B** from poloidal **B**, and small radial motion regenerates the poloidal **B** from the toroidal). Braginsky's dynamos were asymptotic expansions in which the magnetic Reynolds number of the toroidal flow became very large, and the poloidal flow's magnetic Reynolds number became small, but the product became large. No proofs have ever been given that Braginsky's approximate solutions are approximations to solutions, but this situation is the usual one in fluid mechanics, and no one seems to feel a need to work out a proof. Braginsky's algebra is formidable even without any attempt to bound the remainders in his estimates.

About 1968 two new mechanisms began to be treated rigorously, which were alternatives to the poloidal→toroidal→poloidal cycle of Elsasser and Bullard and Gellman. G.O. Roberts (1970) showed how short-wavelength parts of **B** and **u** could combine quadratically in the $\nabla \times (\mathbf{u} \times \mathbf{B})$ term in the dynamo equation. The difference-wavenumbers produced in **B** were so small that their free decay rate was low, and they could be supplied against ohmic decay by the coupling at short wavelengths. The short-wavelength **B** could in turn be regenerated from long-wave **B** via short-wave **u** in the $\nabla \times (\mathbf{u} \times \mathbf{B})$ term.

A related idea had been proposed by E.N. Parker (1955). He proposed that convective upwelling in the earth's core would generate vorticity because of the earth's rotation, and that the resulting helical motion of the rising fluid would twist toroidal lines of force into meridional planes, transferring magnetic energy from toroidal to poloidal free decay modes. This suggestion has since been generalized to the idea that *helicity*, defined as

$$h = \mathbf{u} \cdot (\nabla \times \mathbf{u}), \qquad (6.5.30)$$

is an important if not essential component of dynamo action. Lortz (1968) constructed a particularly simple steady dynamo based on helicity. His dynamo is an infinite, space-filling fluid, but the length scales of both magnetic field and fluid motion are bounded. (One trap in dynamo proofs is to achieve simplicity by working in an infinite fluid. Unless care is taken, length scales can be indefinitely large, leading to $\mathcal{R}_M = \infty$.) There is a potential singularity in the Lortz dynamo along the $\hat{\mathbf{z}}$ axis that he does not analyze. No one seems to have been worried enough to look carefully at whether it exists or not.

6.5.4 Early Numerical Dynamos

A substantial amount of the numerical work on the correct dynamo equation (as opposed to mean field equations; see later) has been done by David Gubbins Gubbins, David when he was at the University of Cambridge in England. We note here a few of the salient results. Using the formalism of Bullard and Gellman, and truncating at the very high spherical harmonic degree permitted by modern computers, he has obtained solutions that appear to become independent of truncation level. By this means he has shown

1. The Bullard and Gellman flow of 1954 is not a dynamo.
2. Axisymmetric fluid motion can generate nonaxisymmetric magnetic fields, although by Cowling's Theorem it cannot generate axisymmetric fields. See Gubbins (1973).
3. Dynamos in bounded conductors appear to be harder to sustain than bounded dynamos in infinite conductors because at the boundary between conductor and insulator a "surface current" boundary layer is produced that is highly dissipative. See Bullard and Gubbins (1977).

6.5.5 Mean Field Dynamos

An important approximation, called *mean field electrodynamics*, was made precise and exploited by Steenbeck, Krause, and Rädler (hereinafter referred to as SKR) (1967; see also Steenbeck and Krause, 1969a, 1969b). The germ of this idea was presented by E.N. Parker, (1955). The application of mean field electrodynamics has been extremely fruitful in the study of the kinematic dynamo and has led to results showing that whole classes of fluid motions will make successful kinematic dynamos. In this section we sketch out the main ideas, filling in some of the mathematical details. We develop one of several kinematic dynamo mechanisms. Before embarking, we must add a note of caution: unlike the situation in fluid turbulence, where there is a large body of experimental data to

check theory, there is no laboratory work to confirm the validity of the
approximations of the mean field dynamos in real physical systems,
nor is the mathematical treatment compellingly rigorous. Whether
these dynamos have anything to do with the earth, or any real-world
dynamo, remains an open question.

The idea is roughly this: SKR argue that the fluid velocity in
the core is probably turbulent, which means that the fluid flow is
extremely complex on short length scales. If $\mathbf{U}(\mathbf{r}, t)$ is the Eulerian
description of this velocity, then

$$\mathbf{U}(\mathbf{r}, t) = \langle \mathbf{U} \rangle (\mathbf{r}, t) + \mathbf{u}(\mathbf{r}, t) \tag{6.5.31}$$

where $\langle \mathbf{U} \rangle$ is some kind of an average and $\mathbf{u}$ is a disorderly, random
component of small scale that we cannot hope to know in detail and
that we make no effort to specify precisely. The magnetic field $\mathbf{B}$
produced by the kinematic dynamo $\mathbf{U}$ will have the same form,

$$\mathbf{B}(\mathbf{r}, t) = \langle \mathbf{B} \rangle (\mathbf{r}, t) + \mathbf{b}(\mathbf{r}, t). \tag{6.5.32}$$

We anticipate an important finding: under appropriate conditions
concerning the symmetry of the turbulent motions, the small-scale
motions generate a large-scale electromotive force parallel to the large-
scale $\langle \mathbf{B} \rangle$ and proportional to it. As we will see, such a field makes
it very easy to obtain dynamo action. Such dynamos have been
called α-effect dynamos, because the coefficient of proportionality is
conventionally written as α.

With the diffusivity η uniform in space, the dynamo equation is

$$\partial_t \mathbf{B} = \nabla \times (\mathbf{U} \times \mathbf{B}) + \eta \nabla^2 \mathbf{B}.$$

Using (6.5.31) and (6.5.32), this is

$$\partial_t \langle \mathbf{B} \rangle + \partial_t \mathbf{b} = \nabla \times [\langle \mathbf{U} \rangle \times \langle \mathbf{B} \rangle + \langle \mathbf{U} \rangle \times \mathbf{b} + \mathbf{u} \times \langle \mathbf{B} \rangle + \mathbf{u} \times \mathbf{b}]$$
$$+ \eta \nabla^2 \langle \mathbf{B} \rangle + \eta \nabla^2 \mathbf{b}.$$

$$\tag{6.5.33}$$

The averaging operation $\langle \ \rangle$ is so far unspecified, and its precise
definition is unimportant as long as it has the following formal

properties: let a, b be any real numbers and f, g any Eulerian descriptions of real physical quantities. Then

$$\langle f \rangle = f \quad \text{if } f \text{ is constant in time and space} \quad (6.5.34)$$
$$\langle af + bg \rangle = a\langle f \rangle + b\langle g \rangle \quad\quad (6.5.35)$$
$$\partial_t \langle f \rangle = \langle \partial_t f \rangle \quad\quad (6.5.36)$$
$$\mathbf{\nabla} \langle f \rangle = \langle \mathbf{\nabla} f \rangle \quad\quad (6.5.37)$$
$$\langle \langle f \rangle g \rangle = \langle f \rangle \langle g \rangle \quad\quad (6.5.38)$$
$$\langle \langle f \rangle \rangle = \langle f \rangle. \quad\quad (6.5.39)$$

Note that (6.5.39) follows from (6.5.38) with $g = 1$ and (6.5.34).

1. Three different candidates for $\langle\ \rangle$ have been proposed: $\langle f \rangle (\mathbf{r}, t)$ can be the average of $f(\mathbf{r}', t)$ over a ball of some radius b centered on $\mathbf{r}$ (the averaging kernel need not be constant in the ball; it can taper to zero on $|\mathbf{r}' - \mathbf{r}| = b$). SKR propose taking the length scale of the turbulence much smaller than the averaging radius, b, and this is turn must be much less than the length scale of the mean fields. For this to be possible, it must be true that

$$\text{length scale of turbulence} << \text{length scale of mean fields.}$$
$$(6.5.40)$$

This is the so-called *two-scale approximation*. This form of averaging $\langle\ \rangle$ satisfies (6.5.31–6.5.37) exactly.

2. A second candidate for $\langle f \rangle$ is a time average instead of a space average. Its mathematical properties (6.5.33) are the same as the space average.

3. A third candidate for the averaging operation $\langle\rangle$ is the Gibbs statistical hypothesis, applied to turbulence in the manner of Kolmogorov and G. I. Taylor. Imagine N identical earths (say, $N \simeq 10^9$ or 10^{23}) subjected to identical experimental conditions as far as we can arrange them. Small errors we necessarily commit in the initial conditions will grow with time, so the actual $\mathbf{U}$ and $\mathbf{B}$ in the earth's core will vary from one earth to another, as will any physical quantity f. If $f_n(\mathbf{r}, t)$ is the value of $f(\mathbf{r}, t)$ observed in the nth experimental earth, we define

$$\langle f \rangle_G (\mathbf{r}, t) = \frac{1}{N} \sum_{n=1}^{N} f_n(\mathbf{r}, t).$$

It is easy to verify that (6.5.34–6.5.38) hold exactly. The problem with $\langle f \rangle_G$ is what it means for the one real earth we live on. People usually hope that $\langle f \rangle_G$ can be approximated by a time average. As far as we know, these questions are still open.

If we have an average satisfying (6.5.34–6.5.38) we can apply it to (6.5.31) and (6.5.32), obtaining

$$\langle \mathbf{u} \rangle = \langle \mathbf{b} \rangle = 0. \tag{6.5.41}$$

Then applying it to (6.5.33) gives

$$\partial_t \langle \mathbf{B} \rangle = \boldsymbol{\nabla} \times [\langle \mathbf{U} \rangle \times \langle \mathbf{B} \rangle + \langle \mathbf{u} \times \mathbf{b} \rangle] + \eta \nabla^2 \langle \mathbf{B} \rangle. \tag{6.5.42}$$

Subtracting this from (6.5.33) gives

$$\partial_t \mathbf{b} = \boldsymbol{\nabla} \times [\langle \mathbf{U} \rangle \times \mathbf{b} + \mathbf{u} \times \langle \mathbf{B} \rangle + \mathbf{u} \times \mathbf{b} - \langle \mathbf{u} \times \mathbf{b} \rangle] + \eta \nabla^2 \mathbf{b},$$

which we prefer to write as

$$\partial_t \mathbf{b} - \boldsymbol{\nabla} \times [\langle \mathbf{U} \rangle \times \mathbf{b} + \mathbf{u} \times \mathbf{b} - \langle \mathbf{u} \times \mathbf{b} \rangle] - \eta \nabla^2 \mathbf{b} = \boldsymbol{\nabla} \times [\mathbf{u} \times \langle \mathbf{B} \rangle]. \tag{6.5.43}$$

Equation (6.5.42) is the dynamo equation for $\langle \mathbf{B} \rangle$ in terms of $\langle \mathbf{U} \rangle$, but with the extra term $\langle \mathbf{u} \times \mathbf{b} \rangle$. The physical meaning of this term can be seen most easily by applying $\langle \ \rangle$ to the pre-Maxwell equations and Ohm's law. The result is

$$\boldsymbol{\nabla} \times \langle \mathbf{E} \rangle = -\partial_t \langle \mathbf{H} \rangle, \qquad \boldsymbol{\nabla} \cdot \langle \mathbf{E} \rangle = \langle \rho \rangle / \epsilon_0$$
$$\boldsymbol{\nabla} \times \langle \mathbf{B} \rangle = \mu_0 \langle \mathbf{J} \rangle, \qquad \boldsymbol{\nabla} \cdot \langle \mathbf{B} \rangle = 0$$
$$\langle \mathbf{J} \rangle = \sigma(\langle \mathbf{U} \rangle \times \langle \mathbf{B} \rangle + \langle \mathbf{u} \times \mathbf{b} \rangle + \langle \mathbf{E} \rangle).$$

In other words, the effect of the interaction of the small-scale $\mathbf{u}$ and $\mathbf{b}$ is to produce an addition to the effective large-scale electric field in Ohm's law, namely

$$\mathcal{E} = \langle \mathbf{u} \times \mathbf{b} \rangle. \tag{6.5.44}$$

How do we calculate $\mathcal{E}$ in terms of $\langle \mathbf{U} \rangle$ and $\langle \mathbf{B} \rangle$ so as to make (6.5.42) an equation for $\langle \mathbf{B} \rangle$ alone? SKR argue that if the length scale of $\mathbf{b}$ really is short, initial conditions will fade away, and (6.5.43) then shows that $\mathbf{b}$ depends linearly on $\langle \mathbf{B} \rangle$ if both are seen as functions. And if the length scale of $\mathbf{b}$ is short compared with that of $\langle \mathbf{B} \rangle$ then $\mathbf{b}(\mathbf{r}, t)$ will depend mainly on the values of $\langle \mathbf{B} \rangle (\mathbf{r}', t')$ with $\mathbf{r}'$ and t' near to $\mathbf{r}$ and t. As a first approximation, $\mathbf{b}(\mathbf{r}, t)$ will depend only on $\langle \mathbf{B} \rangle (\mathbf{r}, t)$. To include effects of nearby points, we can admit dependence of $\mathbf{b}(\mathbf{r}, t)$ on $\partial_t \langle \mathbf{B} \rangle (\mathbf{r}, t)$, $\partial_i \langle \mathbf{B} \rangle (\mathbf{r}, t)$, and higher derivatives at $(\mathbf{r}, t)$. The time derivatives can be omitted, being calculable from the space derivatives via (6.5.42) and (6.5.43). Moreover, from (6.5.43), the dependence of $\mathbf{b}$ on $\langle \mathbf{B} \rangle$ is *linear*. Therefore so is that of $\langle \mathbf{u} \times \mathbf{b} \rangle$ on $\langle \mathbf{B} \rangle$. Thus there must be scalars $\alpha_{ij}(\mathbf{r}, t)$, $\beta_{ijk}(\mathbf{r}, t)$, $\gamma_{ijk\ell}(\mathbf{r}, t)$, etc., such that if $\mathcal{E}$ is (6.5.44) then relative to orthonormal basis $\hat{\mathbf{x}}_1, \hat{\mathbf{x}}_2, \hat{\mathbf{x}}_3$ for R^3,

$$
\begin{aligned}
\mathcal{E}_i(\mathbf{r}, t) = & \alpha_{ij}(\mathbf{r}, t) \langle B_j \rangle (\mathbf{r}, t) + \beta_{ijk}(\mathbf{r}, t) \partial_j \langle B_k \rangle (\mathbf{r}, t) \\
& + \gamma_{ijk\ell} \partial_j \partial_k \langle B_\ell \rangle (\mathbf{r}, t) + \dots .
\end{aligned} \tag{6.5.45}
$$

SKR propose not to calculate $\alpha_{ij}, \beta_{ijk}, \dots$, from (6.5.43) but to regard those coefficients as constitutive quantities describing statistical properties of the turbulence, in the same way that viscosity describes statistically the average effect of *molecular* motion on momentum transfer in a fluid. The coefficients are to be measured empirically, and the measurements are to be simplified as far as possible by symmetry arguments. For example, if the turbulence is stationary (averages are independent of $\mathbf{r}$ and t) then $\alpha_{ij}, \beta_{ijk}, \dots$, will be independent of $\mathbf{r}$ and t.

Considerable work has been done in the fluid mechanics literature on isotropic turbulence, although the earth's rapid rotation makes it almost certain that core turbulence is not isotropic (Busse, 1983). Turbulence is called *fully isotropic* if reports of the statistical properties of the turbulence do not permit the identification of the particular orthonormal basis for R^3 used to make the measurements. Similarly, it is called *skew isotropic* when the statistical properties do not allow the detection of which right-handed orthonormal basis was used.

As an example of these isotropy arguments, suppose all terms in (6.5.45) except the α_{ij} are negligible. Abbreviate $\langle B_j \rangle$ as $\mathcal{B}_j$. Then

$$
\mathcal{E}_i = \alpha_{ij} \mathcal{B}_j . \tag{6.5.46}
$$

Suppose now that this result is to be described in terms of another orthonormal basis $\hat{\mathbf{x}}_1', \hat{\mathbf{x}}_2', \hat{\mathbf{x}}_3'$. Relative to that basis, we will find

$$
\mathcal{E}_k' = \alpha_{k\ell}' \mathcal{B}_\ell' . \tag{6.5.47}
$$

We do not, in fact, have to make new measurements to find the $\alpha'_{k\ell}$. We can calculate them from α_{ij}, which we have already measured. To do so, we must examine carefully how electric and magnetic fields are actually measured. An observer measuring $\mathbf{E}$ and $\mathbf{B}$ must choose a reference frame: an origin and an orthonormal basis for R^3, $\hat{\mathbf{x}}'_1, \hat{\mathbf{x}}'_2, \hat{\mathbf{x}}'_3$. Let h' be the "handedness" of this basis:

$$h' = \hat{\mathbf{x}}'_1 \cdot (\hat{\mathbf{x}}'_2 \times \hat{\mathbf{x}}'_3). \tag{6.5.48}$$

Then $h' = +1$ or -1 for right- or left-handed bases. Suppose the observer has never seen a human, and does not know whether her basis is right-handed. We watch her, and we know. She proceeds exactly as we would if we had chosen a "correct," i.e., right-handed, basis. She defines a cross-product $\times'$ by assuming

$$\hat{\mathbf{x}}'_j \times' \hat{\mathbf{x}}'_k = \epsilon_{ijk}\hat{\mathbf{x}}'_i.$$

Thus she calculates $\mathbf{u} \times' \mathbf{v} = \hat{\mathbf{x}}'_i\epsilon_{ijk}u'_jv'_k$. In fact, her basis satisfies

$$\hat{\mathbf{x}}'_j \times' \hat{\mathbf{x}}'_k = h'\epsilon_{ijk}\hat{\mathbf{x}}'_i,$$

so the relation between her cross-product and the "true" one (ours) is

$$\mathbf{u} \times' \mathbf{v} = h'\mathbf{u} \times \mathbf{v}. \tag{6.5.49}$$

Therefore, when the primed observer measures $\mathbf{E}(\mathbf{r}, t)$ and $\mathbf{B}(\mathbf{r}, t)$ at a point by watching forces on charged particles, she will use the Lorentz force law in the form

$$\mathbf{f} = q(\mathbf{E} + \mathbf{v} \times' \mathbf{B}). \tag{6.5.50}$$

Therefore she will agree with us about $\mathbf{E}$ (assuming she is not moving relative to us) but will disagree about $\mathbf{B}$. She will have to give $\mathbf{B}$ the sign opposite to ours. Therefore

$$\mathcal{E}'_k = (\hat{\mathbf{x}}'_k \cdot \hat{\mathbf{x}}_i)\mathcal{E}_i \tag{6.5.51}$$

$$\mathcal{B}'_\ell = hh'(\hat{\mathbf{x}}'_\ell \cdot \hat{\mathbf{x}}_j)\mathcal{B}_j \tag{6.5.52}$$

where we have put in our own handedness just in case we decided to use left-handed bases too. Then

$$\mathcal{E}'_k = (\hat{\mathbf{x}}'_k \cdot \hat{\mathbf{x}}_i)\alpha_{ij}hh'(\hat{\mathbf{x}}_j \cdot \hat{\mathbf{x}}'_\ell)\mathcal{B}'_\ell.$$

Therefore

$$\alpha'_{k\ell} = hh'(\hat{\mathbf{x}}'_k \cdot \hat{\mathbf{x}}_i)(\hat{\mathbf{x}}'_\ell \cdot \hat{\mathbf{x}}_j)\alpha_{ij}. \qquad (6.5.53)$$

Now suppose the turbulence is skew isotropic. Then none of its statistical properties depend on the basis used to measure them, as long as only bases with the same handedness are used. For such bases, $h = h'$ so $hh' = 1$. Also

$$\alpha'_{k\ell} = \alpha_{k\ell} \qquad (6.5.54)$$

because the values of the α_{ij} are measured statistical properties of the turbulence. Therefore

$$\alpha_{k\ell} = (\hat{\mathbf{x}}'_k \cdot \hat{\mathbf{x}}_i)(\hat{\mathbf{x}}'_\ell \cdot \hat{\mathbf{x}}_j)\alpha_{ij} \qquad (6.5.55)$$

as long as $h = h'$. One way to achieve $h = h'$ is to take

$$\hat{\mathbf{x}}'_1 = \hat{\mathbf{x}}_2, \qquad \hat{\mathbf{x}}'_2 = -\hat{\mathbf{x}}_1, \qquad \hat{\mathbf{x}}'_3 = \hat{\mathbf{x}}_3.$$

Then

$$\begin{aligned}
\alpha_{12} &= (\hat{\mathbf{x}}'_1 \cdot \hat{\mathbf{x}}_i)(\hat{\mathbf{x}}'_2 \cdot \hat{\mathbf{x}}_j)\alpha_{ij} \\
&= -(\hat{\mathbf{x}}_2 \cdot \hat{\mathbf{x}}_i)(\hat{\mathbf{x}}_1 \cdot \hat{\mathbf{x}}_j)\alpha_{ij} = -\delta_{i2}\delta_{j1}\alpha_{ij} = -\alpha_{21}.
\end{aligned}$$

Another way to achieve $h = h'$ is to take

$$\hat{\mathbf{x}}'_1 = \hat{\mathbf{x}}_2, \qquad \hat{\mathbf{x}}'_2 = \hat{\mathbf{x}}_1 \qquad \hat{\mathbf{x}}'_3, = -\hat{\mathbf{x}}_3.$$

Then

$$\alpha_{12} = (\hat{\mathbf{x}}_2 \cdot \hat{\mathbf{x}}_i)(\hat{\mathbf{x}}_1 \cdot \hat{\mathbf{x}}_j) = \alpha_{21} \qquad \text{and} \qquad \alpha_{22} = (\hat{\mathbf{x}}_1 \cdot \hat{\mathbf{x}}_i)(\hat{\mathbf{x}}_1 \cdot \hat{\mathbf{x}}_j) = \alpha_{11}.$$

Thus

$$\alpha_{12} = -\alpha_{21} = \alpha_{21} \quad \text{and} \quad \alpha_{22} = \alpha_{11}.$$

Then $\alpha_{21} = \alpha_{12} = 0$. Similarly,

$$\alpha_{33} = \alpha_{11}, \qquad \alpha_{13} = \alpha_{31} = \alpha_{23} = \alpha_{32} = 0.$$

Thus

$$\alpha_{ij} = \alpha \delta_{ij}. \tag{6.5.56}$$

A similar argument shows that in skew isotropic turbulence

$$\beta_{ijk} = \beta \epsilon_{ijk}. \tag{6.5.57}$$

Now suppose the turbulence is fully isotropic. Then (6.5.44) holds even if the two bases have opposite handedness. One way to achieve that is to take $\hat{\mathbf{x}}'_i = -\hat{\mathbf{x}}_i$. Then (6.5.53) is $\alpha'_{k\ell} = hh'\alpha_{k\ell}$. But if $h' = -h$, this means $\alpha'_{k\ell} = -\alpha_{k\ell}$. With (6.5.54) we get $\alpha_{k\ell} = 0$. In (6.5.56) α must vanish if the turbulence is fully isotropic.

We conclude that in skew isotropic turbulence (6.5.45) is

$$\mathcal{E} = \alpha \langle \mathbf{B} \rangle + \beta \nabla \times \langle \mathbf{B} \rangle + \dots \tag{6.5.58}$$

with $\alpha \neq 0$. The ratio of successive terms in this series is of order (length scale of turbulence/length scale of mean field). Usually all but the first term are neglected. However, in fully isotropic turbulence $\alpha = 0$, and one must start with the second term. A series similar to (6.5.58) exists for many constitutive relations, e.g., viscous stress as a function of strain rate, heat flow as a function of temperature gradient, even Ohm's law in a stationary conductor.

If we keep only the first two terms in (6.5.58), (6.5.42) becomes

$$\partial_t \langle \mathbf{B} \rangle = \nabla \times [\langle \mathbf{U} \rangle \times \langle \mathbf{B} \rangle + \alpha \langle \mathbf{B} \rangle + (\beta - \eta) \nabla \times \langle \mathbf{B} \rangle]. \tag{6.5.59}$$

The effect of the turbulence is to change the magnetic diffusivity η to an "effective" value $\eta - \beta$, and to add an electromotive force $\alpha \langle \mathbf{B} \rangle$. If α, β, η are constant, equation (6.5.59) is

$$\partial_t \langle \mathbf{B} \rangle = \nabla \times (\langle \mathbf{U} \rangle \times \langle \mathbf{B} \rangle) + \alpha \nabla \times \langle \mathbf{B} \rangle + (\eta - \beta) \nabla^2 \langle \mathbf{B} \rangle. \tag{6.5.60}$$

It is now trivial to maintain magnetic fields with fluid motion. We do not even need a mean flow. Suppose $\langle \mathbf{U} \rangle = 0$, and the fluid fills all space. We can find solutions to (6.5.60) in the form

$$\langle \mathbf{B} \rangle (\mathbf{r}, t) = \mathbf{F}(t) e^{i \mathbf{k} \cdot \mathbf{r}} \tag{6.5.61}$$

as long as

$$\partial_t \mathbf{F} = i\alpha \mathbf{k} \times \mathbf{F} - \tilde{\eta} k^2 \mathbf{F} \qquad (6.5.62)$$

where $\tilde{\eta} = \eta - \beta$. Choose $\hat{\mathbf{z}}$ in the $\mathbf{k}$ direction. Then $\nabla \cdot \langle \mathbf{B} \rangle = 0$ requires $\mathbf{k} \cdot \mathbf{F} = 0$, or $F_z = 0$, and (6.5.61) is

$$\partial_t F_x = -i\alpha k F_y - \tilde{\eta} k^2 F_x$$
$$\partial_t F_y = +i\alpha k F_x - \tilde{\eta} k^2 F_y$$
$$\partial_t (F_x \pm i F_y) = (-\tilde{\eta} k^2 \mp \alpha k)(F_x \pm i F_y)$$

$$F_x \pm i F_y = [F_x(0) \pm i F_y(0)] e^{(-\tilde{\eta} k^2 \mp \alpha k)t}. \qquad (6.5.63)$$

There will be dynamo action, signaled by growth of $\langle \mathbf{B} \rangle$, if $|\alpha| k > \tilde{\eta} k^2$, or $|\alpha| + k\beta > k\eta$. Thus a dynamo occurs if

$$\frac{|\alpha|}{k\eta} + \frac{\beta}{\eta} > 1. \qquad (6.5.64)$$

The left-hand side of (6.5.64) is the square of a magnetic Reynolds number. If there is any nonzero α effect, by making k small enough we get a dynamo, called an *α-effect* dynamo.

Estimates of α and β can be found in Moffatt (1978). If k_T is the dominant wavevector in $\mathbf{u}$, so that $2\pi/k_T$ is the size of a typical turbulent eddy, then

$$|\alpha| \approx \frac{\langle u^2 \rangle}{\eta k_T}.$$

Thus, neglecting β,

$$\mathcal{R}_M^2 = \frac{\langle u^2 \rangle}{\eta^2 k k_T} > 1 \qquad (6.5.65)$$

is the condition for dynamo action. The estimate of $|\alpha|$ comes from an estimate of $\mathbf{b}$ via (6.5.43). A similar estimate of $|\beta|$ is

$$|\beta| = \frac{\langle u^2 \rangle}{\eta k_T^2}.$$

If $\alpha = 0$, then the dynamo condition (6.5.64) becomes

$$\frac{\langle u^2 \rangle}{\eta^2 k_T^2} > 1. \qquad (6.5.66)$$

As expected, this is more stringent than (6.5.65).

Paul Roberts (1972) has used the formalism of Bullard and Gellman together with mean field electrodynamics to carry out numerical calculations that exhibit kinematic dynamos in a sphere of conducting fluid in vacuo when there is an isotropic α effect and $\beta = 0$. He permits α to vary with position in ways that are regarded as physically reasonable. For a description of the more recent developments see the article by Roberts and Gubbins (1987).

6.6 The Dynamics of Dynamos

Much less is known about the full dynamo problem than about kinematic dynamos. The following references will give the interested reader a good overview: P.H. Roberts (1971, 1987), Moffatt (1978), Weiss (1971), Busse (1983). We report three items.

6.6.1 The Taylor Theorem

As J. B. Taylor (1963) noted, the correct momentum equation in the earth's core must include the Coriolis term due to the rotation of the earth with angular velocity $\boldsymbol{\Omega} = \Omega \hat{\mathbf{z}}$. When viscous forces are neglected, we saw in section 6.2 that the equation is

$$\breve{\rho}[D_t \mathbf{u} + 2\boldsymbol{\Omega} \times \mathbf{u}] = -\boldsymbol{\nabla}p + \breve{\rho}\mathbf{g} + \mathbf{J} \times \mathbf{B}. \qquad (6.6.1)$$

A "slow" motion is one in which $|D_t \mathbf{u}| \ll |\boldsymbol{\Omega} \times \mathbf{u}|$; essentially this is any motion whose period is much greater than one day. Taylor shows that if the core motion is slow in this sense, then for any cylinder $\mathcal{C}$ with $\varpi = $ constant, whose axis is $\hat{\mathbf{z}}$, and whose top and bottom end at solid boundaries,

$$\int_{\mathcal{C}} dz \, d\lambda \, (\mathbf{J} \times \mathbf{B})_\lambda = 0. \qquad (6.6.2)$$

He also shows that if $\mathbf{B}$ is any magnetic field satisfying (6.6.2), then (6.6.1) with $D_t \mathbf{u} = 0$ has a unique solution $\mathbf{u}$ if $\breve{\rho}\mathbf{g}$ and $\mathbf{B}$ are given. This $\mathbf{u}$ can be expressed by quadrature.

It appears that the dynamics of the core are much easier than the dynamo equation, because Ω is so large compared to the other frequencies present. As far as we know, no one has yet exploited the Taylor Theorem to produce dynamos with a given driving force. Candidates for the driving force all involve buoyancy: favorites at the moment are thermal buoyancy from either radioactive potassium in the body of the core or heat of solidification (freezing) at the inner core–outer core boundary, or chemical buoyancy from differentiation on freezing at the latter boundary.

6.6.2 Bullard Dynamo, Poincaré–Bendixson Theorem, and Chaos

The Bullard disk dynamo with constant torque described at the beginning of the chapter is a dynamo with specified forces; it is albeit a very simple example of a dynamic dynamo. Recall that its most complicated behavior was a periodic solution. Malkus (1972) observed that an inductance across PQ in Figure 6.2.1 would produce reversals and apparently random behavior in Bullard's disk dynamo. A stumbling block in nonlinear dynamics, similar to the Cowling Theorem, has been the Poincaré–Bendixson Theorem (see Coddington and Levinson, 1955, chap. 15), which describes an autonomous differential system

$$\frac{dx_i}{dt} = f_i(x_1, \ldots, x_N), \qquad i = 1, \ldots, N. \qquad (6.6.3)$$

Autonomous means simply that $\partial_t f_i = 0$. The theorem states that, if $N = 2$, all solutions tend to infinity, to a limit, or to a periodic solution of the equations. Clearly, for $N = 2$, things are quite regular and Bullard's dynamic dynamo falls in this class. It was long believed that to produce irregular, chaotic behavior, N had to be very large, and that the chaos of turbulence or thermal motion in a gas came from the large number of degrees of freedom present. Lorenz (1963) found an example of chaos from meteorology with $N = 3$. It turns out that the Bullard disk dynamo with a shunt across PQ is another example of chaos with $N = 3$. It has been analyzed at some length by Robbins (1976). Thus a system only slightly more complex than Bullard's original dynamo can exhibit remarkably erratic behavior, reminiscent of the actual field behavior of the core, while the model possesses only three degrees of freedom.

6.6.3. Data Possibly Relevant to the Dynamics

There is now an increasing body of data concerning the geomagnetic field that may be relevant to the dynamics and could ultimately be used to discriminate in favor of particular kinds of dynamo models. We close with a brief mention of some of these data in anticipation that they will fuel further developments in our understanding of the origin of the geomagnetic field.

It now appears (Morrison, 1979; Jault et al., 1988; Jackson et al., 1992) that variations in the length of day are caused partly by coupling between the core and lower mantle and changes of the flow regime in the core. Neither the nature of this coupling (electromagnetic, thermal, topographic) nor the origin of the changes in flow is understood.

As yet there has been no generally agreed upon explanation of the strength of **B** in the earth. Paleomagnetic observations (Tanaka et al., 1995) indicate that the dipole moment has not exceeded its present value by more than about a factor of 2, and that it drops to about 10–20 percent of its current value during geomagnetic reversals.

The existence of and time scale associated with geomagnetic reversals provide the most robust paleomagnetic data relevant to the geodynamo. There are long periods when the geomagnetic field does not reverse, for example, the 40-million-year normal polarity superchron during the Cretaceous (see Figure 1.4.4). The transition to these *superchrons* follows a gradual decrease in average reversal rate, and a gradual increase in rate follows their termination. These rate variations may involve changes in mantle convection patterns and the resulting changes in heat loss from the core. There has been a long-standing debate about whether the normal and reverse states of the field are equally likely and behave alike (Wilson, 1970; Phillips, 1977; McFadden and Merrill, 1984; Merrill and McFadden, 1994). If the polarity superchrons are excluded from the analysis (on the grounds that the reversal mechanism was somehow turned off during those times), then there is no evidence for differences in stability between the normal and reverse states. There is no explanation for the typical time between reversals, which today stands at around 100,000 years but was several times longer 40 million years ago.

When one considers the actual structure of the field as determined from paleomagnetic data, things become much less certain. The axial dipole remains the dominant first-order feature of the field, but why this should be so is not obvious from models of the geodynamo. The data require models of the time-averaged field with different second-order field contributions during normal and reverse polarity epochs over the last 5 million years (e.g., Merrill and McElhinny, 1977;

Schneider and Kent, 1990; Johnson and Constable, 1995); this may indicate that a longer time interval than one polarity epoch needs to be sampled to get a stable estimate of average field configuration. The typical time taken for a reversal to occur seems to range from about 2000 to 10,000 years; this variability partially reflects inaccuracies in the recording medium, and difficulties in determining what constitutes a transitional field at any given geographical location, but presumably also contains any genuine variations in reversal times for different events. The factors that control the amount of time required for the field to reverse are not understood.

Recent compilations of geomagnetic reversal records suggest the existence of systematic structures recurring in transitional fields (Clement, 1991; Laj et al., 1992). This has been interpreted by some as evidence for spatial variability in the boundary conditions (temperature, topography, or composition) between the earth's outer core and mantle; the spatial variability is then supposed to favor the generation of certain kinds of field configurations (e.g., Laj et al., 1991; Hoffman, 1992). Others attribute their appearance to bias in the geographic distribution of sites from which records are available, and possible systematic biases in recording geomagnetic field directions in rocks when the field intensities are low (Valet et al., 1992; McFadden et al., 1993). The resolution of these issues will require a concerted effort to acquire more and better distributed data from field transitions.

Exercises

1. In a plasma (ionized gas) $\mathbf{B}$ can be so large that the Lorentz force $\mathbf{J} \times \mathbf{B}$ dominates the pressure gradient and all other forces (see subsection 6.2.2). Then the condition for hydrostatic equilibrium in the plasma is simply $\mathbf{J} \times \mathbf{B} = 0$ or $\nabla \times \mathbf{B} \times \mathbf{B} = 0$. A magnetic field with this property is called *force-free*. If $\mathbf{B}$ is force-free, there is a scalar field α such that $\nabla \times \mathbf{B} = \alpha \mathbf{B}$. In force-free fields of minimum magnetic energy α is constant.
   In $S(a, b)$, construct at least one nonzero force-free field $\mathbf{B}$ that has $b_r = 0$ on $S(a)$ and $S(b)$. Show that it has $\mathbf{B} = 0$ there. [Hint: See Abramowitz and Stegun, 1965, chap. 10.]

2. Suppose that all of space is filled by a rigid ohmic conductor with

constant magnetic diffusivity η. $\{\hat{\mathbf{x}}, \hat{\mathbf{y}}, \hat{\mathbf{z}}\}$ is an orthonormal basis for R^3, and position vectors $\mathbf{r}$ are written $\mathbf{r} = x\hat{\mathbf{x}} + y\hat{\mathbf{y}} + z\hat{\mathbf{z}}$. In the conductor there is a magnetic field $\mathbf{B}(\mathbf{r}, t)$.

Suppose $\mathbf{B}(\mathbf{r}, 0) = \hat{\mathbf{z}} \times \nabla\Psi(\mathbf{r}, 0)$ for some function $\Psi(\mathbf{r}, 0)$.

(a) Show that for any later time t there is a function $\Psi(\mathbf{r}, t)$ such that $\mathbf{B}(\mathbf{r}, t) = \hat{\mathbf{z}} \times \nabla\Psi(\mathbf{r}, t)$.

(b) Show that if $\mathbf{B} = \hat{\mathbf{z}} \times \nabla\Psi$ at time t, then at that time the lines of force of $\mathbf{B}$ are the curves of intersection of the planes $z = $ constant and the surfaces $\Psi = $ constant.

(c) Suppose there are positive constants l, m, n such that $\Psi(\mathbf{r}, 0) = (\sin lx)(\sin my)(\cos(nz))$. Sketch the lines of force of $\mathbf{B}(\mathbf{r}, 0)$ that lie in the plane $z = 0$, and show their directions.

(d) If $\mathbf{B}(\mathbf{r}, 0)$ is as in (c) above, find $\mathbf{B}(\mathbf{r}, t)$. Sketch its lines of force, which lie in the plane $z = 0$, and also show their directions.

3. With the infinite rigid conductor of the previous question still in place, now consider a different magnetic field, $\mathbf{B}(\mathbf{r}, t)$.

There are positive constants A and B such that $\mathbf{B}(\mathbf{r}, 0) = A\hat{\mathbf{x}} + B\hat{\mathbf{y}}(\text{sgn}x)$, where $\text{sgn}x = 1$ for $x > 0$, $\text{sgn}x = -1$ for $x < 0$, and $\text{sgn}0 = 0$. At time t in the plane $z = 0$, there is a magnetic line of force whose equation for large $|x|$ is asymptotically $y = (B/A)|x|$. Find $y(0, t)$, the value of y at which this line of force intersects the y axis. Sketch the line of force at time t.

7

APPENDIX: MATHEMATICAL BACKGROUND

7.1 Linear Algebra

The elementary facts of linear algebra as they apply to vectors in three-dimensional space are indispensable tools for the study of geomagnetism. As soon as we meet classical electromagnetism we need to be able to manipulate the various forms of vector dot products, divergences, and curls in an economical way. This section reviews the use of arrays and indices, and the conventions adopted throughout the book.

7.1.1 Arrays

In this section we consider the properties of arrays, with special emphasis on two particularly important arrays, the Kronecker δ and the alternator ϵ. As in the computer language FORTRAN, an *array* is a collection of objects (usually real or complex numbers) indexed by integers. The number of indices is the *order* of the array. Each index takes all integer values between 1 and some positive integer called the *dimension* of the array in that index. For example, if an array A has order 4 then $A_{ijk\ell}$ is the entry in the array whose first index has value i, whose second index has value j, third index k, and fourth index l. An array of order 2 can be thought of in two different ways as a matrix; A_{ij} can be the matrix entry in row i and column j, or in row j and column i. An array of order 1 can be thought of as a row vector or a column vector.

Two arrays will be of most concern to us. The first is the *n-dimensional Kronecker δ*. This array's symbol is $\delta^{(n)}$, or simply δ. It is of second order, and of dimension n in each index; by its definition,

$\delta_{ij}^{(n)} = 0$ if $i \neq j$ and $\delta_{ij}^{(n)} = 1$ if $i = j$. Both the corresponding matrices are the $n \times n$ identity matrix.

The second very important array is the *n-dimensional alternator* ϵ. This array's symbol is $\epsilon^{(n)}$, or simply ϵ. It is of order n, and of dimension n in each index. By its definition, $\epsilon_{i_1 \ldots i_n} = 0$ if any two of the integers $i_1, i_2, \ldots, i_n$ are equal; if not, they are some permutation of the integers $1, 2, \ldots, n$, and $\epsilon_{i_1, \ldots, i_n}$ is $+1$ or -1 according to whether that permutation is even or odd. We will need only $\epsilon^{(2)}$ and $\epsilon^{(3)}$. For $\epsilon^{(2)}$, $\epsilon_{11} = \epsilon_{22} = 0$, $\epsilon_{21} = -1$, $\epsilon_{12} = 1$. For $\epsilon^{(3)}$, $\epsilon_{ijk} = 0$ if any two of i, j, k are the same, while $\epsilon_{123} = \epsilon_{231} = \epsilon_{312} = 1$ and $\epsilon_{213} = \epsilon_{132} = \epsilon_{321} = -1$.

An important pair of properties an array might possess are: *symmetry* and *antisymmetry*. An array A of order 4 (for example) is *antisymmetric* in its first and third indices (for example) if it has the same dimension in those two indices and if for all possible values of i, j, k, ℓ we have

$$A_{ijk\ell} = -A_{kji\ell}.$$

A similar array would be *symmetric* in its first and third indices if for all possible values of i, j, k, ℓ we have

$$A_{ijk\ell} = A_{kji\ell}.$$

7.1.2 Index Conventions

In performing calculations with arrays, a considerable economy can be achieved if we adopt certain conventions regarding the meaning to be attached to an expression when an index appears once or twice in it.

(1) If an index appears exactly once in each term of an array equation, that equation is supposed to be true for all values of the index. Thus, if A, B, C are arrays, the equation $A = B + C$ means that all three arrays have the same order and the same dimension in each separate index, and that each entry in array A is the sum of the corresponding entries in B and C. By the first index convention, if, for example, A, B, C are all $2 \times 3 \times 5 \times 4$ arrays of order 4, then the last condition can be written $A_{ijk\ell} = B_{ijk\ell} + C_{ijk\ell}$. It is understood that this equation holds for $i = 1, 2$, and $j = 1, 2, 3$, and $k = 1, 2, 3, 4, 5$, and $\ell = 1, 2, 3, 4$. The names of the indices are irrelevant. Thus $A_{ij} = E_{ij} + F_{ij}$ means the same thing as $A_{pq} = E_{pq} + F_{pq}$. Another

use of this convention: a fourth-order array A is symmetric in its first and third indices if $A_{ijk\ell} = A_{kji\ell}$; similarly it is antisymmetric when $A_{ijk\ell} = -A_{kji\ell}$. (2) If an index appears exactly *twice* in an array or product of two arrays, it is understood to have the same possible values (dimension) in both places, and a sum is understood to take place over all its possible values even though the symbols for summation are absent. This is called the *Einstein summation convention*, after Albert Einstein, who invented it to save himself effort as he developed his theory of general relativity. Thus A_{iik} stands for $\sum_{i=1}^{n} A_{iik}$, where n is the common dimension of array A in its first and second indices. The name of the summation index is irrelevant. Thus A_{iik} and A_{jjk} stand for the same sum. Using conventions (1) and (2) together, we can write $A_{iik} = A_{jjk}$. As another example of convention (2), $\delta_{ii}^{(n)} = n$. As a third example, suppose A and S are second-order arrays with the same dimension in all their indices and S is symmetric, A antisymmetric. Then $A_{ij}S_{ij} = 0$. To prove this, using conventions (1) and (2), we write $A_{ij}S_{ij} = -A_{ji}S_{ij}$ (from antisymmetry of A) $= -A_{lk}S_{k\ell}$ (changing labels of summation indices) $= -A_{ij}S_{ji}$ (changing labels of summation indices) $= -A_{ij}S_{ij}$ (from symmetry of S). Comparing the two end members of this string of inequalities, we see $A_{ij}S_{ij} = -A_{ij}S_{ij}$, which must $= 0$.

7.1.3 Properties of the Kronecker Delta and the Alternator

We next list and prove a series of properties of the Kronecker delta and the alternator. These will be very useful in establishing results we need in vector calculus. They will also give the reader a chance to practice his or her understanding of the index conventions.

(1) $\delta^{(n)}$ is symmetric in its first and second indices. That is, $\delta_{ij}^{(n)} = \delta_{ji}^{(n)}$.

(2) $\delta_{ii}^{(n)} = n$.

(3) $\delta_{ij}^{(n)} A_{jk\ell} = A_{ik\ell}$ if the dimension of A in its first index is n.

Proof: In $\sum_{j=1}^{n} \delta_{ij} A_{jk\ell}$, all terms are 0 except the one for which $j = i$. That term is $(1)(A_{ik\ell})$. ◁

(4) $\epsilon^{(n)}$ is antisymmetric in each pair of its indices (totally antisymmetric).

Proof: Obvious to the reader who knows about odd and even permutations. Other readers should stick to $n = 2, 3$ and look at the explicit definitions of $\epsilon^{(2)}$ and $\epsilon^{(3)}$. Note that property

7. Appendix: Mathematical Background

(4) for $n = 2$ can be written $\epsilon_{ij} = -\epsilon_{ji}$, while for $n = 3$ it is $\epsilon_{ijk} = \epsilon_{jki} = \epsilon_{kij} = -\epsilon_{jik} = -\epsilon_{ikj} = -\epsilon_{kji}$. ◁

(5) Suppose A is an array of order n, of dimension n in each index, and totally antisymmetric (changes sign under any interchange of a pair of indices). Then $A_{i_1 \ldots i_n} = \alpha \epsilon_{i_1 \ldots i_n}^{(n)}$ for some constant α.

Proof: We give the proof only for $n = 3$. (For $n = 2$, the proof is easier. For $n > 3$, one needs some elementary notation for permutations, which we prefer to omit.) Let $\alpha = A_{123}$. If $i = j$, then $A_{ijk} = A_{jik}$, but $A_{ijk} = -A_{jik}$ by hypothesis. Hence $A_{ijk} = 0$ if $i = j$. Thus $A_{ijk} = \alpha \epsilon_{ijk}$ if $i = j$. Similarly, $A_{ijk} = \alpha \epsilon_{ijk}$ if any of i, j, k are equal. If all are different, we have $\alpha = A_{123} = -A_{213} = A_{312} - -A_{132} = A_{231} = A_{321}$, so $A_{ijk} = \alpha \epsilon_{ijk}$ if i, j, k are all different. ◁

(6) If δ_{ij} is the three-dimensional Kronecker delta,

$$\epsilon_{ijk}\epsilon_{\ell mn} = \delta_{i\ell}\delta_{jm}\delta_{kn} + \delta_{im}\delta_{jn}\delta_{k\ell} + \delta_{in}\delta_{j\ell}\delta_{km}$$
$$- \delta_{im}\delta_{j\ell}\delta_{kn} - \delta_{i\ell}\delta_{jn}\delta_{km} - \delta_{in}\delta_{jm}\delta_{k\ell}. \qquad (7.1.1)$$

(Mnemonic device: ijk appear in the same order in all terms. All six permutations of ℓmn appear, and the sign of any term is $+$ or $-$ according to whether the permutation of ℓ, m, n is even or odd.)

Proof: Let the right-hand side of the above equation be denoted by $R_{ijk\ell mn}$. Inspection shows that $R_{ijk\ell mn}$ is antisymmetric in any pair of indices among $\{\ell, m, n\}$. If we fix i, j, k, it follows that there is a constant S_{ijk} such that $R_{ijk\ell mn} = S_{ijk}\epsilon_{\ell mn}$. Moreover, $S_{ijk} = S_{ijk}\epsilon_{123} = R_{ijk123}$. Inspection shows that $R_{ijk\ell mn}$ is antisymmetric in any pair of indices among $\{i, j, k\}$. It follows that there is a constant α such that $S_{ijk} = \alpha \epsilon_{ijk}$. Thus $R_{ijk\ell mn} = \alpha \epsilon_{ijk}\epsilon_{\ell mn}$. In particular, $\alpha = \alpha \epsilon_{123}\epsilon_{123} = R_{123123}$, and by inspection $R_{123123} = 1$. Thus $\alpha = 1$. ◁

(7) If δ_{ij} is the three-dimensional Kronecker delta then

$$\epsilon_{ijk}\epsilon_{imn} = \delta_{jm}\delta_{kn} - \delta_{jn}\delta_{km} \qquad (7.1.2)$$

$$\epsilon_{ijk}\epsilon_{ijn} = 2\delta_{kn} \qquad (7.1.3)$$

$$\epsilon_{ijk}\epsilon_{ijk} = 6. \qquad (7.1.4)$$

Proof: To prove (7.1.2), set $\ell = i$ in (7.1.1) and sum over $i = 1, 2, 3$, obtaining

$$\epsilon_{ijk}\epsilon_{imn} = \delta_{ii}\delta_{jm}\delta_{kn} + \delta_{im}\delta_{jn}\delta_{ki} + \delta_{in}\delta_{ji}\delta_{km}$$
$$- \delta_{im}\delta_{ji}\delta_{kn} - \delta_{ii}\delta_{jn}\delta_{km} - \delta_{in}\delta_{jm}\delta_{ki}$$
$$= 3\delta_{jm}\delta_{kn} + \delta_{km}\delta_{jn} + \delta_{nj}\delta_{km}$$
$$- \delta_{jm}\delta_{kn} - 3\delta_{jn}\delta_{km} - \delta_{kn}\delta_{jm}$$
$$= \delta_{jm}\delta_{kn} - \delta_{jn}\delta_{km}.$$

To prove (7.1.3), set $j = m$ in (7.1.2) and sum over $j = 1, 2, 3$. To prove (7.1.4), set $k = n$ in (7.1.3) and sum over $k = 1, 2, 3$. Equation (7.1.4) is also obvious from the definition of $\epsilon^{(3)}$ given in subsection 7.1.1. ◁

(8) A 3×3 array S_{ij} is symmetric (a) if, and (b) only if, $\epsilon_{ijk}S_{jk} = 0$. *Proof:* (a) For any fixed i, ϵ_{ijk} is antisymmetric in j and k. By hypothesis, S_{jk} is symmetric in j and k. Recall the result for antisymmetric arrays A that $A_{ij}S_{ij} = 0$ shown in the paragraph on index conventions; thus $\epsilon_{ijk}S_{jk} = 0$. (b) By hypothesis $\epsilon_{ijk}S_{jk} = 0$. Multiply by $\epsilon_{i\ell m}$ and sum over $i = 1, 2, 3$, obtaining $\epsilon_{i\ell m}\epsilon_{ijk}S_{jk} = 0$. By (7.1.2), this is $(\delta_{\ell j}\delta_{mk} - \delta_{\ell k}\delta_{mj})S_{jk} = 0$ or by (3) $S_{\ell m} - S_{m\ell} = 0$. Thus S is symmetric. ◁

(9) If δ_{ij} is the three-dimensional Kronecker delta,

$$\delta_{ij}\epsilon_{k\ell m} = \delta_{ik}\epsilon_{j\ell m} + \delta_{i\ell}\epsilon_{kjm} + \delta_{im}\epsilon_{k\ell j}. \tag{7.1.5}$$

(Mnemonic device: i appears first in all terms. On the right, j is interchanged successively with each of k, ℓ, m.)
Proof: Let $R_{ijk\ell m}$ denote the right-hand side of (7.1.5). It is clearly antisymmetric in each pair (k, ℓ), (k, m), (ℓ, m). Thus, for any fixed i, j there is a constant S_{ij} such that $R_{ijk\ell m} = S_{ij}\epsilon_{k\ell m}$. Multiply this equation by $\epsilon_{k\ell m}$ and sum k, ℓ, and m from 1 to 3. Then

$$S_{ij}\epsilon_{k\ell m}\epsilon_{k\ell m} = (\delta_{ik}\epsilon_{j\ell m} + \delta_{i\ell}\epsilon_{kjm} + \delta_{im}\epsilon_{k\ell j})\epsilon_{k\ell m}.$$

By (7.1.3) and (7.1.4), this is

$$6S_{ij} = 2\delta_{ik}\delta_{jk} + 2\delta_{i\ell}\delta_{j\ell} + 2\delta_{im}\delta_{jm}.$$

By (3) of the delta and alternator properties $6S_{ij} = 6\delta_{ij}$, so $S_{ij} = \delta_{ij}$. $\triangleleft$

(10) If A_{ij} is any 3×3 array, and $\det A$ is its determinant

$$(\det A)\epsilon_{ijk} = A_{i\ell}A_{jm}A_{kn}\epsilon_{\ell mn}. \qquad (7.1.6)$$

Let R_{ijk} be the right-hand side of (7.1.6). It is antisymmetric in i, j, k. For example,

$$R_{jik} = A_{j\ell}A_{im}A_{kn}\epsilon_{\ell mn} = A_{jm}A_{i\ell}A_{kn}\epsilon_{m\ell n}$$
$$= -A_{i\ell}A_{jm}A_{kn}\epsilon_{\ell mn} = -R_{ijk}.$$

Thus there is a constant R such that $R_{ijk} = R\epsilon_{ijk}$. Then $R = R_{123}$, so

$$R = A_{1\ell}A_{2m}A_{3n}\epsilon_{\ell mn}$$
$$= A_{11}A_{22}A_{33} + A_{12}A_{23}A_{31} + A_{13}A_{21}A_{32}$$
$$- A_{12}A_{21}A_{33} - A_{11}A_{23}A_{32} - A_{13}A_{22}A_{31} = \det A.$$

Note that in general, for an $n \times n$ array A_{ij}, $(\det A)\epsilon^{(n)}_{i_1 \ldots i_n} = A_{i_1 j_1} \ldots A_{i_n j_n}\epsilon^{(n)}_{j_1 \ldots j_n}$.

7.1.4 Applications of Delta and the Alternator to Vector Algebra

As promised, we apply the foregoing material to derive some simple relations in vector algebra. The reader is certainly familiar with many of these results, but may find the practice in going through the material helpful. In the following, and throughout the book, we adopt these conventions: R^n is real n-dimensional Euclidean space. Position vectors in R^n are written $\mathbf{r}, \mathbf{x}, \mathbf{u}$, *etc.* Unit vectors are written with carets, as $\hat{\mathbf{r}}$, while $\sqrt{\mathbf{r} \cdot \mathbf{r}}$, which is the length of the vector $\mathbf{r}$, is written $|\mathbf{r}|$ or simply r. Thus, if $\mathbf{r} \neq \mathbf{0}$, $\hat{\mathbf{r}} = \mathbf{r}/r$ is a unit vector, called the *direction* of $\mathbf{r}$.

(1) A basis $\mathbf{x}_1, \ldots, \mathbf{x}_n$ for R^n is called *orthonormal* if $\mathbf{x}_i \cdot \mathbf{x}_j = \delta_{ij}$. It follows that each vector $\mathbf{x}_i$ of the basis is a unit vector, that is: $\mathbf{x}_i = \hat{\mathbf{x}}_i$.

(2) If $\hat{\mathbf{x}}_1, \ldots, \hat{\mathbf{x}}_n$ is an orthonormal basis for R^n, and $\mathbf{r}$ is any vector in R^n, then

$$\mathbf{r} = \hat{\mathbf{x}}_i r_i \quad \text{where} \quad r_i = \hat{\mathbf{x}}_i \cdot \mathbf{r}. \tag{7.1.7}$$

The numbers r_i are the components of $\mathbf{r}$ relative to the basis $\hat{\mathbf{x}}_1, \ldots, \hat{\mathbf{x}}_n$.

(3) If $\hat{\mathbf{x}}_1, \ldots, \hat{\mathbf{x}}_n$ is an orthonormal basis for R^n and $\mathbf{r} = \hat{\mathbf{x}}_i r_i$, $\mathbf{s} = \hat{\mathbf{x}}_j s_j$ then $\mathbf{r} \cdot \mathbf{s} = r_i s_i$. For $\mathbf{r} \cdot \mathbf{s} = r_i s_j \, \hat{\mathbf{x}}_i \cdot \hat{\mathbf{x}}_j = r_i s_j \delta_{ij} = r_i s_i$.

(4) An orthonormal basis $\hat{\mathbf{x}}_1, \hat{\mathbf{x}}_2, \hat{\mathbf{x}}_3$ for R^3 is right-handed if and only if

$$\hat{\mathbf{x}}_i \times \hat{\mathbf{x}}_j = \epsilon_{ijk}\hat{\mathbf{x}}_k \,, \quad \text{or} \quad (\hat{\mathbf{x}}_i \times \hat{\mathbf{x}}_j) \cdot \hat{\mathbf{x}}_k = \epsilon_{ijk}. \tag{7.1.8}$$

Proof: The two equations (7.1.8) are equivalent, as we see by setting $\mathbf{r} = \hat{\mathbf{x}}_i \times \hat{\mathbf{x}}_j$ in (7.1.7) and changing the summation index in (7.1.7) to k. If (7.1.8) holds, then $\hat{\mathbf{x}}_1 \times \hat{\mathbf{x}}_2 = \epsilon_{12k}\hat{\mathbf{x}}_k = \epsilon_{123}\hat{\mathbf{x}}_3 = \hat{\mathbf{x}}_3$, so the basis is right-handed. Conversely, if the basis is right-handed, $(\hat{\mathbf{x}}_i \times \hat{\mathbf{x}}_j) \cdot \hat{\mathbf{x}}_k$ is $+1$ or -1 when ijk is an even or odd permutation of 123, and is 0 otherwise. Thus it is ϵ_{ijk}. ◁

(5) Let $\hat{\mathbf{x}}_1, \hat{\mathbf{x}}_2, \hat{\mathbf{x}}_3$ be a right-handed orthonormal basis for R^3. Let $\mathbf{u} = u_i\hat{\mathbf{x}}_i$, $\mathbf{v} = v_j\hat{\mathbf{x}}_j$, $\mathbf{w} = w_k\mathbf{x}_k$. Then

$$(\mathbf{v} \times \mathbf{w})_i = \epsilon_{ijk}v_j w_k \tag{7.1.9}$$

$$\mathbf{u} \cdot (\mathbf{v} \times \mathbf{w}) = \epsilon_{ijk}u_i v_j w_k. \tag{7.1.10}$$

Proof: $\mathbf{v} \times \mathbf{w} = (v_j\hat{\mathbf{x}}_j) \times (w_k\hat{\mathbf{x}}_k) = v_j w_k \mathbf{x}_j \times \mathbf{x}_k = v_j w_k \epsilon_{ijk}\hat{\mathbf{x}}_i$. And $\mathbf{u} \cdot (\mathbf{v} \times \mathbf{w}) = u_i(\mathbf{v} \times \mathbf{w})_i$. ◁

From (7.1.10) we obtain the well-known identities $\mathbf{v} \times \mathbf{w} = -\mathbf{w} \times \mathbf{v}$ and

$$\mathbf{u} \cdot (\mathbf{v} \times \mathbf{w}) = \mathbf{v} \cdot (\mathbf{w} \times \mathbf{u}) = \mathbf{w} \cdot (\mathbf{u} \times \mathbf{v}) = (\mathbf{u} \times \mathbf{v}) \cdot \mathbf{w}. \tag{7.1.11}$$

And from (7.1.9) we obtain the notorious

$$\mathbf{u} \times (\mathbf{v} \times \mathbf{w}) = (\mathbf{u} \cdot \mathbf{w})\mathbf{v} - (\mathbf{u} \cdot \mathbf{v})\mathbf{w}. \tag{7.1.12}$$

Proof:

$$\begin{aligned}
[\mathbf{u} \times (\mathbf{v} \times \mathbf{w})]_i &= \epsilon_{ijk}u_j(\mathbf{v} \times \mathbf{w})_k = \epsilon_{ijk}u_j\epsilon_{k\ell m}v_\ell w_m \\
&= (\delta_{i\ell}\delta_j m - \delta_{im}\delta_{j\ell})u_j v_\ell w_m = u_j v_i w_j - u_j v_j w_i \\
&= (\mathbf{u} \cdot \mathbf{w})v_i - (\mathbf{u} \cdot \mathbf{v})w_i = [(\mathbf{u} \cdot \mathbf{w})\mathbf{v} - (\mathbf{u} \cdot \mathbf{v})\mathbf{w}]_i. \quad ◁
\end{aligned}$$

7.2 Vector Analysis: Differential Calculus

The subject of this lengthy section is formal definition of vector and scalar fields and the construction of operators to act on them; the most important are the differential operators that are generalizations of the familiar ∇. We will also examine under what circumstances it is permitted to have such operators act one after the other, and when this is allowed what happens. These results find applications in Chapters 3 and 6.

7.2.1 Scalar and Vector Fields

Let W be any subset of R^3; recall R^3 is ordinary Euclidean 3 space, with an origin chosen, so that points are described by position vectors $\mathbf{r}$. A *scalar field* on W is a function f that assigns to each point $\mathbf{r}$ in W a scalar (a real or complex number), written $f(\mathbf{r})$. A *vector field* on W is a function $\mathbf{v}$ that assigns to each point $\mathbf{r}$ in W a vector $\mathbf{v}(\mathbf{r})$. Suppose $f, g, f_1, \ldots, f_n$ are scalar fields on W and $\mathbf{u}, \mathbf{v}, \mathbf{v}_1, \ldots, \mathbf{v}_n$ are vector fields on W and $a_1, \ldots, a_n$ are constant scalars. The scalar fields on W, which are written $fg, a_\nu f_\nu$ (remember to sum on ν from 1 to n), $\mathbf{u} \cdot \mathbf{v}$, and the vector fields on W, which are written $f\mathbf{v}$ or $\mathbf{v}f$, $a_\nu \mathbf{v}_n u$, $\mathbf{u} \times \mathbf{v}$ are defined by requiring for every $\mathbf{r}$ in W that

$$(fg)(\mathbf{r}) = f(\mathbf{r})g(\mathbf{r}) \tag{7.2.1}$$

$$(a_\nu f_\nu)(\mathbf{r}) = a_\nu [f_\nu(\mathbf{r})] \tag{7.2.2}$$

$$(\mathbf{u} \cdot \mathbf{v})(\mathbf{r}) = \mathbf{u}(\mathbf{r}) \cdot \mathbf{v}(\mathbf{r}) \tag{7.2.3}$$

$$(f\mathbf{v})(\mathbf{r}) = (\mathbf{v}f)(\mathbf{r}) = f(\mathbf{r})\mathbf{v}(\mathbf{r}) = \mathbf{v}(\mathbf{r})f(\mathbf{r}) \tag{7.2.4}$$

$$(a_\nu \mathbf{v}_\nu)(\mathbf{r}) = a_\nu [\mathbf{v}_\nu(\mathbf{r})] \tag{7.2.5}$$

$$(\mathbf{u} \times \mathbf{v})(\mathbf{r}) = \mathbf{u}(\mathbf{r}) \times \mathbf{v}(\mathbf{r}) \tag{7.2.6}$$

where the Einstein summation convention continues to operate in equations (7.2.2) and (7.2.5).

If $\mathbf{v}$ is a vector field on W and $\hat{\mathbf{x}}_1, \hat{\mathbf{x}}_2, \hat{\mathbf{x}}_3$ is an orthonormal basis for R^3, the three scalar fields $\hat{\mathbf{x}}_i \cdot \mathbf{v}$ are called the component fields of $\mathbf{v}$ relative to $\hat{\mathbf{x}}_1, \hat{\mathbf{x}}_2, \hat{\mathbf{x}}_3$, and one often abbreviates $\hat{\mathbf{x}}_i \cdot \mathbf{v}$ as v_i, so $\mathbf{v} = \hat{\mathbf{x}}_i v_i = v_i \hat{\mathbf{x}}_i$.

Given a subset W of R^3 and a collection D of scalar fields on W, D is a *linear space* of scalar fields if it is a linear space under

(7.2.2); that is, whenever $f_1, \ldots, f_n$ are members of D and $a_1, \ldots, a_n$ are scalars, $a_\nu f_\nu$ is also a member of D. Similarly, D is a linear space of vector fields if it is a linear space under (7.2.5).

Example 1: Let W be an open set and let D be the collection of all continuously differentiable scalar fields on W. At the risk of stating the obvious, we say that the reason open sets are employed so often in this subject is that then every point is surrounded in every direction by other points of the set, making differentials and other limiting operations simple to define in a uniform way; a point in the boundary is more difficult to accommodate and so we just omit such awkward elements.   ◁

Example 2: Let W be an open set and let D be the collection of all continuously differentiable vector fields on W. It is important to note that a vector field $\mathbf{v}$ is called continuously differentiable relative to an orthonormal basis $\hat{\mathbf{x}}_1, \hat{\mathbf{x}}_2, \hat{\mathbf{x}}_3$ if the scalar fields $v_i = \hat{\mathbf{x}}_i \cdot \mathbf{v}$ are continuously differentiable. Continuous differentiability does not depend on the basis. To see this, suppose that vector field $\mathbf{v}$ is continuously differentiable relative to orthonormal basis $\hat{\mathbf{x}}_1, \hat{\mathbf{x}}_2, \hat{\mathbf{x}}_3$. Let $\hat{\mathbf{x}}'_1, \hat{\mathbf{x}}'_2, \hat{\mathbf{x}}'_3$ be any other orthonormal basis. Then $v_i = \hat{\mathbf{x}}_i \cdot \mathbf{v}$, $v'_j = \hat{\mathbf{x}}'_j \cdot \mathbf{v}$, and $\mathbf{v} = \hat{\mathbf{x}}_i v_i$, so $v'_j = \hat{\mathbf{x}}'_j \cdot (\hat{\mathbf{x}} \cdot v_i)$, or

$$v'_j = (\hat{\mathbf{x}}'_j \cdot \hat{\mathbf{x}}_i) v_i. \tag{7.2.7}$$

Thus if the v_i are continuously differentiable so are the v'_j.   ◁

A *function* is an ordered triple (f, D, R). D and R are sets, and f is a rule that assigns to each object d in D a unique object $f(d)$ in R. The object $f(d)$ is called the value of f at d. Every member of R is $f(d)$ for at least one d in D. The set D is called the *domain* of the function, and the set R is called its *range*. If R is a subset of another set S, we write $f : D \to S$, and say that f maps D into S. Often (f, D, R) is abbreviated as f, and we speak of the function f instead of the function (f, D, R).

7.2.2 Scalar Linear Operators

We now consider the idea of transforming one scalar field into another. We restrict our attention, however, to a particularly simple kind of transformation. A *scalar linear operator* on a subset W of R^3 is a function L with these properties:

(1) the domain of L is a *linear space* D_L of *scalar* fields on W;

(2) the range of L is a set R_L of *scalar* fields on some subset W' of R^3 (usually $W' = W$, but not always);

(3) L is linear. That is, if $a_1, \ldots, a_n$ are scalar constants and $f_1, \ldots, f_n$ are scalar fields in D_L, then

$$L(a_\nu f_\nu) = a_\nu L(f_\nu). \tag{7.2.8}$$

We often write $L(f)$ as Lf.

The range R_L of a scalar linear operator L is a linear space of scalar fields on W', but this requires a proof.

Proof: Suppose $g_1, \ldots, g_n$ are in R_L and $a_1, \ldots, a_n$ are scalars. Then there are scalar fields $f_1, \ldots, f_n$ in D_L such that $g_\nu = L(f_\nu)$. Then $a_\nu g_\nu = a_\nu L(f_\nu) = L(a_\nu f_\nu)$ by (7.2.8). But $a_\nu f_\nu$ is in D_L because D_L is a linear space. Hence $a_\nu g_\nu$ is $L(f)$ for some f in D_L; that is, $a_\nu g_\nu$ is in R_L.   ◁

We provide next a number of examples illustrating the perhaps unexpected variety of objects that can fall under the classification of scalar linear operators.

Example 1: W is an open subset of R^3. D_L is the set of all continuously differentiable scalar fields on W. $\hat{\mathbf{x}}_1, \hat{\mathbf{x}}_2, \hat{\mathbf{x}}_3$ is an orthonormal basis for R^3. Position vectors $\mathbf{r}$ are written $\mathbf{r} = \hat{\mathbf{x}}_i r_i$, and $\partial/\partial r_i$ is abbreviated ∂_i. For any f in D_L, we take $L_1 f$ to be $\partial_1 f$. Then $\partial_1 f$ is a scalar field on W, but is not necessarily itself continuously differentiable (it is, however, continuous). Thus the range of L_1 is different from its domain. Similarly, ∂_2 and ∂_3 are scalar linear operators on W.   ◁

Example 2: W and D_L as in Example 1. Let u_1, u_2, u_3 be any smooth curvilinear coordinate system in W. Let $\partial/\partial u_i$ be abbreviated ∂_i, Then ∂_1, ∂_2, and ∂_3 are scalar linear operators on W.   ◁

Example 3: W is any subset of R^3. D_L is any linear space of scalar fields on W, and g is any particular scalar field on W. For any f in D_L, we define $Lf = gf$. Then L is a scalar linear operator on W, called the "multiplication operator generated or defined by g." We will denote such an operator by g^M. Thus for every scalar field f in D_L, $g^M f = gf$.   ◁

Example 4: W and D_L as in Example 3. $\mathbf{r}_0$ is any particular position vector in R^3. For any f in D_L, define Lf by requiring

$$(Lf)(\mathbf{r}) = f(\mathbf{r} - \mathbf{r}_0).$$

The scalar field Lf is defined not on W but on the set W' of all position vectors of the form $\mathbf{r} + \mathbf{r}_0$, with $\mathbf{r}$ in W. This set is usually written $\mathbf{r}_0 + W$ or $W + \mathbf{r}_0$ and is obtained by translating W rigidly without rotation through the displacement $\mathbf{r}_0$. Thus R_L is a linear space of scalar fields defined on $W + \mathbf{r}_0$, not W. ◁

7.2.3 Sums and Products of Scalar Linear Operators

The next topic for development is a kind of arithmetic between the scalar linear operators. We can give meaning to the concepts of the sum of two operators and to the product. An interesting fact is that the resultant operators retain the property of linearity.

Suppose that $a_1, \ldots, a_n$ are scalar constants, and $L_1, \ldots, L_n$ are scalar linear operators with the same domain D. Then $a_\nu L_\nu$ (don't forget the summation convention) is defined as the linear operator whose domain is D and which assigns to any scalar field f in D the scalar field $(a_\nu L_\nu)f$ defined by

$$(a_\nu L_\nu)f = a_\nu(L_\nu f). \tag{7.2.9}$$

Note, for example, that (7.2.9) *defines* what is meant by the operator $L_1 + L_2$. Suppose that L and M are scalar linear operators and $R_M \subseteq D_L$, which means merely R_M is a subset of D_L. Then the scalar linear operator LM is defined as the operator whose domain is D_M and which assigns to any scalar field f in D_M the scalar field $(LM)f$ defined by

$$(LM)f = L(Mf). \tag{7.2.10}$$

No one can prevent us from defining $a_\nu L_\nu$ and LM by (7.2.9) and (7.2.10), but it must be proved that $a_\nu L_\nu$ and LM are *linear* operators. To prove this, suppose $b_1, \ldots, b_m$ are constant scalars and $f_1, \ldots, f_m$ are in D_M. Then

$$(a_\nu L_\nu)(b_\mu f_\mu) = a_\nu[L_\nu(b_\mu f_\mu)] = a_\nu[b_\mu L_\nu(f_\mu)]$$
$$= a_\nu b_\mu L_\nu(f_\mu) = b_\mu[a_\nu L_\nu(f_\mu)] = b_\mu[(a_\nu L_\nu)(f_\mu)],$$

so $a_\nu L_\nu$ is linear. And

$$(LM)(b_\mu f_\mu) = L[M(b_\mu f_\mu)] = L[b_\mu M(f_\mu)]$$
$$= b_\mu L[M f_\mu] = b_\mu[(LM)f_\mu],$$

so LM is linear.

From the definitions (7.2.9) and (7.2.10), we verify that scalar linear operators satisfy the following arithmetic rules:

$$L + M = M + L \tag{7.2.11}$$

$$L + (M + N) = (L + M) + N \tag{7.2.12}$$

$$L(M + N) = LM + LN \tag{7.2.13}$$

$$(M + N)L = ML + NL \tag{7.2.14}$$

$$L(MN) = (LM)N \tag{7.2.15}$$

$$Lc^M = c^M L \tag{7.2.16}$$

if c^M is the operator of multiplication determined by a scalar constant c. In the relations above with additions, we appeal for meaning to the obvious interpretation of (7.2.9). In general,

$$LM \neq ML,$$

and so we can not reverse the order of operators unless this has been shown to be allowed. In these equations, ranges and domains are supposed related so as to make sense. Thus in (7.2.13) we must have $D_M = D_N$ and R_M, R_N, R_{M+N} must be subsets of D_L.

Proofs:

(7.2.11): $(L + M)f = Lf + Mf = Mf + Lf = (M + L)f$ because $g + h = h + g$ when g and h are scalar fields. This is true because of (7.2.2) and its truth for scalars.

(7.2.12): The argument is the same as in (7.2.11).

(7.2.13): $[L(M + N)]f = L[(M + N)f] = L(Mf + Nf)$. But L is linear, so this is $L(Mf) + L(Nf)$, which is $(LM)f + (LN)f = (LM + LN)f$.

(7.2.14): $[(M + N)L]f = (M + N)(Lf) = M(LF) + N(Lf) = (ML)f + (NL)f = (ML + NL)f$. Note here that the property of linearity is not used.

(7.2.15): $[L(MN)]f = L[(MN)f] = L[M(Nf)] = (LM)(Nf) = [(LM)N]f$.

(7.2.16): $(Lc^M)f = L(cf)$. But L is linear, so $L(cf) = c(Lf) = c^M[Lf] = (c^M L)f$. ◁

In all these proofs we have used the definition of a function: Two scalar linear operators L and M are equal if they have the same domain and if $Lf = Mf$ for every f in the domain.

We commented earlier that in general $LM \neq ML$. We return to this important fact in subsection 7.2.11; it is shown there, for example, that, if $W, D, \hat{\mathbf{x}}_1, \hat{\mathbf{x}}_2, \hat{\mathbf{x}}_3$ are as in Example 1, subsection 7.2.2 and ∂_i is the operator of partial differentiation with respect to r_i, and if r_j^M is the operator of multiplication by the scalar field $\hat{\mathbf{x}}_j \cdot \mathbf{r}$, then

$$\partial_i r_j^M = r_j^M \partial_i + \delta_{ij}^M. \tag{7.2.17}$$

7.2.4 Scalar Linear Operators Acting on Vector Fields

So far scalar linear operators have been restricted to act on scalar fields. A scalar linear operator L can be applied to vector fields as follows: suppose L is a scalar linear operator on some subset W of R^3, and $\mathbf{v}$ is a vector field on R^3. Suppose that relative to some particular orthonormal basis $\hat{\mathbf{x}}_1, \hat{\mathbf{x}}_2, \hat{\mathbf{x}}_3$, the component scalar fields v_i are in D_L, so that Lv_i are well-defined scalar fields. Then we define $L\mathbf{v}$ by

$$L\mathbf{v} = \hat{\mathbf{x}}_i(Lv_i). \tag{7.2.18}$$

Some comments about this definition are needed. If $\hat{\mathbf{x}}_1', \hat{\mathbf{x}}_2', \hat{\mathbf{x}}_3'$ is any other orthonormal basis of R^3, we have (7.2.7). But D_L is a linear space, so v_1', v_2', v_3' are in D_L, and Lv_j' is a well-defined scalar field. What if we prefer to define $L\mathbf{v} = \hat{\mathbf{x}}_j'(Lv_j')$? In fact all is well. We will get the same answer as (7.2.18). To see this, recall that L is linear, so (7.2.7) implies $Lv_j' = (\hat{\mathbf{x}}_j' \cdot \hat{\mathbf{x}}_i)Lv_i$. Then

$$\hat{\mathbf{x}}_j'(Lv_j') = \hat{\mathbf{x}}_j'[(\hat{\mathbf{x}}_j' \cdot \hat{\mathbf{x}}_i)Lv_i] = [\hat{\mathbf{x}}_j'(\hat{\mathbf{x}}_j' \cdot \hat{\mathbf{x}}_i)]Lv_i.$$

But $\mathbf{u} = \hat{\mathbf{x}}_j'(\hat{\mathbf{x}}_j' \cdot \mathbf{u})$ for any $\mathbf{u}$. So $\hat{\mathbf{x}}_j'(\hat{\mathbf{x}}_j' \cdot \hat{\mathbf{x}}_2) = \hat{\mathbf{x}}_i$, and

$$\hat{\mathbf{x}}_j'(Lv_j') = \hat{\mathbf{x}}_i(Lv_i).$$

Another, equivalent, way to write the definition (7.2.18) is

$$(L\mathbf{v})_i = Lv_i. \tag{7.2.19}$$

7.2.5 Vector Linear Operators

A *vector linear operator* $\mathbf{L}$ is defined in exactly the same way as a scalar linear operator (see subsection 7.2.2) except that $R_\mathbf{L}$, the range of $\mathbf{L}$, is a set of *vector* fields on some subset W' of R^3 (usually $W' = W$). Thus, if f is a scalar field in the domain $D_\mathbf{L}$ then $\mathbf{L}f$ is a *vector* field. Some authors would call these *vector-valued* linear operators.

Given an orthonormal basis $\hat{\mathbf{x}}_1, \hat{\mathbf{x}}_2, \hat{\mathbf{x}}_3$ for R^3, and a vector linear operator $\mathbf{L}$, the components of $\mathbf{L}$ relative to this basis are the three scalar linear operators L_i defined by requiring for each f in $D_\mathbf{L}$ that

$$L_i f = \hat{\mathbf{x}}_i \cdot (\mathbf{L}f). \tag{7.2.20}$$

The domain of L_i is $D_\mathbf{L}$, and L_i is also written $\hat{\mathbf{x}}_i \cdot \mathbf{L}$. Thus, simply because of this choice of notation, $(\hat{\mathbf{x}}_i \cdot \mathbf{L})f = \hat{\mathbf{x}}_i \cdot (\mathbf{L}f)$. Another, equivalent, way to write the definition (7.2.20) is

$$L_i f = (\mathbf{L}f)_i. \tag{7.2.21}$$

Note that (7.2.19) defines $L\mathbf{v}$, while (7.2.21) defines $L_i f$.

For any vector $\mathbf{u}$, $\mathbf{u} = \hat{\mathbf{x}}_i u_i = u_i \hat{\mathbf{x}}_i$, so (7.2.21) implies

$$\mathbf{L}f = \hat{\mathbf{x}}_i(L_i f) = (L_i f)\hat{\mathbf{x}}_i \tag{7.2.22}$$

relative to any orthonormal basis for R^3. It follows from (7.2.22) that if $\mathbf{L}$ and $\mathbf{N}$ are vector linear operators and there is even one basis relative to which $L_i = N_i$, then $\mathbf{L} = \mathbf{N}$.

This observation implies that if we know the component operators L_i of $\mathbf{L}$ relative to one orthonormal basis $\hat{\mathbf{x}}_1, \hat{\mathbf{x}}_2, \hat{\mathbf{x}}_3$, then $\mathbf{L}$ is determined, so we should be able to calculate the component operators L'_j relative to any other orthonormal basis $\hat{\mathbf{x}}'_1, \hat{\mathbf{x}}'_2, \hat{\mathbf{x}}'_3$. The calculation is easy. For any scalar field f, $L'_j f = \hat{\mathbf{x}}'_j \cdot (\mathbf{L}f) = \hat{\mathbf{x}}'_j \cdot [\hat{\mathbf{x}}_i(L_i f)] = (\hat{\mathbf{x}}'_j \cdot \hat{\mathbf{x}}_i)(L_i f) = [(\hat{\mathbf{x}}'_j \cdot \hat{\mathbf{x}}_i)L_i]f$. Since this holds for every f in $D_\mathbf{L}$,

$$L'_j = (\hat{\mathbf{x}}'_j \cdot \hat{\mathbf{x}}_i)L_i. \tag{7.2.23}$$

This is the same as (7.2.7). The components of a vector linear operator transform like the components of a vector when the orthonormal basis is changed.

We come now to two very important examples of vector linear operators.

Example 1: Let W, D, $\hat{\mathbf{x}}_1, \hat{\mathbf{x}}_2, \hat{\mathbf{x}}_3$ be as in Example 1 of subsection 7.2.2. For any f in D, define $\mathbf{L}f = \hat{\mathbf{x}}_i(\partial_i f)$. The vector linear operator $\mathbf{L}$ is the gradient operator ∇. From its definition,

$$\nabla f = \hat{\mathbf{x}}_i(\partial_i f) \qquad (7.2.24)$$

there appears to be a ∇ operator for every orthonormal basis. In fact, all these operators are the same. To see this, let $\hat{\mathbf{x}}'_1, \hat{\mathbf{x}}'_2, \hat{\mathbf{x}}'_3$ be any other orthonormal basis. The position vector is

$$\mathbf{r} = \hat{\mathbf{x}}_i r_i \quad \text{so} \quad r'_j = \hat{\mathbf{x}}'_j \cdot \mathbf{r} = \hat{\mathbf{x}}'_j \cdot (\hat{\mathbf{x}}_i r_i) = (\hat{\mathbf{x}}'_j \cdot \hat{\mathbf{x}}_i)r_i,$$

a special case of (7.2.7). Then

$$\frac{\partial r'_j}{\partial r_i} = (\hat{\mathbf{x}}'_j \cdot \hat{\mathbf{x}}_i) \qquad (7.2.25)$$

By the chain rule,

$$\partial_i f = \frac{\partial f}{\partial r_i} = \frac{\partial r'_j}{\partial r_i}\frac{\partial f}{\partial r'_j} = (\hat{\mathbf{x}}'_j \cdot \hat{\mathbf{x}}_i)\partial'_j f,$$

so

$$\hat{\mathbf{x}}_i(\partial_i f) = [\hat{\mathbf{x}}_i(\hat{\mathbf{x}}'_j \cdot \hat{\mathbf{x}}_i)](\partial'_j f) = \hat{\mathbf{x}}'_j(\partial'_j f)$$

and

$$\nabla f = \hat{\mathbf{x}}_i(\partial_i f) = \hat{\mathbf{x}}'_j(\partial'_j f). \qquad (7.2.26)$$

What are the component scalar linear operators ∇_i relative to the orthonormal basis $\hat{\mathbf{x}}_1, \hat{\mathbf{x}}_2, \hat{\mathbf{x}}_3$? By definition, for any scalar field f, $\nabla_j f = \hat{\mathbf{x}}_j \cdot (\nabla f)$, so $\nabla_j f = \hat{\mathbf{x}}_j \cdot [\hat{\mathbf{x}}_i(\partial_i f)] = \delta_{ij}\partial_i f = \partial_j f$. Thus, relative to any orthonormal basis,

$$\nabla_i = \partial_i. \qquad \lhd$$

Example 2: Let W, D be as in Example 3 of subsection 7.2.2. Let $\mathbf{v}$ be any particular vector field on W. For any f in D, define $\mathbf{L}f = \mathbf{v}f$. The vector linear operator $\mathbf{L}$ is called the multiplication operator defined by $\mathbf{v}$. We will write it $\mathbf{v}^M$. Thus for any scalar field f in D, $\mathbf{v}^M f = \mathbf{v}f$. Relative to any orthonormal basis $\hat{\mathbf{x}}_1, \hat{\mathbf{x}}_2, \hat{\mathbf{x}}_3$, we have for scalar fields f that

$$(\mathbf{v}^M)_i f = \hat{\mathbf{x}}_i \cdot (\mathbf{v}^M f) = \hat{\mathbf{x}}_i \cdot (\mathbf{v}f) = (\hat{\mathbf{x}}_i \cdot \mathbf{v})f = v_i f = (v_i)^M f.$$

That is, $(\mathbf{v}^M)_i = (v_i)^M$. In other words, the ith component of the vector operator $\mathbf{v}^M$ is the multiplication operator generated by v_i, the ith component of the vector field $\mathbf{v}$.   ◁

7.2.6 Linear Combinations of Vector Linear Operators

Linear combinations of vector linear operators are defined as for scalar linear operators. Suppose $a_1, \ldots, a_n$ are scalar constants and $\mathbf{L}_1, \ldots, \mathbf{L}_n$ are vector linear operators with the same domain D. Then $a_\nu \mathbf{L}_\nu$ denotes a vector linear operator whose domain is D and that assigns to any scalar field f in D a vector field $(a_\nu \mathbf{L}_\nu)f$ according to the rule

$$(a_\nu \mathbf{L}_\nu)f = a_\nu (\mathbf{L}_\nu f) \tag{7.2.27}$$

where the summation convention remains in force as always. The component operators of $a_\nu \mathbf{L}_\nu$ relative to an orthonormal basis $\hat{\mathbf{x}}_1, \hat{\mathbf{x}}_2, \hat{\mathbf{x}}_3$ are obtained as follows:

$$[\hat{\mathbf{x}}_i \cdot (a_\nu \mathbf{L}_\nu)]f = \hat{\mathbf{x}}_i \cdot [(a_\nu \mathbf{L}_\nu)f] = \hat{\mathbf{x}}_i \cdot [a_\nu (\mathbf{L}_\nu f)]$$

$$= a_\nu \hat{\mathbf{x}}_i \cdot (\mathbf{L}_\nu f) = a_\nu [(\hat{\mathbf{x}}_i \cdot \mathbf{L}_\nu)f] = [a_\nu (\hat{\mathbf{x}}_i \cdot \mathbf{L}_\nu)]f.$$

Hence

$$\hat{\mathbf{x}}_i \cdot (a_\nu \mathbf{L}_\nu) = a_\nu (\hat{\mathbf{x}}_i \cdot \mathbf{L}_\nu). \tag{7.2.28}$$

In particular, if a is a scalar constant,

$$(a\mathbf{L})_i = aL_i \tag{7.2.29}$$

$$(\mathbf{L} + \mathbf{M})_i = L_i + M_i. \tag{7.2.30}$$

7.2.7 Products of Vector Linear Operators

Let K, M be scalar linear operators and $\mathbf{L}$, $\mathbf{N}$ be vector linear operators. We have already defined KM; see (7.2.10). We would also like to define $K\mathbf{L}$ and $\mathbf{L}K$ (which we expect may be different) and $\mathbf{LN}$. But $\mathbf{LN}$ is a second-order-tensor linear operator, so we will define only the scalar and vector linear operators that can be extracted from it, namely $\mathbf{L} \cdot \mathbf{N}$ and $\mathbf{L} \times \mathbf{N}$. We return to these products of two vector linear operators in the next subsection.

To define $K\mathbf{L}$ and $\mathbf{L}M$, we require for any scalar field f in $D_\mathbf{L}$ and any scalar field g in D_M that

$$(K\mathbf{L})f = K(\mathbf{L}f) \tag{7.2.31}$$

$$(\mathbf{L}M)g = \mathbf{L}(Mg). \tag{7.2.32}$$

$\mathbf{L}M$ is defined only if $R_M \subseteq D_\mathbf{L}$, and its domain is D_M. $K\mathbf{L}$ is defined only if $R_{L_i} \subseteq D_K$, where L_i are the component fields of $\mathbf{L}$; and the domain of $K\mathbf{L}$ is $D_\mathbf{L}$. Note that the right-hand side of (7.2.31) is defined by (7.2.18) or (7.2.19). The component operators of $K\mathbf{L}$ and $\mathbf{L}M$ relative to any orthonormal basis $\hat{\mathbf{x}}_1, \hat{\mathbf{x}}_2, \hat{\mathbf{x}}_3$ are easy to calculate. We have

$$(K\mathbf{L})_i f = [(K\mathbf{L})f]_i = [K(\mathbf{L}f)]_i = K(\mathbf{L}f)_i = K[L_i f] = (KL_i)f$$

where the equalities are justified by equations (7.2.21), (7.2.31), (7.2.19), (7.2.21), and (7.2.10) respectively. Since this holds for all f in $D_\mathbf{L}$,

$$(K\mathbf{L})_i = KL_i \tag{7.2.33}$$

and $(\mathbf{L}M)_i f = [(\mathbf{L}M)f]_i = [\mathbf{L}(Mf)]_i = L_i(Mf) = (L_iM)f$, so

$$(\mathbf{L}M)_i = L_iM. \tag{7.2.34}$$

Equations (7.2.33, 7.2.34) hold relative to every orthonormal basis $\hat{\mathbf{x}}_1, \hat{\mathbf{x}}_2, \hat{\mathbf{x}}_3$.

Another useful consequence of (7.2.32) is

$$\mathbf{L} = \hat{\mathbf{x}}_i^M L_i. \tag{7.2.35}$$

The proof is simple. If f is in $D_\mathbf{L}$,

$$\mathbf{L}f = \hat{\mathbf{x}}_i(L_i f) = \hat{\mathbf{x}}_i^M(L_i f) = (\hat{\mathbf{x}}_i^M L_i)f.$$

Here we have called upon (7.2.22), the definition of the multiplication operator and (7.2.32). This holds for all possible f, which gives (7.2.35).

The arithmetic rules obeyed by scalar and vector linear operators are the following. Here c is a constant scalar and $\mathbf{b}$ a constant vector, while K and L are scalar linear operators and $\mathbf{N}$, $\mathbf{P}$, $\mathbf{Q}$ are vector linear operators. Then (compare subsection 7.2.3)

$$\mathbf{N} + \mathbf{P} = \mathbf{P} + \mathbf{N} \qquad (7.2.36)$$
$$\mathbf{N} + (\mathbf{P} + \mathbf{Q}) = (\mathbf{N} + \mathbf{P}) + \mathbf{Q} \qquad (7.2.37)$$
$$\mathbf{N}(K + L) = \mathbf{N}K + \mathbf{N}L \qquad (7.2.38)$$
$$(K + L)\mathbf{N} = K\mathbf{N} + L\mathbf{N} \qquad (7.2.39)$$
$$L(\mathbf{N} + \mathbf{P}) = L\mathbf{N} + L\mathbf{P} \qquad (7.2.40)$$
$$(\mathbf{N} + \mathbf{P})L = \mathbf{N}L + \mathbf{P}L \qquad (7.2.41)$$
$$\mathbf{N}(KL) = (\mathbf{N}K)L \qquad (7.2.42)$$
$$K(\mathbf{N}L) = (K\mathbf{N})L \qquad (7.2.43)$$
$$K(L\mathbf{N}) = (KL)\mathbf{N} \qquad (7.2.44)$$
$$c^M \mathbf{N} = \mathbf{N}c^M \qquad (7.2.45)$$
$$K\mathbf{b}^M = \mathbf{b}^M K \qquad (7.2.46)$$

but in general $L\mathbf{N} \neq \mathbf{N}L$. Of these rules, only (7.2.45) and (7.2.46) are not trivial consequences of the definitions. To prove (7.2.45), note that for any scalar field f,

$$(\mathbf{N}c^M)f = \mathbf{N}(c^M f) = \mathbf{N}(cf).$$

But $\mathbf{N}$ is linear, so

$$\mathbf{N}(cf) = c(\mathbf{N}f) = c^M(\mathbf{N}f) = (c^M\mathbf{N})f.$$

To prove (7.2.46), use components relative to any orthonormal basis. We have

$$(K\mathbf{b}^M)_i = K(\mathbf{b}^M)_i = K(\mathbf{b}_i)^M = (\mathbf{b}_i)^M K = (\mathbf{b}^M)_i K = (\mathbf{b}^M K)_i,$$

which proves (7.2.46), where the third equality follows from (7.2.16). From (7.2.35, 7.2.46) we can rewrite (7.2.35) as

$$\mathbf{L} = L_i \hat{\mathbf{x}}_i^M. \tag{7.2.47}$$

7.2.8 Dot and Cross Products of Vector Linear Operators

Finally, we define $\mathbf{L} \cdot \mathbf{N}$ and $\mathbf{L} \times \mathbf{N}$ for vector linear operators $\mathbf{L}$ and $\mathbf{N}$. Let $\hat{\mathbf{x}}_1, \hat{\mathbf{x}}_2, \hat{\mathbf{x}}_3$ be an orthonormal basis for R^3, and let L_i and N_i be the component operators of $\mathbf{L}$ and $\mathbf{N}$ relative to this basis. Define

$$\mathbf{L} \cdot \mathbf{N} = L_i N_i \tag{7.2.48}$$
$$\mathbf{L} \times \mathbf{N} = (\hat{\mathbf{x}}_j \times \hat{\mathbf{x}}_k)^M L_j N_k. \tag{7.2.49}$$

To show that $\mathbf{L} \cdot \mathbf{N}$ and $\mathbf{L} \times \mathbf{N}$ do not depend on the choice of orthonormal basis, let $\mathbf{x}'_1, \mathbf{x}'_2, \mathbf{x}'_3$ be any other orthonormal basis and let $L'_j = \mathbf{x}'_j \cdot \mathbf{L}$, $N'_j = \hat{\mathbf{x}}'_j \cdot \mathbf{N}$. Then, on account of (7.2.23), we have

$$L'_j N'_j = [(\hat{\mathbf{x}}'_j \cdot \hat{\mathbf{x}}_i) L_i][(\hat{\mathbf{x}}'_j \cdot \hat{\mathbf{x}}_k) N_k].$$

Using the arithmetic rules in subsection 7.2.3, we find

$$L'_j N'_j = [(\hat{\mathbf{x}}'_j \cdot \hat{\mathbf{x}}_i)(\hat{\mathbf{x}}'_j \cdot \hat{\mathbf{x}}_k)] L_i N_k.$$

But $(\hat{\mathbf{x}}'_j \cdot \hat{\mathbf{x}}_i)(\hat{\mathbf{x}}'_j \cdot \hat{\mathbf{x}}_k) = [(\hat{\mathbf{x}}_i \cdot \hat{\mathbf{x}}'_j)\hat{\mathbf{x}}'_j] \cdot \hat{\mathbf{x}}_k = \hat{\mathbf{x}}_i \cdot \hat{\mathbf{x}}_k = \delta_{ik}$, so

$$L'_j N'_j = \delta_{ik} L_i N_k = L_i N_i.$$

Thus $\mathbf{L} \cdot \mathbf{N}$ is independent of the orthonormal basis used to calculate it. For equations (7.2.23) and (7.2.49) imply

$$L'_\ell N'_m = [(\hat{\mathbf{x}}'_\ell \cdot \hat{\mathbf{x}}_j) L_j][(\hat{\mathbf{x}}'_m \cdot \hat{\mathbf{x}}_k) N_k]$$

and the arithmetic rules of subsection 7.2.3 give

$$L'_\ell N'_m = (\hat{\mathbf{x}}'_\ell \cdot \hat{\mathbf{x}}_j)(\hat{\mathbf{x}}'_m \cdot \hat{\mathbf{x}}_k) L_j N_k.$$

Then

$$(\hat{\mathbf{x}}'_\ell \times \hat{\mathbf{x}}'_m)^M L'_\ell N'_m f = (\hat{\mathbf{x}}'_\ell \times \hat{\mathbf{x}}'_m)(\hat{\mathbf{x}}'_\ell \cdot \hat{\mathbf{x}}_j)(\hat{\mathbf{x}}'_m \cdot \hat{\mathbf{x}}_k) L_j N_k f$$

for any scalar f. But from the arithmetic of the ordinary vector cross product,

$$(\hat{\mathbf{x}}'_\ell \times \hat{\mathbf{x}}'_m)(\hat{\mathbf{x}}'_\ell \cdot \hat{\mathbf{x}}_j)(\hat{\mathbf{x}}'_m \cdot \hat{\mathbf{x}}_k) = [\hat{\mathbf{x}}'_\ell(\hat{\mathbf{x}}'_\ell \cdot \hat{\mathbf{x}}_j)] \times [\hat{\mathbf{x}}'_m(\hat{\mathbf{x}}'_m \cdot \hat{\mathbf{x}}_k)] = \hat{\mathbf{x}}_j \times \hat{\mathbf{x}}_k,$$

so

$$(\hat{\mathbf{x}}'_\ell \times \hat{\mathbf{x}}'_m)^M L'_\ell N'_m f = (\hat{\mathbf{x}}_j \times \hat{\mathbf{x}}_k) L_j N_k f = (\hat{\mathbf{x}}_j \times \hat{\mathbf{x}}_k)^M L_j N_k f.$$

Since this is true for every possible scalar f,

$$(\hat{\mathbf{x}}'_\ell \times \hat{\mathbf{x}}'_m)^M L'_\ell N'_m = (\hat{\mathbf{x}}_j \times \hat{\mathbf{x}}_k)^M L_j N_k.$$

From (7.2.34), a definition of $\mathbf{L} \times \mathbf{N}$ equivalent to (7.2.49) is

$$(\mathbf{L} \times \mathbf{N})_i = [\hat{\mathbf{x}}_i \cdot (\hat{\mathbf{x}}_j \times \hat{\mathbf{x}}_k)] L_j N_k. \tag{7.2.50}$$

If $\hat{\mathbf{x}}_1, \hat{\mathbf{x}}_2, \hat{\mathbf{x}}_3$ is a right-handed orthonormal basis, this becomes

$$(\mathbf{L} \times \mathbf{N})_i = \epsilon_{ijk} L_j N_k. \tag{7.2.51}$$

The arithmetic rules for the dot and cross product of vector linear operators are like those for ordinary vectors, except that there are some complications because products of scalar component operators do not always commute $(N_i P_j \neq P_j N_i)$. In what follows, K and L are scalar linear operators, $\mathbf{N}, \mathbf{P}, \mathbf{Q}$ are vector linear operators, $\mathbf{b}$ is a constant vector, and $*$ stands for either $\cdot$ or $\times$, since most of the rules hold for both the dot and the cross product.

$$\mathbf{L} \cdot \mathbf{N} = \mathbf{N} \cdot \mathbf{L} \qquad \text{if } \mathbf{L} = \mathbf{b}^M \tag{7.2.52}$$

$$\mathbf{L} \times \mathbf{N} = -\mathbf{N} \times \mathbf{L} \qquad \text{if } \mathbf{L} = \mathbf{b}^M \tag{7.2.53}$$

$$\mathbf{N} * (\mathbf{P} + \mathbf{Q}) = \mathbf{N} * \mathbf{P} + \mathbf{N} * \mathbf{Q} \tag{7.2.54}$$

$$(\mathbf{P} + \mathbf{Q}) * \mathbf{N} = \mathbf{P} * \mathbf{N} + \mathbf{Q} * \mathbf{N} \tag{7.2.55}$$

$$K(\mathbf{N} * \mathbf{P}) = (K\mathbf{N}) * \mathbf{P} \tag{7.2.56}$$

$$\mathbf{N} * (K\mathbf{P}) = (\mathbf{N}K) * \mathbf{P} \tag{7.2.57}$$

$$\mathbf{N} * (\mathbf{P}K) = (\mathbf{N} * \mathbf{P})K \tag{7.2.58}$$

$$\mathbf{N} \cdot (\mathbf{P} \times \mathbf{Q}) = (\mathbf{N} \times \mathbf{P}) \cdot \mathbf{Q} \tag{7.2.59}$$

$$\mathbf{N} \times (\mathbf{P} \times \mathbf{Q}) = N_j \mathbf{P} Q_j - (\mathbf{N} \cdot \mathbf{P})\mathbf{Q} \tag{7.2.60}$$

$$(\mathbf{N} \times \mathbf{P}) \times \mathbf{Q} = N_j \mathbf{P} Q_j - \mathbf{N}(\mathbf{P} \cdot \mathbf{Q}). \tag{7.2.61}$$

In (7.2.60) and (7.2.52), N_j and Q_j are the component operators of $\mathbf{N}$ and $\mathbf{Q}$ relative to any single orthonormal basis for R^3. Parentheses here are unnecessary because, according to (7.2.31), $(N_j\mathbf{P})Q_j = N_j(\mathbf{P}Q_j)$.

The proofs are most conveniently carried out by using components relative to some fixed right-handed orthonormal basis for R^3. The only tools needed are the rules (7.2.29), (7.2.30), (7.2.33), (7.2.34), (7.2.48), (7.2.51) for finding the components of vector operators, and the arithmetic rules (7.2.11–7.2.16) for scalar linear operators. As an example, we prove (7.2.61). We have

$$[\mathbf{N} \times (\mathbf{P} \times \mathbf{Q})]_i = \epsilon_{ijk} N_j (\mathbf{P} \times \mathbf{Q})_k = \epsilon_{ijk} N_j (\epsilon_{k\ell m} P_\ell Q_m)$$

$$= \epsilon_{ijk} \epsilon_{k\ell m} N_j (P_\ell Q_m) = \epsilon_{kij} \epsilon_{k\ell m} N_j P_\ell Q_m$$

$$= (\delta_{i\ell} \delta_{jm} - \delta_{im} \delta_{j\ell}) N_j P_\ell Q_m$$

$$= N_j P_i Q_j - N_j P_j Q_i = N_j (\mathbf{P} Q_j)_i - (\mathbf{N} \cdot \mathbf{P}) Q_i$$

$$= [N_j \mathbf{P} Q_j]_i - [(\mathbf{N} \cdot \mathbf{P})\mathbf{Q}]_i = [N_j \mathbf{P} Q_j - (\mathbf{N} \cdot \mathbf{P})\mathbf{Q}]_i.$$

Thus the vector linear operators constituting the left and right sides of (7.2.60) have the same components relative to one orthonormal basis. Therefore they are equal.

The proofs of (7.2.52–7.2.61) are like the proofs for ordinary vectors and scalars, except that for the latter we have one additional rule of scalar arithmetic, $KL = LK$, which is not available for scalar operators. Therefore in (7.2.54–7.2.61) the left–right order of the individual operator in multiplications must be the same throughout the whole equation. We cannot, for example, assume $\mathbf{P}Q_j = Q_j\mathbf{P}$, and simplify (7.2.60) by writing $N_j\mathbf{P}Q_j = N_jQ_j\mathbf{P} = (\mathbf{N} \cdot \mathbf{Q})\mathbf{P}$; that calculation is permitted for vectors and vector fields, but not for vector linear operators.

7.2.9 FODOs

The left–right order preservation leads to an interesting observation. We can insert a fence anywhere in the ordered lineup of the symbols; for example, in (7.2.59) we can have $|\mathbf{N}, \mathbf{P}, \mathbf{Q}$, or $\mathbf{N}, |\mathbf{P}, \mathbf{Q}$, or $\mathbf{N}, \mathbf{P}, |\mathbf{Q}$ or $\mathbf{N}, \mathbf{P}, \mathbf{Q}|$. If we interpret all symbols left of the fence as linear operators and all symbols right of the fence as fields, the equations make sense; all their terms are well defined. Are the equations still true? The proofs require the rules (7.2.29), (7.2.30), (7.2.33), (7.2.34), (7.2.48), (7.2.51) for finding components of vectors. These work wherever we put the fence. The proofs also require the arithmetic rules (7.2.11–7.2.16) for scalars. These work for all positions of the fence in (7.2.54) and (7.2.55). Rule (7.2.15), $(LM)N = L(MN)$, works for the fence positions $|L, M, N$, and $L, M, |N$, and $L, M, N|$, but *not* for $L|, M, N$. It is not true, for example, that if f and g are scalar fields, $\partial_i(fg) = (\partial_i f)g$. Therefore (7.2.54–7.2.61) are true if all symbols are operators, if all symbols are fields, or if the *last* symbol on the right is a field and the preceding symbols are operators. Equations (7.2.56) through (7.2.61) fail if the last two symbols on the right are fields and the first symbol is an operator.

This gap can be filled for one kind of operator, a *first-order differential operator*, or FODO, although the rules (7.2.54–7.2.61) need to be modified appropriately.

A scalar or vector linear operator, D or $\mathbf{D}$, on an open subset W of R^3, is a FODO if it has these three properties (they are the same for D and $\mathbf{D}$, so we give them only for $\mathbf{D}$):

i) $D_{\mathbf{D}}$, the domain of $\mathbf{D}$, consists of all continuously differentiable scalar fields on W.

ii) for any fixed $\mathbf{r}_0$ in W, if f is any scalar field in $D_{\mathbf{D}}$ whose gradient vanishes at $\mathbf{r}_0$, then the field $\mathbf{D}f$ also vanishes at $\mathbf{r}_0$.

iii) for any scalar fields in $D_{\mathbf{D}}$, $\mathbf{D}(fg) = (\mathbf{D}f)g + f(\mathbf{D}g)$.

Example 1: Let u_1, u_2, u_3 be a smooth coordinate system on W, either curvilinear or not. Abbreviate $\partial/\partial u_i$ as ∂_i. Then ∂_i is a scalar FODO.   ◁

Example 2: The gradient operator ∇ is a vector FODO.   ◁

Example 3: Let $\mathbf{v}$ be a vector field on W. Then $\mathbf{v}^M \cdot \nabla$ is a scalar FODO.   ◁

Example 4: Any linear combination of FODOs is a FODO.   ◁

Example 5: If f and $\mathbf{v}$ are a scalar field and a vector field, and D and $\mathbf{D}$ are a scalar and a vector FODO, then $f^M D, f^M \mathbf{D}, v^M D,$ $\mathbf{v}^M \cdot \mathbf{D}$, and $\mathbf{v}^M \times \mathbf{D}$ are all FODOs. ◁

These examples include all FODOs, as we can see from the following theorem and its corollary.

Theorem: If D is a scalar FODO, there is a unique vector field $\mathbf{v}$ such that $D = \mathbf{v}^M \cdot \boldsymbol{\nabla}$.

Proof: To demonstrate uniqueness, let $\hat{\mathbf{x}}_1, \hat{\mathbf{x}}_2, \hat{\mathbf{x}}_3$ be an orthonormal basis for R^3 and write position vectors as $\mathbf{r} = \hat{\mathbf{x}}_i r_i$. We assume $\mathbf{v}^M \cdot \boldsymbol{\nabla} = \mathbf{w}^M \cdot \boldsymbol{\nabla}$, and want to prove $\mathbf{v} = \mathbf{w}$. For any scalar field f, $(\mathbf{v}^M \cdot \boldsymbol{\nabla})f = (\mathbf{w}^M \cdot \boldsymbol{\nabla})f$, so $\mathbf{v} \cdot (\boldsymbol{\nabla} f) = \mathbf{w} \cdot (\boldsymbol{\nabla} f)$. Let $f = r_i$. Then $\boldsymbol{\nabla} f = \hat{\mathbf{x}}_i$, so $\mathbf{v} \cdot \hat{\mathbf{x}}_i = \mathbf{w} \cdot \hat{\mathbf{x}}_i$. Hence $\mathbf{v} = \mathbf{w}$.

To show that the vector field must exist, we proceed as follows: We are given a scalar FODO D, and we want to construct a vector field $\mathbf{v}$ such that $D = \mathbf{v}^M \cdot \boldsymbol{\nabla}$. We take a hint from the uniqueness proof. For any scalar field f, we know the scalar field Df. Therefore we know the three scalar fields Dr_i. We define $\mathbf{v} = \hat{\mathbf{x}}_i(Dr_i)$. Is it true that $Df = (\mathbf{v}^M \cdot \boldsymbol{\nabla})f$ for all f? It suffices to prove that at each point $\mathbf{r}_0$, the field Df and the field $(\mathbf{v}^M \cdot \boldsymbol{\nabla})f$ have the same numerical value, whatever scalar field f we use. To see this, choose a particular $\mathbf{r}_0$. Define the scalar field g by $g = f - r_i\, \partial_i f(\mathbf{r}_0)$. Then by construction, $\boldsymbol{\nabla} g$ vanishes at $\mathbf{r}_0$; hence by (ii) in the definition of a FODO, Dg vanishes there. But $Dg = Df - \partial_i f(\mathbf{r}_0)Dr_i$, since D is linear and $\partial_i f(\mathbf{r}_0)$ is a constant scalar. Therefore,

$$Dg = Df - v_i(\partial_i f(\mathbf{r}_0)) = Df - \mathbf{v} \cdot [\boldsymbol{\nabla} f(\mathbf{r}_0)].$$

Since $Dg = 0$ at $\mathbf{r} = \mathbf{r}_0$,

$$Df = \mathbf{v} \cdot [\boldsymbol{\nabla} f(\mathbf{r}_0)]$$

at $\mathbf{r} = \mathbf{r}_0$. ◁

Corollary 1: Let $\mathbf{D}$ be any vector FODO, and let $\hat{\mathbf{x}}_1, \hat{\mathbf{x}}_2, \hat{\mathbf{x}}_3$ be any orthonormal basis for R^3. Then there are unique *vector* fields $\mathbf{v}_1, \mathbf{v}_2, \mathbf{v}_3$ such that $\mathbf{D} = \hat{\mathbf{x}}_i(\mathbf{v}_i^M \cdot \boldsymbol{\nabla})$.

Proof: The uniqueness follows at once from the relations: $D_j = \hat{\mathbf{x}}_j \cdot \mathbf{D} = (\hat{\mathbf{x}}_j \cdot \hat{\mathbf{x}}_i)(\mathbf{v}_i^M \cdot \boldsymbol{\nabla}) = \delta_{ij}(\mathbf{v}_i^M \cdot \boldsymbol{\nabla}) = \mathbf{v}_j^M \cdot \boldsymbol{\nabla}$. Since D_j is uniquely determined by $\mathbf{D}$ and $\hat{\mathbf{x}}_1, \hat{\mathbf{x}}_2, \hat{\mathbf{x}}_3$, so is $\mathbf{v}_j$.

To demonstrate existence, we note that D_j is a scalar FODO, so there is a vector field $\mathbf{v}_j$ such that $D_j = \mathbf{v}_j^M \cdot \nabla$. But $\mathbf{D} = \hat{\mathbf{x}}_j D_j$.◁

Corollary 2: The chain rule. Suppose f is a function of a single real variable and g is a scalar field. Define the scalar field $f{\circ}g$ by requiring

$$(f{\circ}g)(\mathbf{r}) = f[g(\mathbf{r})].$$

Suppose D is a scalar or vector FODO. Then

$$D(f{\circ}g) = \frac{df}{dg} Dg.$$

Proof: This holds if $D = \partial/\partial r_i$, from the chain rule. Therefore it holds for $D = \nabla$. Therefore it holds for any $D = \mathbf{v} \cdot \nabla$. Therefore, by Corollary 1, it holds for any vector FODO as well.   ◁

7.2.10 Arithmetic with FODOs

Now we return to equations (7.2.56–7.2.61), and place the fence so there are two fields to its right and a linear operator to its left; however, now we assume that this linear operator is a FODO. As an example, consider again (7.2.60), but now with $\mathbf{N} = \mathbf{D}$, a FODO, and $\mathbf{P}$ and $\mathbf{Q}$ being vector fields. We have

$$\begin{aligned}
[\mathbf{D} \times (\mathbf{P} \times \mathbf{Q})]_i &= \epsilon_{ijk} D_j (\mathbf{P} \times \mathbf{Q})_k \\
&= \epsilon_{ijk} D_j [\epsilon_{k\ell m} P_\ell Q_m] \\
&= \epsilon_{kij}\epsilon_{k\ell m} D_j (P_\ell Q_m) \\
&= (\delta_{i\ell}\delta_{jm} - \delta_{im}\delta_{j\ell}) D_j (P_\ell Q_m) \\
&= D_j(P_i Q_j) - D_j(P_j Q_i) \\
&= (D_j P_i)Q_j + P_i(D_j Q_j) - (D_j P_j)Q_i - P_j(D_j Q_i) \\
&= Q_j(D_j P_i) + P_i(D_j Q_j) - Q_i(D_j P_j) - P_j(D_j Q_i).
\end{aligned}$$

Here we have used that $D_j P_i$ and Q_j are scalar fields, so $(D_j P_i)Q_j =$

$Q_j(D_j P_i)$. Now we have

$$[\mathbf{D} \times (\mathbf{P} \times \mathbf{Q})]_i$$
$$= (Q_j^M D_j)P_i + P_i(\mathbf{D} \cdot \mathbf{Q}) - Q_i(\mathbf{D} \cdot \mathbf{P}) - (P_j^M D_j)Q_i$$
$$= (\mathbf{Q}^M \cdot \mathbf{D})P_i + (\mathbf{D} \cdot \mathbf{Q})P_i - (\mathbf{P}^M \cdot \mathbf{D})Q_i - (\mathbf{D} \cdot \mathbf{P})Q_i$$
$$= [\mathbf{Q}^M \cdot \mathbf{D} + (\mathbf{D} \cdot \mathbf{Q})^M]P_i - [\mathbf{P}^M \cdot \mathbf{D} + (\mathbf{D} \cdot \mathbf{P})^M]Q_i$$
$$= \left\{ [\mathbf{Q}^M \cdot \mathbf{D} + (\mathbf{D} \cdot \mathbf{Q})^M]\mathbf{P} - [\mathbf{P}^M \cdot \mathbf{D} + (\mathbf{D} \cdot \mathbf{P})^M]\mathbf{Q} \right\}_i.$$

Therefore (7.2.60) is replaced by

$$\mathbf{D} \times (\mathbf{P} \times \mathbf{Q}) = [\mathbf{Q}^M \cdot \mathbf{D} + (\mathbf{D} \cdot \mathbf{Q})^M]\mathbf{P} - [\mathbf{P}^M \cdot \mathbf{D} + (\mathbf{D} \cdot \mathbf{P})^M]\mathbf{Q}. \quad (7.2.62)$$

The appearance of this equation can be simplified with a useful and accepted convention. In the term $\mathbf{Q}^M \cdot \mathbf{D}$, the vector field $\mathbf{Q}$ is obviously acting as a multiplication operator, so we can drop the M without confusion. And $(\mathbf{D} \cdot \mathbf{Q})^M \mathbf{P} = (\mathbf{D} \cdot \mathbf{Q})\mathbf{P}$, since $\mathbf{P}$ is a vector field. Therefore we can omit all ^{M} superscripts in (7.2.62).

The complete list of rules for differentiating products, the rules that replace (7.2.56–7.2.61), can be obtained in the same way. In these rules, given below, D and $\mathbf{D}$ are a scalar and vector FODO, f and g are scalar fields, and $\mathbf{u}$ and $\mathbf{v}$ are vector fields. Throughout any equation, $*$ can be either $\cdot$ or $\times$. We do not repeat the defining rules $D(fg) = (Df)g + f(Dg)$ and $\mathbf{D}(fg) = (\mathbf{D}f)g + f(\mathbf{D}g)$.
Scalar FODO D:

$$D(f\mathbf{u}) = (Df)\mathbf{u} + f(D\mathbf{u}) \quad (7.2.63)$$
$$D(\mathbf{u} * \mathbf{v}) = (D\mathbf{u}) * \mathbf{v} + \mathbf{u} * (D\mathbf{v}). \quad (7.2.64)$$

Vector FODO D:

$$\mathbf{D} * (f\mathbf{u}) = (\mathbf{D}f) * \mathbf{u} + f(\mathbf{D} * \mathbf{u}) \quad (7.2.65)$$
$$\mathbf{D}(\mathbf{u} \cdot \mathbf{v}) = (\mathbf{D}u_i)v_i + (\mathbf{u} \cdot \mathbf{D})\mathbf{v} \quad (7.2.66)$$
$$\mathbf{D} \cdot (\mathbf{u} \times \mathbf{v}) = \mathbf{v} \cdot (\mathbf{D} \times \mathbf{u}) - \mathbf{u} \cdot (\mathbf{D} \times \mathbf{v}) \quad (7.2.67)$$
$$\mathbf{D} \times (\mathbf{u} \times \mathbf{v}) = (\mathbf{D} \cdot \mathbf{v} + \mathbf{v} \cdot \mathbf{D})\mathbf{u} - (\mathbf{D} \cdot \mathbf{u} + \mathbf{u} \cdot \mathbf{D})\mathbf{v}. \quad (7.2.68)$$

We have just introduced the convention that we omit the ^{M} on a multiplication operator if the context makes clear that it is an operator, not a field. Thus $\mathbf{v}^M \cdot \boldsymbol{\nabla}$ will be written $\mathbf{v} \cdot \boldsymbol{\nabla}$. There is no danger of error, because vector and scalar multiplication operators behave among themselves like vector and scalar fields. If f and g are scalar fields and $\mathbf{u}$ and $\mathbf{v}$ are vector fields, we have

$$
\begin{aligned}
f^M + g^M &= (f+g)^M & \mathbf{u}^M + \mathbf{v}^M &= (\mathbf{u}+\mathbf{v})^M \\
f^M g^M &= (fg)^M & f^M \mathbf{v}^M &= (f\mathbf{v})^M \\
\mathbf{u}^M \cdot \mathbf{v}^M &= (\mathbf{u}\cdot\mathbf{v})^M & \mathbf{u}^M \times \mathbf{v}^M &= (\mathbf{u}\times\mathbf{v})^M .
\end{aligned}
$$

The first four rules are trivial consequences of definitions, and the last two are easily proved using components relative to an orthonormal basis.

7.2.11 Commutation

In our list (7.2.52–7.2.61), the two rules (7.2.52) and (7.2.53) were given only for $\mathbf{L}$ a multiplication operator by a constant-vector field. It is of some interest to know the other situations in which these rules are valid. One common situation in which they hold is that $\mathbf{L}$ and $\mathbf{N}$ commute. Vector linear operators $\mathbf{L}$ and $\mathbf{N}$ are said to *commute* relative to orthonormal basis $\hat{\mathbf{x}}_1, \hat{\mathbf{x}}_2, \hat{\mathbf{x}}_3$, if all their scalar components relative to that basis commute, i.e., $L_i N_j = N_j L_i$. From (7.2.23) and the linearity of $\mathbf{L}$ and $\mathbf{N}$, it is clear that if they commute relative to one orthonormal basis, they commute relative to all such bases. Therefore we will say simply that they commute. (Tensor enthusiasts should note that $\mathbf{L}$ and $\mathbf{N}$ commute not when $\mathbf{LN} = \mathbf{NL}$, but when $\mathbf{LN} = (\mathbf{NL})^T$, where T is the transpose operation on second-order tensors.)

A scalar linear operator K always commutes with itself, since $KK = KK$. A vector linear operator $\mathbf{L}$ may not commute with itself. It may not be true that $L_i L_j = L_j L_i$. Any two multiplication operators commute, since their components are ordinary scalar fields.

Commutation simplifies two other rules in subsection 7.2.8 as well as generalizing (7.2.52) and (7.2.53). If $\mathbf{N}$ and $\mathbf{P}$ commute, then $N_j \mathbf{P} = \mathbf{P} N_j$, and (7.2.60) and (7.2.61) simplify to

$$\mathbf{N} \times (\mathbf{P} \times \mathbf{Q}) = \mathbf{P}(\mathbf{N} \cdot \mathbf{Q}) - (\mathbf{N} \cdot \mathbf{P})\mathbf{Q} \qquad (7.2.69)$$

$$(\mathbf{N} \times \mathbf{P}) \times \mathbf{Q} = \mathbf{P}(\mathbf{N} \cdot \mathbf{Q}) - \mathbf{N}(\mathbf{P} \cdot \mathbf{Q}). \qquad (7.2.70)$$

If $\mathbf{N} = \mathbf{P} = \nabla$, these equations work. The relationship (7.2.70) is uninteresting, on account of (7.2.81). However, (7.2.69) becomes the well-known

$$\nabla \times (\nabla \times \mathbf{Q}) = \nabla(\nabla \cdot \mathbf{Q}) - \nabla^2 \mathbf{Q}. \qquad (7.2.71)$$

Moreover, if $\mathbf{N} = \mathbf{P} = \nabla$, (7.2.59) and (7.2.81) give

$$\nabla \cdot (\nabla \times \mathbf{Q}) = 0. \qquad (7.2.72)$$

Equations (7.2.71) and (7.2.72) hold whether $\mathbf{Q}$ is a vector linear operator or a vector field.

Example: $\mathbf{L} = \nabla$, $\mathbf{N} = \mathbf{r}^M$. Then $L_i = \partial_i$ so $L_i L_j = \partial_i \partial_j = \partial_j \partial_i = L_j L_i$. Thus $\mathbf{L}$ commutes with itself. So does $\mathbf{N}$, since $\mathbf{N}$ is a multiplication operator. However, $\mathbf{L}$ and $\mathbf{N}$ do not commute. We have

$$L_i N_j - N_j L_i = \partial_i r_j^M - r_j^M \partial_i = \delta_{ij}^M$$

because

$$\left(\partial_i r_j^M \right) f = \partial_i (r_j f) = (\partial_i r_j) f + r_j \partial_i f = \delta_{ij} f + \left(r_j^M \partial_i \right) f$$
$$= \left(\delta_{ij}^M + r_j^M \partial_i \right) f.$$

Thus

$$\partial_i r_j^M - r_j^M \partial_i = \delta_{ij}^M. \qquad (7.2.73)$$

7.2.12 An Important FODO and Its Commutation Properties

We define

$$\Lambda = \mathbf{r}^M \times \nabla. \qquad (7.2.74)$$

Equation (7.2.74) may also be written $\Lambda = \mathbf{r} \times \nabla$. Then in components,

$$\Lambda_i = \epsilon_{ijk} r_j \partial_k.$$

The following are easy to verify, in the same manner as (7.2.73):

$$\Lambda_i r_\ell^M = r_\ell^M \Lambda_i + \epsilon_{ij\ell} r_j^M = r_\ell^M \Lambda_i - \epsilon_{i\ell j} r_j^M \qquad (7.2.75)$$

$$\Lambda_i \partial_\ell = \partial_\ell \Lambda_i - \epsilon_{i\ell k} \partial_k. \qquad (7.2.76)$$

Moreover,

$$\Lambda_i \Lambda_\ell = \epsilon_{ijk} r_j \partial_k (\epsilon_{\ell mn} r_m \partial_n) = \epsilon_{ijk} \epsilon_{\ell mn} r_j \partial_k (r_m \partial_n)$$
$$= \epsilon_{ijk} \epsilon_{\ell mn} r_j (r_m \partial_k + \delta_{km}) \partial_n$$
$$= \epsilon_{ijk} \epsilon_{\ell mn} r_j r_m \partial_k \partial_n + \epsilon_{ijk} \epsilon_{\ell kn} r_j \partial_n.$$

But

$$\epsilon_{ijk} \epsilon_{\ell kn} = \epsilon_{kij} \epsilon_{kn\ell} = \delta_{in} \delta_{j\ell} - \delta_{i\ell} \delta_{jn},$$

so

$$\Lambda_i \Lambda_\ell = [\epsilon_{ijk} \epsilon_{\ell mn} r_j r_m \partial_k \partial_n - \delta_{i\ell} r_j \partial_j] + r_\ell \partial_i. \qquad (7.2.77)$$

Since the term in square brackets is symmetric in i and l,

$$\Lambda_i \Lambda_\ell - \Lambda_\ell \Lambda_i = r_l \partial_i - r_i \partial_\ell.$$

Now

$$\epsilon_{ijk} \Lambda_k = \epsilon_{ijk} \epsilon_{k\ell m} r_\ell \partial_m = (\delta_{i\ell} \delta_{jm} - \delta_{im} \delta_{j\ell}) r_l \partial_m = r_i \partial_j - r_j \partial_i$$

so

$$\Lambda_i \Lambda_\ell - \Lambda_\ell \Lambda_i = -\epsilon_{i\ell k} \Lambda_k.$$

We will later want all three commutation relations for Λ, so we collect them here:

$$\Lambda_i r_j^M - r_j^M \Lambda_i = -\epsilon_{ijk} r_k^M \qquad (7.2.78)$$

$$\Lambda_i \partial_j - \partial_j \Lambda_i = -\epsilon_{ijk} \partial_k \qquad (7.2.79)$$

$$\Lambda_i \Lambda_j - \Lambda_j \Lambda_i = -\epsilon_{ijk} \Lambda_k. \qquad (7.2.80)$$

It is clear from (7.2.78, 7.2.79, 7.2.80) that Λ does not commute with $\mathbf{r}^M$ or with ∇ or with itself.

Now if $\mathbf{L}$ and $\mathbf{N}$ commute with each other, it is obvious from (7.2.48) and (7.2.51) that $\mathbf{L} \cdot \mathbf{N} = \mathbf{N} \cdot \mathbf{L}$ and $\mathbf{L} \times \mathbf{N} = -\mathbf{N} \times \mathbf{L}$, as with ordinary vectors. In particular, if $\mathbf{L}$ commutes with itself, $\mathbf{L} \times \mathbf{L} = -\mathbf{L} \times \mathbf{L} = \mathbf{0}$. If not, $\mathbf{L} \times \mathbf{L}$ may not be zero. Thus we have

$$\nabla \times \nabla = \mathbf{0}. \qquad (7.2.81)$$

However, from (7.2.80),

$$2(\mathbf{\Lambda} \times \mathbf{\Lambda})_i = 2\epsilon_{ijk}\Lambda_j\Lambda_k = \epsilon_{ijk}\Lambda_j\Lambda_k + \epsilon_{ikj}\Lambda_k\Lambda_j$$
$$= \epsilon_{ijk}(\Lambda_j\Lambda_k - \Lambda_k\Lambda_j) = -\epsilon_{ijk}\epsilon_{jk\ell}\Lambda_\ell = -2\delta_{i\ell}\Lambda_\ell = -2\Lambda_i,$$

so

$$\mathbf{\Lambda} \times \mathbf{\Lambda} = -\mathbf{\Lambda}. \tag{7.2.82}$$

Both $\boldsymbol{\nabla}$ and $\mathbf{\Lambda}$ are vector FODOs, so a vector FODO may or may not satisfy $\mathbf{D} \times \mathbf{D} = \mathbf{0}$.

In order to have $\mathbf{L} \cdot \mathbf{N} = \mathbf{N} \cdot \mathbf{L}$, it is not *necessary* that $\mathbf{L}$ and $\mathbf{N}$ commute; only that, relative to some orthonormal basis, $L_i N_i = N_i L_i$. (If this holds relative to one orthonormal basis, then $\mathbf{L} \cdot \mathbf{N} = \mathbf{N} \cdot \mathbf{L}$, so $L'_j N'_j = N'_j L'_j$ relative to any other orthonormal basis.) For example, from (7.2.78) and (7.2.79) we have $\mathbf{\Lambda} \cdot \mathbf{r}^M = \mathbf{r}^M \cdot \mathbf{\Lambda}$ and $\mathbf{\Lambda} \cdot \boldsymbol{\nabla} = \boldsymbol{\nabla} \cdot \mathbf{\Lambda}$. These particular examples are somewhat special because we actually have

$$\mathbf{\Lambda} \cdot \mathbf{r}^M = \mathbf{r}^M \cdot \mathbf{\Lambda} = 0 \tag{7.2.83}$$

$$\mathbf{\Lambda} \cdot \boldsymbol{\nabla} = \boldsymbol{\nabla} \cdot \mathbf{\Lambda} = 0. \tag{7.2.84}$$

To prove (7.2.83), note that $\mathbf{r}^M \cdot \mathbf{\Lambda} = 0$ from (7.2.59) and the definition (7.2.74). To prove (7.2.84) note that $\boldsymbol{\nabla} \cdot \mathbf{\Lambda} = \partial_i \Lambda_i = \partial_i \epsilon_{ijk} r_j \partial_k = \epsilon_{ijk}[\delta_{ij} + r_j \partial_i]\partial_k = (\epsilon_{ijk}\delta_{ij})\partial_k + r_j\epsilon_{ijk}\partial_i\partial_k$. Since δ_{ik} and $\partial_i\partial_k$ are both symmetric in i and k, $\partial_i\Lambda_i = 0$.

In order to have $\mathbf{L} \times \mathbf{N} = -\mathbf{N} \times \mathbf{L}$, it is not necessary that $\mathbf{L}$ and $\mathbf{N}$ commute. All that is required is that relative to one orthonormal basis (and hence all orthonormal bases) $\epsilon_{ijk}L_j N_k = -\epsilon_{ijk}N_j L_k$. But

$$-\epsilon_{ijk}N_j L_k = -\epsilon_{ikj}N_k L_j = \epsilon_{ijk}N_k L_j,$$

so $\mathbf{L} \times \mathbf{N} = -\mathbf{N} \times \mathbf{L}$ if and only if $\epsilon_{ijk}(L_j N_k - N_k L_j) = 0$, that is, if and only if $L_j N_k - N_k L_j$ is symmetric in j and k. From this and (7.2.73) we have

$$\mathbf{r}^M \times \boldsymbol{\nabla} = -\boldsymbol{\nabla} \times \mathbf{r}^M. \tag{7.2.85}$$

That is, $\mathbf{\Lambda} = -\boldsymbol{\nabla} \times \mathbf{r}^M$ as well as $\mathbf{r}^M \times \boldsymbol{\nabla}$.

Although $\mathbf{L} \times \mathbf{L}$ may not be $-\mathbf{L} \times \mathbf{L}$, $\mathbf{L} \cdot \mathbf{L}$ is always $\mathbf{L} \cdot \mathbf{L}$ (interchanging the two $\mathbf{L}$'s as if they were vectors always works for the dot product, but not for the cross product when linear operators are involved). The scalar linear operator $\mathbf{L} \cdot \mathbf{L}$ is usually written L^2.

Thus $\boldsymbol{\nabla}\cdot\boldsymbol{\nabla} = \nabla^2$, $\boldsymbol{\Lambda}\cdot\boldsymbol{\Lambda} = \Lambda^2$. The operator ∇^2 is called the *Laplacian*. Relative to any orthonormal basis it is

$$\nabla^2 = \partial_i \partial_i. \qquad (7.2.86)$$

We will encounter Λ^2 again, so we calculate it also relative to an orthonormal basis. Setting $i = \ell$ and summing from 1 to 3 in (7.2.77) gives

$$
\begin{aligned}
\Lambda^2 &= (\delta_{jm}\delta_{kn} - \delta_{jn}\delta_{km})r_j r_m \partial_k \partial_n - 3r_j\partial_j + r_i\partial_i \\
&= r^2 \partial_k \partial_k - r_j r_k \partial_k \partial_j - 2r_j\partial_j \\
&= r^2\nabla^2 - r_k(r_j\partial_k)\partial_j - 2r_j\partial_j \\
&= r^2\nabla^2 - r_k(-\delta_{jk} + \partial_k r_j)\partial_j - 2r_j\partial_j.
\end{aligned}
$$

So

$$\Lambda^2 = r^2\nabla^2 - r_k\partial_k r_j\partial_j - r_j\partial_j. \qquad (7.2.87)$$

As it happens, (7.2.87) does not really need an orthonormal basis. We have $r_j\partial_j = \mathbf{r}\cdot\boldsymbol{\nabla}$, so

$$\Lambda^2 = r^2\nabla^2 - (\mathbf{r}\cdot\boldsymbol{\nabla})^2 - (\mathbf{r}\cdot\boldsymbol{\nabla}) = r^2\nabla^2 - (\mathbf{r}\cdot\boldsymbol{\nabla})(\mathbf{r}\cdot\boldsymbol{\nabla}+1^M). \qquad (7.2.88)$$

7.2.13 Curvilinear Coordinates and $\boldsymbol{\nabla}$

Suppose u_1, u_2, u_3 are curvilinear coordinates in an open subset W of R^3. Each coordinate triple (u_1, u_2, u_3) determines a position vector $\mathbf{r}(u_1, u_2, u_3)$, and each position $\mathbf{r}$ in W has unique coordinate values $u_1(\mathbf{r}), u_2(\mathbf{r}), u_3(\mathbf{r})$. The coordinates are scalar fields on W. Abbreviate $\partial/\partial u_i$ as ∂_i, so define

$$\partial_i = \partial/\partial u_i. \qquad (7.2.89)$$

We claim that

$$\partial_i = (\partial_i\mathbf{r})\cdot\boldsymbol{\nabla} \qquad (7.2.90)$$

and

$$\boldsymbol{\nabla} = (\boldsymbol{\nabla} u_i)\partial_i. \qquad (7.2.91)$$

To prove these results, choose a particular orthonormal basis $\hat{\mathbf{x}}_1, \hat{\mathbf{x}}_2, \hat{\mathbf{x}}_3$ and let $r_i = \hat{\mathbf{x}}_i \cdot \mathbf{r}$. For any scalar field f, the chain rule implies

$$\partial_i f = \partial f/\partial u_i = (\partial r_j/\partial u_i)(\partial f/\partial r_j)$$
$$= (\partial r_j/\partial u_i)\hat{\mathbf{x}}_j \cdot \nabla f = \partial(r_j\hat{\mathbf{x}}_j)/\partial u_i \cdot \nabla f$$
$$= (\partial_i \mathbf{r}) \cdot \nabla f = [(\partial_i \mathbf{r}) \cdot \nabla]f.$$

This gives (7.2.90). The chain rule also implies

$$\partial f/\partial r_j = (\partial u_i/\partial r_j)(\partial f/\partial u_i),$$

so

$$\nabla f = \hat{\mathbf{x}}_j(\partial f/\partial r_j) = [\hat{\mathbf{x}}_j(\partial u_i/\partial r_j)]\partial_i f = (\nabla u_i)(\partial_i f) = [(\nabla u_i)\partial_i]f.$$

Hence (7.2.91).

Note that (7.2.90) is an example of the theorem of subsection 7.2.9. The scalar linear operator ∂_i is a FODO, and (7.2.90) exhibits it explicitly in the form $\mathbf{v} \cdot \nabla$.

Usually $\partial_i \mathbf{r}$ is easier to compute than ∇u_i, and in (7.2.91) one usually wants ∇u_i expressed as a function of u_1, u_2, u_3 rather than $\mathbf{r}$. Therefore it is useful to know the algebra for computing ∇u_i directly from $\partial_i \mathbf{r}$. The computation is based on the following identity:

$$\partial_i \mathbf{r} \cdot \nabla u_j = \delta_{ij}. \tag{7.2.92}$$

To obtain this identity, apply (7.2.90) to the scalar field u_j. To use (7.2.92) we exploit the notion of dual basis. Suppose $(\mathbf{b}_1, \mathbf{b}_2, \mathbf{b}_3)$ is any triple of vectors in R^3. Another triple of vectors, $(\boldsymbol{\beta}_1, \boldsymbol{\beta}_2, \boldsymbol{\beta}_3)$, is said to be *dual* to $(\mathbf{b}_1, \mathbf{b}_2, \mathbf{b}_3)$ if

$$\mathbf{b}_i \cdot \boldsymbol{\beta}_j = \delta_{ij}. \tag{7.2.93}$$

Any two vector triples that satisfy (7.2.93) are each linearly independent and hence each is a basis for R^3. To see this, suppose there are scalars a_1, a_2, a_3 such that $a_j\boldsymbol{\beta}_j = \mathbf{0}$, Then

$$0 = \mathbf{b}_i \cdot (a_j\boldsymbol{\beta}_j) = a_j(\mathbf{b}_i \cdot \boldsymbol{\beta}_j) = a_j\delta_{ij} = a_i.$$

Any basis $(\mathbf{b}_1, \mathbf{b}_2, \mathbf{b}_3)$ for R^3 is dual to exactly one triple of vectors. This triple, $(\boldsymbol{\beta}_1, \boldsymbol{\beta}_2, \boldsymbol{\beta}_3)$, being also a basis for R^3, is called the basis dual to $(\mathbf{b}_1, \mathbf{b}_2, \mathbf{b}_3)$. To prove uniqueness, suppose $(\mathbf{b}_1, \mathbf{b}_2, \mathbf{b}_3)$ is given. From (7.2.93), $\boldsymbol{\beta}_1$ is perpendicular to $\mathbf{b}_2$ and $\mathbf{b}_3$, so $\boldsymbol{\beta}_1 = C_1 \mathbf{b}_2 \times \mathbf{b}_3$. Since $\mathbf{b}_1 \cdot \boldsymbol{\beta}_1 = 1$, $C_1 = [\mathbf{b}_1 \cdot (\mathbf{b}_2 \times \mathbf{b}_3)]^{-1}$, so $\boldsymbol{\beta}_1 = \mathbf{b}_2 \times \mathbf{b}_3 / Q$ where $Q = \mathbf{b}_1 \cdot (\mathbf{b}_2 \times \mathbf{b}_3)$. Similarly,

$$\boldsymbol{\beta}_2 = \mathbf{b}_3 \times \mathbf{b}_1 / [\mathbf{b}_2 \cdot (\mathbf{b}_3 \times \mathbf{b}_1)] \quad \text{and} \quad \boldsymbol{\beta}_3 = \mathbf{b}_1 \times \mathbf{b}_2 / [\mathbf{b}_3 \cdot (\mathbf{b}_1 \times \mathbf{b}_2)].$$

But

$$\mathbf{b}_1 \cdot (\mathbf{b}_2 \times \mathbf{b}_3) = \mathbf{b}_2 \cdot (\mathbf{b}_3 \times \mathbf{b}_1) = \mathbf{b}_3 \cdot (\mathbf{b}_1 \times \mathbf{b}_2),$$

so

$$\left. \begin{aligned} \boldsymbol{\beta}_1 &= (\mathbf{b}_2 \times \mathbf{b}_3)/Q \\ \boldsymbol{\beta}_2 &= (\mathbf{b}_3 \times \mathbf{b}_1)/Q \\ \boldsymbol{\beta}_3 &= (\mathbf{b}_1 \times \mathbf{b}_2)/Q \end{aligned} \right\}, \tag{7.2.94}$$

where

$$Q = \mathbf{b}_1 \cdot (\mathbf{b}_2 \times \mathbf{b}_3).$$

To prove existence, given a basis $\mathbf{b}_1, \mathbf{b}_2, \mathbf{b}_3$, simply define $\boldsymbol{\beta}_1, \boldsymbol{\beta}_2, \boldsymbol{\beta}_3$ by (7.2.94). Then clearly $\mathbf{b}_i \cdot \boldsymbol{\beta}_j = \delta_{ij}$.

There are two ways to find ∇u_i from $\partial_j \mathbf{r}$. First, we can set $\mathbf{b}_i = \partial_i \mathbf{r}$, and then ∇u_i is the $\boldsymbol{\beta}_i$ given by (7.2.94). Even this computation can be avoided if we are lucky. If we can guess a $\boldsymbol{\beta}_1, \boldsymbol{\beta}_2, \boldsymbol{\beta}_3$ satisfying

$$\partial_i \mathbf{r} \cdot \boldsymbol{\beta}_j = \delta_{ij} \tag{7.2.95}$$

then we know from the uniqueness of the dual basis of $\partial_1 \mathbf{r}, \partial_2 \mathbf{r}, \partial_3 \mathbf{r}$ that $\nabla u_i = \boldsymbol{\beta}_i$.

Notice that by applying this second technique, we can immediately deduce that the basis dual to an *orthonormal* basis $\hat{\mathbf{x}}_1, \hat{\mathbf{x}}_2, \hat{\mathbf{x}}_3$ is that basis itself. For one triple satisfying $\hat{\mathbf{x}}_i \cdot \boldsymbol{\beta}_j = \delta_{ij}$ is $\boldsymbol{\beta}_i = \hat{\mathbf{x}}_i$, and we know there is only one dual basis for $\hat{\mathbf{x}}_1, \hat{\mathbf{x}}_2, \hat{\mathbf{x}}_3$. An orthonormal basis is said to be self-dual.

7.2.14 Spherical Polar Coordinates

Our version of spherical polar coordinates is shown in Figure 7.2.1. Each position vector $\mathbf{r}$ is described by a radius r, a colatitude θ, and a longitude λ, the description being

$$\mathbf{r}(r, \theta, \lambda) = \hat{\mathbf{x}} r \sin \theta \cos \lambda + \hat{\mathbf{y}} r \sin \theta \sin \lambda + \hat{\mathbf{z}} r \cos \theta . \tag{7.2.96}$$

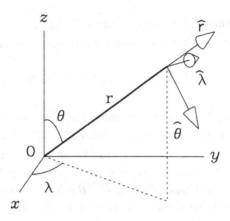

Figure 7.2.1: Spherical polar coordinates, as described by (7.2.96).

The scalar fields $r(\mathbf{r})$, $\theta(\mathbf{r})$, $\lambda(\mathbf{r})$ are given by

$$\left.\begin{array}{l} r(\mathbf{r}) = |\mathbf{r}| \\[2mm] \theta(\mathbf{r}) = \arctan\left[|\mathbf{r}|^2/(\hat{\mathbf{z}} \cdot \mathbf{r})^2 - 1\right]^{\frac{1}{2}} \\[2mm] \lambda(\mathbf{r}) = \arctan\left(\dfrac{\hat{\mathbf{y}} \cdot \mathbf{r}}{\hat{\mathbf{x}} \cdot \mathbf{r}}\right) \end{array}\right\}. \tag{7.2.97}$$

We make no use of (7.2.97) except to observe, as is geometrically obvious, that the scalar fields θ and λ depend on which orthonormal basis $\hat{\mathbf{x}}, \hat{\mathbf{y}}, \hat{\mathbf{z}}$ we choose. The scalar field r does not depend on the basis. In order to get most of the advantages of spherical coordinates without choosing a particular basis, it is useful to think of $\mathbf{r}$ as determined by a radial coordinate r and a direction $\hat{\mathbf{r}}$ defined by

$$\hat{\mathbf{r}} = \mathbf{r}/r \tag{7.2.98}$$

instead of r and θ and λ.

From (7.2.96), we see immediately that

$$\left.\begin{array}{l} \partial_r\mathbf{r} = \hat{\mathbf{x}} \sin\theta \cos\lambda + \hat{\mathbf{y}} \sin\theta \sin\lambda + \hat{\mathbf{z}} \cos\theta = \hat{\mathbf{r}} \\[2mm] \partial_\theta\mathbf{r} = \hat{\mathbf{x}} r \cos\theta \cos\lambda + \hat{\mathbf{y}} r \cos\theta \sin\lambda - \hat{\mathbf{z}} r \sin\theta \\[2mm] \partial_\lambda\mathbf{r} = -\hat{\mathbf{x}} r \sin\theta \sin\lambda + \hat{\mathbf{y}} r \sin\theta \cos\lambda \end{array}\right\}. \tag{7.2.99}$$

The lengths of these vectors are $|\partial_r \mathbf{r}| = 1$, $|\partial_\theta \mathbf{r}| = r$, $|\partial_\lambda \mathbf{r}| = r\sin\theta$, so at each location $\mathbf{r}$ there are unit vectors $\hat{\boldsymbol{\theta}}(\mathbf{r})$ and $\hat{\boldsymbol{\lambda}}(\mathbf{r})$ such that

$$\partial_r \mathbf{r} = \hat{\mathbf{r}} \qquad \partial_\theta \mathbf{r} = \hat{\boldsymbol{\theta}} r \qquad \partial_\lambda \mathbf{r} = \hat{\boldsymbol{\lambda}} r \sin\theta. \qquad (7.2.100)$$

The vectors $\hat{\mathbf{r}}(\mathbf{r}), \hat{\boldsymbol{\theta}}(\mathbf{r}), \hat{\boldsymbol{\lambda}}(\mathbf{r})$ are called the unit vectors at $\mathbf{r}$ in the directions of increasing r, θ, and λ. From (7.2.97), clearly $\hat{\mathbf{r}} \cdot \hat{\boldsymbol{\theta}} = \hat{\mathbf{r}} \cdot \hat{\boldsymbol{\lambda}} = \hat{\boldsymbol{\theta}} \cdot \hat{\boldsymbol{\lambda}} = 0$, so at each position $\mathbf{r}$, the three vectors $\hat{\mathbf{r}}(\mathbf{r}), \hat{\boldsymbol{\theta}}(\mathbf{r}), \hat{\boldsymbol{\lambda}}(\mathbf{r})$ constitute an orthonormal basis for R^3. (In fact $\hat{\mathbf{r}}, \hat{\boldsymbol{\theta}}, \hat{\boldsymbol{\lambda}}$ is a right-handed orthonormal basis. That is, $\hat{\mathbf{r}} = \hat{\boldsymbol{\theta}} \times \hat{\boldsymbol{\lambda}}$. See Figure 7.2.1 or use (7.2.99).) It is easy to guess a basis dual to (7.2.100), namely $\hat{\mathbf{r}}, \hat{\boldsymbol{\theta}}/r$ and $\hat{\boldsymbol{\lambda}}/r\sin\theta$. The uniqueness of the dual basis then implies

$$\nabla r = \hat{\mathbf{r}} \qquad \nabla\theta = \hat{\boldsymbol{\theta}}/r \qquad \nabla\lambda = \hat{\boldsymbol{\lambda}}/r\sin\theta,$$

so from (7.2.91)

$$\nabla = \hat{\mathbf{r}}\partial_r + \frac{\hat{\boldsymbol{\theta}}}{r}\partial_\theta + \frac{\hat{\boldsymbol{\lambda}}}{r\sin\theta}\partial_\lambda. \qquad (7.2.101)$$

The vector fields $\hat{\boldsymbol{\theta}}$ and $\hat{\boldsymbol{\lambda}}$ depend on which orthonormal basis $\hat{\mathbf{x}}, \hat{\mathbf{y}}, \hat{\mathbf{z}}$ we choose to establish our spherical coordinate system. The same is therefore true of the FODOs $\partial_\theta = r\hat{\boldsymbol{\theta}} \cdot \nabla$ and $\partial_\lambda = r\sin\theta\hat{\boldsymbol{\lambda}} \cdot \nabla$. The vector field $\hat{\mathbf{r}}$ does *not* depend on which spherical coordinates we use, and therefore neither does the differential operator

$$\partial_r = \hat{\mathbf{r}} \cdot \nabla. \qquad (7.2.102)$$

We can define ∂_r without introducing any particular system of spherical coordinates.

It follows that the FODO ∇_1, defined by

$$\nabla_1 = r\nabla - \mathbf{r}\partial_r, \qquad (7.2.103)$$

does not depend on any choice of spherical coordinates. However, we can always express it in spherical coordinates. From (7.2.101), relative to every system of spherical coordinates, ∇_1 has the form

$$\nabla_1 = \hat{\boldsymbol{\theta}}\partial_\theta + \frac{1}{\sin\theta}\hat{\boldsymbol{\lambda}}\partial_\lambda. \qquad (7.2.104)$$

And, of course,

$$\boldsymbol{\nabla} = \hat{\mathbf{r}}\partial_r + \frac{1}{r}\boldsymbol{\nabla}_1. \tag{7.2.105}$$

If a vector field **v** is given in the form

$$\mathbf{v(r)} = \hat{\mathbf{r}}v_r + \hat{\boldsymbol{\theta}}v_\theta + \hat{\boldsymbol{\lambda}}v_\lambda \tag{7.2.106}$$

we can use (7.2.101) to calculate $\boldsymbol{\nabla} \cdot \mathbf{v}$ or $\boldsymbol{\nabla} \times \mathbf{v}$ if we know the derivatives of $\hat{\mathbf{r}}(r,\theta,\lambda)$, $\hat{\boldsymbol{\theta}}(r,\theta,\Lambda)$, $\hat{\boldsymbol{\lambda}}(r,\theta,\lambda)$ with respect to r,θ,λ. From (7.2.99) and the definitions of $\hat{\mathbf{r}}, \hat{\boldsymbol{\theta}}, \hat{\boldsymbol{\lambda}}$, clearly

$$\left. \begin{array}{l} \hat{\mathbf{r}} = \hat{\mathbf{x}}\sin\theta\cos\lambda + \hat{\mathbf{y}}\sin\theta\sin\lambda + \hat{\mathbf{z}}\cos\theta \\[4pt] \hat{\boldsymbol{\theta}} = \hat{\mathbf{x}}\cos\theta\cos\lambda + \hat{\mathbf{y}}\cos\theta\sin\lambda - \hat{\mathbf{z}}\sin\theta \\[4pt] \hat{\boldsymbol{\lambda}} = -\hat{\mathbf{x}}\sin\lambda + \hat{\mathbf{y}}\cos\lambda \end{array} \right\}. \tag{7.2.107}$$

It is straightforward differentiation to show from (7.2.107) that

$$\begin{array}{lll} \partial_r\hat{\mathbf{r}} = \mathbf{0} & \partial_\theta\hat{\mathbf{r}} = \hat{\boldsymbol{\theta}} & \partial_\lambda\hat{\mathbf{r}} = \hat{\boldsymbol{\lambda}}\sin\theta \\[4pt] \partial_r\hat{\boldsymbol{\theta}} = \mathbf{0} & \partial_\theta\hat{\boldsymbol{\theta}} = -\hat{\mathbf{r}} & \partial_\lambda\hat{\boldsymbol{\theta}} = \hat{\boldsymbol{\lambda}}\cos\theta \\[4pt] \partial_r\hat{\boldsymbol{\lambda}} = \mathbf{0} & \partial_\theta\hat{\boldsymbol{\lambda}} = \mathbf{0} & \partial_\lambda\hat{\boldsymbol{\lambda}} = -\hat{\mathbf{r}}\sin\theta - \hat{\boldsymbol{\theta}}\cos\theta. \end{array} \tag{7.2.108}$$

These relations are also visible directly from the geometry of Figure 7.2.1. From them it follows that

$$\begin{aligned} \partial_r\mathbf{v} &= \hat{\mathbf{r}}(\partial_r v_r) + \hat{\boldsymbol{\theta}}(\partial_r v_\theta) + \hat{\boldsymbol{\lambda}}(\partial_r v_\lambda) \\ \partial_\theta\mathbf{v} &= \hat{\mathbf{r}}(\partial_\theta v_r - v_\theta) + \hat{\boldsymbol{\theta}}(\partial_\theta v_\theta + v_r) + \hat{\boldsymbol{\lambda}}(\partial_\theta v_\lambda) \\ \partial_\lambda\mathbf{v} &= \hat{\mathbf{r}}(\partial_\lambda v_r - \sin\theta v_\lambda) + \hat{\boldsymbol{\theta}}(\partial_\lambda v_\theta - \cos\theta v_\lambda) \\ &\quad + \hat{\boldsymbol{\lambda}}(\partial_\lambda v_\lambda + v_r\sin\theta + v_\theta\cos\theta). \end{aligned} \tag{7.2.109}$$

Then by (7.2.57)

$$\boldsymbol{\nabla} \cdot \mathbf{v} = \left(\hat{\mathbf{r}}\partial_r + \frac{\hat{\boldsymbol{\theta}}}{r}\partial_\theta + \frac{\hat{\boldsymbol{\lambda}}}{r\sin\theta}\partial_\lambda \right) \cdot \mathbf{v}$$

$$\boldsymbol{\nabla} \cdot \mathbf{v} = \hat{\mathbf{r}} \cdot (\partial_r\mathbf{v}) + \frac{1}{r}\hat{\boldsymbol{\theta}} \cdot (\partial_\theta\mathbf{v}) + \frac{1}{r\sin\theta}\hat{\boldsymbol{\lambda}} \cdot (\partial_\lambda\mathbf{v}),$$

so, from (7.2.109)

$$\boldsymbol{\nabla} \cdot \mathbf{v} = \partial_r v_r + \frac{2}{r} v_r + \frac{1}{r\sin\theta} \partial_\theta (\sin\theta v_\theta) + \frac{1}{r\sin\theta}(\partial_\lambda v_\lambda). \quad (7.2.110)$$

In particular, if $\mathbf{v} = \boldsymbol{\nabla} f$, then $v_r = \partial_r f$, $v_\theta = \partial_\theta f/r$, $v_\lambda = \partial_\lambda f/r\sin\theta$, so (7.2.110) gives

$$\boldsymbol{\nabla} \cdot \boldsymbol{\nabla} f = \left[\partial_r^2 + \frac{2}{r}\partial_r + \frac{1}{r^2\sin\theta}\partial_\theta\sin\theta\partial_\theta + \frac{1}{r^2\sin\theta}\partial_\lambda^2 \right] f$$

or

$$r^2\nabla^2 = r^2\partial_r^2 + 2r\partial_r + \frac{1}{\sin\theta}\partial_\theta\sin\theta\partial_\theta + \frac{1}{\sin^2\theta}\partial_\lambda^2. \quad (7.2.111)$$

From (7.2.110) we can also calculate $\boldsymbol{\nabla}_1 \cdot \mathbf{v}$. From (7.2.105),

$$\boldsymbol{\nabla} \cdot \mathbf{v} = (\mathbf{r}\partial_r) \cdot v + \frac{1}{r}\boldsymbol{\nabla}_1 \cdot \mathbf{v} = \hat{\mathbf{r}} \cdot (\partial_r \mathbf{v}) + \frac{1}{r}\boldsymbol{\nabla}_1 \cdot v$$

$$= \partial_r v_r + \frac{1}{r}\boldsymbol{\nabla}_1 \cdot \mathbf{v}.$$

Comparing this with (7.2.110) gives

$$\boldsymbol{\nabla}_1 \cdot \mathbf{v} = 2v_r + \frac{1}{\sin\theta}\partial_\theta(\sin\theta v_\theta) + \frac{1}{\sin\theta}(\partial_\lambda v_\lambda). \quad (7.2.112)$$

Now if $\mathbf{v} = \boldsymbol{\nabla}_1 f$, we have $v_r = 0$, $v_\theta = \partial_\theta f$, $v_\lambda = \partial_\lambda f/\sin\theta$, so

$$\boldsymbol{\nabla}_1 \cdot \boldsymbol{\nabla}_1 f = \left(\frac{1}{\sin\theta}\partial_\theta\sin\theta\partial_\theta + \frac{1}{\sin^2\theta}\partial_\lambda^2 \right) f.$$

Therefore

$$\nabla_1^2 = \frac{1}{\sin\theta}\partial_\theta\sin\theta\partial_\theta + \frac{1}{\sin^2\theta}\partial_\lambda^2. \quad (7.2.113)$$

This result can also be obtained directly from (7.2.104), (7.2.108); and (7.2.112) can be obtained from (7.2.103), (7.2.108), and (7.2.106). Comparing (7.2.111) and (7.2.113) we see that

$$\left. \begin{aligned} r^2\nabla^2 &= r^2\partial_r^2 + 2r\partial_r + \nabla_1^2 \\ &= r\partial_r^2 r + \nabla_1^2 \\ &= \partial_r r^2 \partial_r + \nabla_1^2 \\ &= r\partial_r(r\partial_r + 1) + \nabla_1^2 \end{aligned} \right\}. \quad (7.2.114)$$

It will also be useful to discuss $\mathbf{\Lambda} = \mathbf{r} \times \mathbf{\nabla}$ in spherical polar coordinates. From (7.2.105),

$$\mathbf{\Lambda} = \hat{\mathbf{r}} \times \mathbf{\nabla}_1. \tag{7.2.115}$$

From (7.2.104),

$$\mathbf{\Lambda} = \hat{\boldsymbol{\lambda}}\partial_\theta - \frac{\hat{\boldsymbol{\theta}}\partial_\lambda}{\sin\theta}. \tag{7.2.116}$$

Then $\mathbf{\Lambda} \cdot \mathbf{v} = \hat{\boldsymbol{\lambda}} \cdot (\partial_\theta \mathbf{v}) - \hat{\boldsymbol{\theta}} \cdot (\partial_\lambda \mathbf{v})/\sin\theta$. From (7.2.109),

$$\mathbf{\Lambda} \cdot \mathbf{v} = \frac{1}{\sin\theta}\partial_\theta(\sin\theta v_\lambda) - \frac{1}{\sin\theta}(\partial_\lambda v_\theta). \tag{7.2.117}$$

In particular, if $\mathbf{v} = \mathbf{\Lambda}f$, (7.2.116, 7.2.117) give

$$\mathbf{\Lambda} \cdot \mathbf{\Lambda}f = \frac{1}{\sin\theta}\partial_\theta\sin\theta\partial_\theta f + \frac{1}{\sin^2\theta}\partial_\lambda^2 f.$$

Thus comparing this with (7.2.113) we find

$$\mathbf{\Lambda} \cdot \mathbf{\Lambda} = \mathbf{\nabla}_1 \cdot \mathbf{\nabla}_1 \quad \text{or} \quad \Lambda^2 = \nabla_1^2. \tag{7.2.118}$$

If $\mathbf{v} = \mathbf{\Lambda}f$, (7.2.116) and (7.2.112) give $\mathbf{\nabla}_1 \cdot \mathbf{\Lambda}f = 0$, so

$$\mathbf{\nabla}_1 \cdot \mathbf{\Lambda} = 0. \tag{7.2.119}$$

If $\mathbf{v} = \mathbf{\nabla}_1 f$, (7.2.104) and (7.2.117) give $\mathbf{\Lambda} \cdot \mathbf{\nabla}_1 f = 0$, so

$$\mathbf{\Lambda} \cdot \mathbf{\nabla}_1 = 0. \tag{7.2.120}$$

By our definition of components of operators,

$$\Lambda_x = \hat{\mathbf{x}} \cdot \mathbf{\Lambda} = (\hat{\mathbf{x}} \cdot \hat{\boldsymbol{\lambda}})\partial_\theta - \frac{(\hat{\mathbf{x}} \cdot \hat{\boldsymbol{\theta}})}{\sin\theta}\partial_\lambda = -\sin\lambda\partial_\theta - \cot\theta\cos\lambda\partial_\lambda$$

$$\Lambda_y = \hat{\mathbf{y}} \cdot \mathbf{\Lambda} = (\hat{\mathbf{y}} \cdot \hat{\boldsymbol{\lambda}})\partial_\theta - \frac{(\hat{\mathbf{y}} \cdot \hat{\boldsymbol{\theta}})}{\sin\theta}\partial_\lambda = \cos\lambda\partial_\theta - \cot\theta\sin\lambda\partial_\lambda$$

$$\Lambda_z = \hat{\mathbf{z}} \cdot \mathbf{\Lambda} = (\hat{\mathbf{z}} \cdot \hat{\boldsymbol{\lambda}})\partial_\theta - \frac{(\hat{\mathbf{z}} \cdot \hat{\boldsymbol{\theta}})}{\sin\theta}\partial_\lambda = \partial_\lambda.$$

The expressions for Λ_x and Λ_y can be made almost as simple as Λ_z by combining them. Clearly

$$\left. \begin{array}{l} \Lambda_x + i\Lambda_y = e^{i\lambda}[i\partial_\theta - \cot\theta\partial_\lambda] \\ \Lambda_x - i\Lambda_y = e^{-i\lambda}[-i\partial_\theta - \cot\theta\partial_\lambda] \\ \Lambda_z = \partial_\lambda \end{array} \right\} . \qquad (7.2.121)$$

Finally, we note that the three vectors $\hat{\mathbf{r}}$, $\boldsymbol{\nabla}_1 f$, and $\boldsymbol{\Lambda} f$ form a local orthogonal triple. In view of (7.2.119, 7.2.120) we can see that the *operators* $\hat{\mathbf{r}}^M$, $\boldsymbol{\nabla}_1$, and $\boldsymbol{\Lambda}$ also act as orthogonal linear operators.

7.3 Vector Analysis: Integral Calculus

In this section we set out the well-known theorems of Gauss and Stokes and then derive less familiar statements of them. We go on to use them to prove the uniqueness of the representation of a vector field in terms of its sources.

7.3.1 The Theorems of Stokes and Gauss

First we require the definition of several standard terms. We define the notion of differentiability at a point as follows: A scalar field f on subset V of R^3 is *differentiable* at $\mathbf{r}_0$ in V if a vector $\boldsymbol{\nabla} f(\mathbf{r}_0)$ exists such that

$$\lim_{\mathbf{h} \to 0} \frac{f(\mathbf{r}_0 + \mathbf{h}) - f(\mathbf{r}_0) - \mathbf{h} \cdot \boldsymbol{\nabla} f(\mathbf{r}_0)}{|\mathbf{h}|} = 0.$$

Of course this limit also serves to define the gradient vector $\boldsymbol{\nabla} f(\mathbf{r}_0)$. If $\boldsymbol{\nabla} f$ exists everywhere in V, and is continuous there, and if f is differentiable everywhere in V, the scalar field f is said to be *continuously differentiable* in V.

The idea of continuously differentiable can easily be extended to vector fields: a vector field $\mathbf{v}$ on set V is continuously differentiable if there is an orthonormal basis $\hat{\mathbf{x}}_1, \hat{\mathbf{x}}_2, \hat{\mathbf{x}}_3$ such that the scalar fields $v_i = \hat{\mathbf{x}}_i \cdot \mathbf{v}$ are continuously differentiable. Note that if $\hat{\mathbf{x}}_1', \hat{\mathbf{x}}_2', \hat{\mathbf{x}}_3'$ is

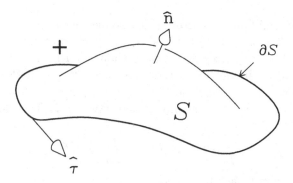

Figure 7.3.1: An oriented surface S. Only the positive side is visible.

any other orthonormal basis then the scalar fields $v_j{'} = (\hat{\mathbf{x}}_j' \cdot \hat{\mathbf{x}}_i)v_i$ are also continuously differentiable.

The third piece of terminology we need is the oriented surface (see Figure 7.3.1). An *oriented surface* is a smooth two-sided surface S, one of whose sides has been chosen and labeled positive. The other side of S is then called negative. The unit normal on the positive side of S is the positive unit normal. The positive unit tangent vector to ∂S, the boundary curve of S, is the unit tangent vector $\hat{\boldsymbol{\tau}}$ whose direction makes $\hat{\mathbf{n}} \times \hat{\boldsymbol{\tau}}$ point into S.

With these concepts defined we are in a position to state the celebrated integral theorems of vector calculus. The proofs will not be supplied but can be found in any standard text on advanced calculus, for example: Thomas and Finney (1984).

Theorem: (Stokes) Suppose $\mathbf{v}$ is a vector field continuously differentiable in open subset V of R^3. Suppose S is an oriented surface in V, with positive unit normal $\hat{\mathbf{n}}$ and positive boundary unit tangent vector $\hat{\boldsymbol{\tau}}$. Then

$$\int_S d^2\mathbf{r}\, \hat{\mathbf{n}} \cdot (\boldsymbol{\nabla} \times \mathbf{v}) = \int_{\partial S} d\ell\, \hat{\boldsymbol{\tau}} \cdot \mathbf{v} \qquad (7.3.1)$$

where $d\ell$ is the element of length on ∂S.

Theorem: (Gauss) Let V be an open subset of R^3 with smooth boundary ∂V and outward unit normal $\hat{\mathbf{n}}$. Let $\bar{V}$ be the closure of V, i.e., V together with ∂V. Suppose $\mathbf{v}$ is a continuously differentiable vector field in $\bar{V}$. Then

$$\int_V d^3\mathbf{r}\, \boldsymbol{\nabla} \cdot \mathbf{v} = \int_{\partial V} d^2\mathbf{r}\, \hat{\mathbf{n}} \cdot \mathbf{v}. \qquad (7.3.2)$$

Stokes' and Gauss' Theorems are sometimes easier to work with in array notation. Let f be a scalar field that is continuously differentiable. Let $\hat{x}_1, \hat{x}_2, \hat{x}_3$ be an orthonormal basis for R^3. Then $v = \hat{x}_\ell f$ is a continuously differentiable vector field, whether ℓ is 1, 2, or 3. Moreover, $v_k = \delta_{\ell k} f$, so $(\nabla \times v)_i = \epsilon_{ijk} \partial_j v_k = \epsilon_{ijk}(\partial_j f)\delta_{\ell k} = \epsilon_{ij\ell}\partial_j f$. Thus $\hat{n} \cdot \nabla \times v = \epsilon_{ij\ell} n_i \partial_j f$ and $\hat{\tau} \cdot v = \tau_\ell f$. Therefore (7.3.1) gives

$$\int_S d^2r \; \epsilon_{ij\ell} n_i \partial_j f = \int_{\partial S} d\ell \; \tau_\ell f.$$

Changing index ℓ back to k gives

$$\int_{\partial S} d\ell \; \tau_k f = \epsilon_{ijk} \int_S d^2r \; n_i \partial_j f. \tag{7.3.3}$$

This is an alternative form of Stokes' Theorem. It implies (7.3.1), because we can apply (7.3.3) to each of the scalar fields $f = v_k$, thus recovering (7.3.1). We may rewrite (7.3.3) in vector notation. Multiply (7.3.3) by $\hat{x}_k$ and sum over k. Since $\hat{x}_k$ is constant, it can be taken inside the integral sign, so (7.3.3) becomes

$$\int_{\partial S} d\ell \; \hat{\tau} f = \int_S d^2r \; \hat{n} \times \nabla f. \tag{7.3.4}$$

Similarly, Gauss' Theorem (7.3.2) can be recast into a relation between vectors. Setting $v = f\hat{x}_i$ in (7.3.2) gives

$$\int_V d^3r \; \partial_i f = \int_{\partial V} d^2r \; n_i f. \tag{7.3.5}$$

This alternative statement of Gauss' Theorem implies (7.3.2), because we can set $f = v_i$ in (7.3.5), thus recovering (7.3.2). To put (7.3.5) into vector notation we multiply (7.3.4) by $\hat{x}_i$ and sum over i, and then

$$\int_{\partial V} d^2r \; \hat{n} f = \int_V d^3r \; \nabla f. \tag{7.3.6}$$

Parenthetically, by taking $f = 1$ in (7.3.5), we obtain the geometrically obvious fact that $\int_{\partial S} d\ell \; \hat{\tau} = 0$ for any closed curve ∂S. By taking $f = 1$ in (7.3.6) we obtain the possibly less obvious fact that $\int_{\partial V} d^2r \; \hat{n} = 0$ for any closed surface ∂V.

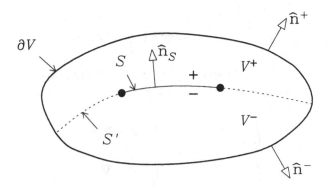

Figure 7.3.2: Section through the region V showing the surface S and its extension S'.

7.3.2 Jump Discontinuities

The scalar field for Gauss' Theorem (7.3.6) is required to be continuously differentiable; we now obtain the modification to the theorem when certain simple discontinuities are permitted.

Let S be an oriented surface in an open subset V of R^3. Let f be a scalar field on $V \setminus S$, which means V with S removed. Suppose that at every point s on S, $\lim_{\mathbf{r} \to \mathbf{s}} f(\mathbf{r})$ exists as long as $\mathbf{r}$ approaches $\mathbf{s}$ from one side of S. Call the limit on the positive side $f^+(\mathbf{s})$ and the limit on the negative side $f^-(\mathbf{s})$. If $f^+(\mathbf{s})$ and $f^-(\mathbf{s})$ exist for every $\mathbf{s}$ on S, and are continuous functions of position on S, then f is said to have a *jump discontinuity* across S. The difference $f^+(\mathbf{s}) - f^-(\mathbf{s})$ is written $[f]_-^+(\mathbf{s})$ and is called the *jump* across S. If the jump is 0 everywhere on S, the jump discontinuity is called *removable*.

A vector field has a jump discontinuity on S if its components relative to one orthonormal basis do so. Then its components relative to every orthonormal basis do so.

Theorem: (Gauss' Theorem with jump discontinuity) Let V be an open subset of R^3 with boundary ∂V and outward unit normal $\hat{\mathbf{n}}$. Let S be a smooth oriented surface in V, with positive unit normal $\hat{\mathbf{n}}_s$. Let f be a scalar field in $V \setminus S$ that is continuously differentiable in $\overline{V}$ except for jump discontinuities in f and $\boldsymbol{\nabla} f$ across S. Then

$$\int_{\partial V} d^2\mathbf{r}\, n_i f = \int_V d^3\mathbf{r}\, \partial_i f + \int_S d^2\mathbf{r}\, (\hat{n}_s)_i [f]_-^+ \qquad (7.3.7)$$

or

$$\int_{\partial V} d^2\mathbf{r}\ \hat{\mathbf{n}}f = \int_V d^3\mathbf{r}\ \boldsymbol{\nabla}f + \int_S d^2\mathbf{r}\ \hat{\mathbf{n}}_s[f]_-^+. \qquad (7.3.8)$$

Proof: Extend S to a surface S' that cuts V into two parts, V^+ and V^-. Then, by the ordinary Gauss Theorem (7.3.6),

$$\int_{\partial V^+} d^2\mathbf{r}\ \hat{\mathbf{n}}^+ f = \int_{V^+} d^3\mathbf{r}\ \boldsymbol{\nabla}f$$

$$\int_{\partial V^-} d^2\mathbf{r}\ \hat{\mathbf{n}}^- f = \int_{V^-} d^3\mathbf{r}\ \boldsymbol{\nabla}f.$$

Now ∂V^+ is part of ∂V, together with S'. On S', $\hat{\mathbf{n}}^+ = -\hat{\mathbf{n}}_S$. And ∂V^- is part of ∂V, together with S'. On S', $\hat{\mathbf{n}}^- = \hat{\mathbf{n}}_S$. Thus adding the two equations above gives

$$\int_{\partial V} d^2\mathbf{r}\ \hat{\mathbf{n}}f - \int_{S'} d^2\mathbf{r}\ \hat{\mathbf{n}}_S f^+ + \int_{S'} d^2\mathbf{r}\ \hat{\mathbf{n}}_S f^- = \int_V d^3\mathbf{r}\ \boldsymbol{\nabla}f.$$

This is (7.3.8).   ◁

7.3.3 Sources of a Vector Field

A vector field $\mathbf{v}$ in R^3 is uniquely determined by its divergence and curl, its behavior at infinity, and its jump discontinuities.

A more precise statement of this result is as follows:

Theorem: Suppose that vector field $\mathbf{v}$ is continuous in all of R^3, and continuously differentiable in R^3 except on one or more smooth oriented surfaces S, across which the partial derivatives with respect to a Cartesian coordinate system, $\partial_i\mathbf{v}$, may have jump discontinuities. Suppose that there is a constant M such that $r^2|\mathbf{v}| \le M$ for all $\mathbf{r}$. Suppose that $\boldsymbol{\nabla}\cdot\mathbf{v} = \boldsymbol{\nabla}\times\mathbf{v} = 0$ everywhere off S. Then $\mathbf{v} = 0$ everywhere.

Proof: The hypotheses assure us that $\oint_C d\ell\,\hat{\boldsymbol{\tau}}\cdot\mathbf{v} = \oint_C d\mathbf{r}\cdot\mathbf{v} = 0$ for every closed curve C and that $\int_{\mathbf{r}_0}^\infty d\mathbf{r}\cdot\mathbf{v}$ is the same for all curves leading from $\mathbf{r}_0$ to infinity. Therefore, there is a well-defined single-valued function $\phi(\mathbf{r})$ given by the line integral

$$\phi(\mathbf{r}) = \int_{\mathbf{r}}^\infty d\mathbf{s}\cdot\mathbf{v}(\mathbf{s}).$$

The path to infinity doesn't matter. Moreover, $r|\phi| \leq M$. Furthermore, $\mathbf{v} = \boldsymbol{\nabla}\phi$ so $\nabla^2\phi = 0$ and $|\mathbf{v}|^2 = |\boldsymbol{\nabla}\phi|^2 = \boldsymbol{\nabla}\cdot(\phi\boldsymbol{\nabla}\phi) - \phi\nabla^2\phi = \boldsymbol{\nabla}\cdot(\phi\boldsymbol{\nabla}\phi) = \boldsymbol{\nabla}\cdot(\phi\mathbf{v})$. Let $B(a)$ be the ball of radius a centered on $\mathbf{0}$, i.e., the set of all $\mathbf{r}$ with $|\mathbf{r}| < a$.

$$\int_{B(a)} d^3\mathbf{r}\,|\mathbf{v}|^2 = \int_{\partial B(a)} d^2\mathbf{r}\,\hat{\mathbf{n}}\cdot\mathbf{v}\phi.$$

But on $\partial B(a)$, $|\mathbf{v}\phi| \leq M^2/a^3$ and the area of $\partial B(a)$ is $4\pi a^2$, so

$$\int_{B(a)} d^3\mathbf{r}\,|\mathbf{v}|^2 \leq 4\pi M^2/a.$$

Letting $a \to \infty$, we have $\int_{B(a)} d^3\mathbf{r}\,|\mathbf{v}|^2 = 0$, so $|\mathbf{v}| = 0$, so $\mathbf{v} = \mathbf{0}$. ◁

In fact we do not need the condition $r^2|\mathbf{v}| \leq M$, only that $v_r(r\hat{\mathbf{r}}) \to 0$ uniformly in $\hat{\mathbf{r}}$ as $r \to \infty$. But the proof is considerably harder. We give a sketch of the more difficult proof, but an appeal must be made to some results of potential theory, which we do not prove.

Proof: Now we must define $\phi(\mathbf{r}) = \int_0^\mathbf{r} d\mathbf{s}\cdot\mathbf{v}(\mathbf{s})$. We cannot be sure at the outset that $\int_\mathbf{r}^\infty d\mathbf{s}\cdot\mathbf{v}(\mathbf{s})$ converges, much less is path-independent. Now we have $\boldsymbol{\nabla}\phi = \mathbf{v}$ so $\nabla^2\phi = 0$ except on S. We must appeal to the theory of harmonic functions (Kellogg, 1953). The singularities on S are removable, so ϕ is harmonic in R^3, i.e., it is infinitely differentiable and satisfies $\nabla^2\phi = 0$. Therefore (Kellogg pp. 251–253) there are constants ϕ_ℓ such that

$$\phi(\mathbf{r}) = \sum_{\ell=0}^\infty \phi_\ell h_\ell(\mathbf{r}) \tag{7.3.9}$$

where $h_\ell(\mathbf{r})$ is a homogeneous polynomial of degree ℓ in $\mathbf{r}$ that satisfies $\nabla^2 h_\ell = 0$; these polynomials should be familiar from Chapter 3. Every partial derivative of (7.3.9) to any order converges uniformly inside every ball $B(a)$; and

$$\frac{1}{4\pi}\int_{S(1)} d^2\hat{\mathbf{r}}\,h_\ell(\hat{\mathbf{r}})h_k(\hat{\mathbf{r}}) = \delta_{\ell k}$$

where $S(1) = \partial B(1)$ and $d^2\hat{\mathbf{r}}$ is the element of area on $S(1)$. Since h_ℓ is homogeneous of degree ℓ,

$$h_\ell(\mathbf{r}) = h_\ell(r\hat{\mathbf{r}}) = r^\ell h_\ell(\hat{\mathbf{r}})$$

so

$$\partial_r h_\ell(\mathbf{r}) = \ell r^{\ell-1} h_\ell(\hat{\mathbf{r}}).$$

Differentiating (7.3.9) term by term gives

$$\partial_r \phi = v_r = \sum_{\ell=0}^{\infty} \ell r^{\ell-1} \phi_\ell h_\ell(\hat{\mathbf{r}}).$$

Multiply by $h_k(\hat{\mathbf{r}})$ and integrate over $S(1)$. Then, dividing by 4π,

$$\frac{1}{4\pi} \int_{S(1)} d^2\hat{\mathbf{r}} \; v_r(r\hat{\mathbf{r}}) h_k(\hat{\mathbf{r}}) = k r^{k-1} \phi_k.$$

The left side of this equation $\to 0$ as $r \to \infty$, so the same must be true of the right side. Hence $\phi_k = 0$ if $k \geq 1$. Thus $\phi = \phi_0 =$ constant, and $\mathbf{v} = \nabla\phi = \mathbf{0}$.   ◁

7.4 Scalar and Vector Fields on Orientable Surfaces

In this final section we discuss scalar and vector fields that are confined to a surface and we describe the new versions of the operators and integral theorems of Stokes and Gauss that apply in this case. For geophysicists the obvious surface to which this theory applies is the surface of the earth.

7.4.1 Projection of a Vector onto a Plane

We begin with the simplest of oriented surfaces, the plane. Let S be any oriented plane surface. Let $\hat{\mathbf{n}}_S = \hat{\mathbf{n}}$ be its positive unit normal. Let $\mathbf{v}$ be any fixed vector.

There is a unique scalar v_n and a unique vector $\mathbf{v}_S$ such that $\mathbf{v} = v_n\hat{\mathbf{n}} + \mathbf{v}_S$ and $\hat{\mathbf{n}} \cdot \mathbf{v}_S = 0$. The idea is to *decompose* the arbitrary vector into a part lying in the plane, and a part normal to the plane.

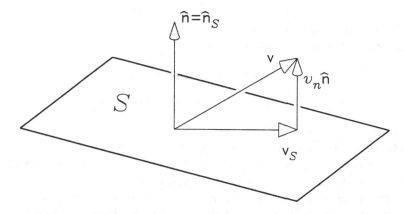

Figure 7.4.1: Projection of a vector onto a plane oriented surface.

The part in the plane $\mathbf{v}_S$ is called the *orthogonal projection* of $\mathbf{v}$ onto S; and v_n is called the *normal component*.

Proof: All we need to do is to note that $\hat{\mathbf{n}} \cdot \mathbf{v} = v_n$, and $\mathbf{v}_S = \mathbf{v} - v_n \hat{\mathbf{n}}$. The second part insures a successful reconstruction of the vector:

$$\mathbf{v} = v_n \hat{\mathbf{n}} + \mathbf{v}_S. \tag{7.4.1}$$

The following calculation verifies the required property $\hat{\mathbf{n}} \cdot \mathbf{v}_S = 0$:

$$\hat{\mathbf{n}} \cdot \mathbf{v}_S = \hat{\mathbf{n}} \cdot (\mathbf{v} - v_n \hat{\mathbf{n}}) = \hat{\mathbf{n}} \cdot \mathbf{v} - v_n \hat{\mathbf{n}} \cdot \hat{\mathbf{n}}$$

$$= v_n - v_n = 0.$$

This establishes the existence of the decomposition for any vector $\mathbf{v}$. The uniqueness of the decomposition is shown by assuming there is another representation, say, $\mathbf{v} = a\hat{\mathbf{n}} + \mathbf{w}$ with $\hat{\mathbf{n}} \cdot \mathbf{w} = 0$. Equating the alternative with (7.4.1)

$$a\hat{\mathbf{n}} + \mathbf{w} = \mathbf{v} = v_n \hat{\mathbf{n}} + \mathbf{v}_S.$$

Now if this equation is dotted with $\hat{\mathbf{n}}$ we see at once that $a = v_n$. Then by subtracting $v_n \hat{\mathbf{n}}$ or $a\hat{\mathbf{n}}$ from $\mathbf{v}$ we obtain $\mathbf{w} = \mathbf{v}_S$. ◁

Here are a number of simple properties of such decompositions all but the last of which are left as exercises to prove. In the following $\mathbf{u}$ and $\mathbf{v}$ are both fixed vectors.

(1) $\mathbf{u} \cdot \mathbf{v} = u_n v_n + \mathbf{u}_S \cdot \mathbf{v}_S$.

(2) For any scalars a and b

$$(a\mathbf{u} + b\mathbf{v})_S = a\mathbf{u}_S + b\mathbf{v}_S$$

$$(a\mathbf{u} + b\mathbf{v})_n = a\mathbf{u}_n + b\mathbf{v}_n.$$

(3) $(\hat{\mathbf{n}} \times \mathbf{v})_S = \hat{\mathbf{n}} \times \mathbf{v}_S$.

(4) $\mathbf{v}_S = -\hat{\mathbf{n}} \times (\hat{\mathbf{n}} \times \mathbf{v}) = -\hat{\mathbf{n}} \times (\hat{\mathbf{n}} \times \mathbf{v}_S)$.

(5) $\mathbf{v}_S = \mathbf{0}$ (a) if, and (b) only if, $\mathbf{v} \cdot \hat{\boldsymbol{\tau}} = 0$ for every unit vector $\hat{\boldsymbol{\tau}}$ that is perpendicular to $\hat{\mathbf{n}}$, that is, for which $\hat{\mathbf{n}} \cdot \hat{\boldsymbol{\tau}} = 0$.

Proof: (a) $\mathbf{v} \cdot \hat{\boldsymbol{\tau}} = v_n \hat{\boldsymbol{\tau}} \cdot \hat{\mathbf{n}} + \mathbf{v}_S \cdot \hat{\boldsymbol{\tau}} = v_n 0 + 0 \cdot \hat{\boldsymbol{\tau}} = 0$. (b) If $\mathbf{v}_S \neq \mathbf{0}$, then $\mathbf{v}_S / |\mathbf{v}_S|$ is a unit vector perpendicular to $\hat{\mathbf{n}}$. Hence by hypothesis $\mathbf{v}_S \cdot \mathbf{v}_S / |\mathbf{v}_S| = 0$, i.e., $|\mathbf{v}_S| = 0$, which is a contradiction. Thus $\mathbf{v}_S = \mathbf{0}$.   ◁

7.4.2 Vector Fields on an Oriented Surface

For a plane oriented surface the unit normal is a fixed vector. More generally, the normal varies from place to place in S. We now generalize the decomposition of the previous section to deal with vector fields on general smooth surfaces as follows.

If S is an oriented surface, $\hat{\mathbf{n}}$ is its positive unit normal, and $\mathbf{v}$ is a vector field on S, then we define $v_n = \hat{\mathbf{n}} \cdot \mathbf{v}$ to be the *normal part* of $\mathbf{v}$; further, we define $\mathbf{v}_S = \mathbf{v} - \hat{\mathbf{n}} v_n$ as the *tangential part* of $\mathbf{v}$. Finally, if $v_n = 0$ everywhere on S, i.e., if $\mathbf{v} = \mathbf{v}_S$, then $\mathbf{v}$ is called a *tangent vector field* on S.

Suppose $\mathbf{v}$ is continuous on S and $\int_C d\ell \, \hat{\boldsymbol{\tau}} \cdot \mathbf{v} = 0$ for every curve C lying on S. Here $d\ell$ is the element of arc length on C, and $\hat{\boldsymbol{\tau}}$ is the unit vector tangent to C. Then the tangential part of $\mathbf{v}$ vanishes on S, that is, $\mathbf{v}_S = \mathbf{0}$ on S.

Proof: Choose any point $\mathbf{s}_0$ on S. Let $\hat{\boldsymbol{\tau}}$ be any unit vector satisfying $\hat{\boldsymbol{\tau}} \cdot \hat{\mathbf{n}}(\mathbf{s}_0) = 0$. Then there is a curve C in S passing through $\mathbf{s}_0$ whose unit tangent vector at $\mathbf{s}_0$ is $\hat{\boldsymbol{\tau}}$. Let C' be any short piece of C containing $\mathbf{s}_0$. Then $\int_{C'} d\ell \, \hat{\boldsymbol{\tau}} \cdot \mathbf{v} = 0$, so, letting the length of C' shrink to zero. $\hat{\boldsymbol{\tau}} \cdot \mathbf{v}(\mathbf{s}_0) = 0$. By property (5) of subsection 7.4.1, $\mathbf{v}_S(\mathbf{s}_0) = \mathbf{0}$. Since $\mathbf{s}_0$ can be any point on S, we have $\mathbf{v}_S = \mathbf{0}$.   ◁

7.4.3 Surface Gradient and Normal Derivative

Now we get to something interesting: the idea of differential operators confined to the surface. Most of the work of this section will be

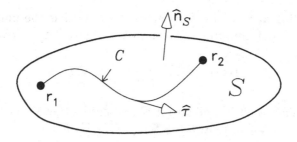

Figure 7.4.2: A line integral in the oriented surface S.

devoted to showing that there is a surface gradient operator with all the properties one would expect of the analog of $\boldsymbol{\nabla}$ in ordinary space.

Let V be any open set containing an oriented surface S. Let f be any scalar field continuously differentiable in V. Define, for every $\mathbf{r}$ on S,

$$\partial_n f = \hat{\mathbf{n}} \cdot \boldsymbol{\nabla} f$$

$$\boldsymbol{\nabla}_S f = (\boldsymbol{\nabla} f)_S = \boldsymbol{\nabla} f - \hat{\mathbf{n}}\partial_n f.$$

Thus $\partial_n f$ is a scalar field defined only on S, and $\boldsymbol{\nabla}_S f$ is a vector field defined only on S. Clearly $\boldsymbol{\nabla}_S f$ is a tangent vector field on S. We call $\boldsymbol{\nabla}_S f$ the *surface gradient* of f.

Suppose a and b are constant scalars and f and g are continuously differentiable scalar fields in V. Then

$$\partial_n(af + bg) = a\partial_n f + b\partial_n g$$

$$\boldsymbol{\nabla}_S(af + bg) = a\boldsymbol{\nabla}_S f + b\boldsymbol{\nabla}_S g. \tag{7.4.2}$$

That is, ∂_n and $\boldsymbol{\nabla}_S$ are linear operators on V. Moreover,

$$\partial_n(fg) = (\partial_n f)g + f(\partial_n g)$$

$$\boldsymbol{\nabla}_S(fg) = (\boldsymbol{\nabla}_S f)g + f(\boldsymbol{\nabla}_S g). \tag{7.4.3}$$

That is, ∂_n and $\boldsymbol{\nabla}_S$ are FODOs.

Next we show the surface gradient acts like an ordinary gradient when we integrate it along lines in S. Suppose f is continuously differentiable in an open set containing S, and C is any curve lying in S, starting at $\mathbf{r}_1$ and ending at $\mathbf{r}_2$. Suppose $\hat{\boldsymbol{\tau}}$ is the unit tangent vector on C, pointing from $\mathbf{r}_1$ to $\mathbf{r}_2$. (See Figure 7.4.2.) Then

$$f(\mathbf{r}_2) - f(\mathbf{r}_1) = \int_C d\ell \, \hat{\boldsymbol{\tau}} \cdot \boldsymbol{\nabla}_S f. \tag{7.4.4}$$

Proof: The result is standard if $\boldsymbol{\nabla}_S$ is replaced by $\boldsymbol{\nabla}$. But $\hat{\boldsymbol{\tau}} \cdot \hat{\mathbf{n}} = 0$ on C, so $\hat{\boldsymbol{\tau}} \cdot \boldsymbol{\nabla} f = \hat{\boldsymbol{\tau}} \cdot [\hat{\mathbf{n}} \partial_n f + \boldsymbol{\nabla}_S f] = \hat{\boldsymbol{\tau}} \cdot \boldsymbol{\nabla}_S f.$ ◁

Another property that is intuitively satisfying is the following: if $\boldsymbol{\nabla}_S f = \mathbf{0}$ on S, f is constant on S, and conversely.

Proof: If $\boldsymbol{\nabla}_S f = \mathbf{0}$ on S, (7.4.4) shows $f(\mathbf{r}_2) = f(\mathbf{r}_1)$ for any $\mathbf{r}_1$ and $\mathbf{r}_2$ on S. Conversely, if f is constant on S, (7.4.4) shows that the vector field $\mathbf{v} = \boldsymbol{\nabla}_S f$ satisfies the hypothesis of the remark at the end of subsection 7.4.2. Hence, by that remark, $\mathbf{v}_S = \mathbf{0}$. But $\mathbf{v}_S = \mathbf{v} = \boldsymbol{\nabla}_S f.$ ◁

In defining $\boldsymbol{\nabla}_S f$, we have required that f be defined not just on S but on an open set V containing S in order to define a derivative normal to the surface; and yet $\boldsymbol{\nabla}_S f$ is defined only on S. In fact, f also need be defined only on S, as we see from the following comments. But first we need some more definitions.

Suppose S is any subset of set V, and g is any function whose domain is V. Then $g|S$ stands for that function f whose domain is S and for which $f(s) = g(s)$ for all s in S. The function $f = g|S$ is the *restriction of g to S* and g is said to be an *extension of f to V*.

Suppose S is a smooth surface in R^3 and f is a scalar field on S. We say f is *continuously differentiable on S* if $f = g|S$ for a least one function g defined and continuously differentiable on an open set V containing S.

Suppose f is continuously differentiable on the smooth surface S. We define $\boldsymbol{\nabla}_S f$ to be $\boldsymbol{\nabla}_S g$ where g is any function of the sort described in the previous paragraph.

Of course we must show that if V_1 and V_2 are two open sets containing S, and g_i is continuously differentiable on V_i, and $f = g_i|S$, for $i = 1, 2$, then $\boldsymbol{\nabla}_S g_1 = \boldsymbol{\nabla}_S g_2$. But $g_1 = f$ and $g_2 = f$ on S, so $g_2 - g_1 = 0$ on S. Therefore, by (7.4.4) $\boldsymbol{\nabla}_S(g_2 - g_1) = \mathbf{0}$, and so $\boldsymbol{\nabla}_S g_2 = \boldsymbol{\nabla}_S g_1$.

It follows that $\boldsymbol{\nabla}_S$ is a FODO on S. It assigns to each continuously differentiable scalar field f on S a tangent vector field $\boldsymbol{\nabla}_S f$ on S.

The last comment needed to show that ∇_S is the surface analog of ∇ is the following: Suppose S is a smooth *simply connected* surface in R^3, and $\mathbf{v}_S$ is a continuous tangent vector field on S. Suppose that for every closed curve C on S, $\oint_C d\ell \hat{\boldsymbol{\tau}} \cdot \mathbf{v}_S = 0$. Then there is a scalar field f on S such that $\mathbf{v}_S = \nabla_S f$. The field f is continuously differentiable and is unique up to a constant.

Proof: Pick a fixed $\mathbf{r}_0$ on S, and define $f(\mathbf{r}) = \int_{\mathbf{r}_0}^{\mathbf{r}} d\ell \, \hat{\boldsymbol{\tau}} \cdot \mathbf{v}_S$, the integral being along any curve C that starts at $\mathbf{r}_0$ and ends at $\mathbf{r}$. Then

$$f(\mathbf{r}_2) - f(\mathbf{r}_1) = \int_{\mathbf{r}_1}^{\mathbf{r}_2} d\ell \, \hat{\boldsymbol{\tau}} \cdot \mathbf{v}_S = \int_{\mathbf{r}_1}^{\mathbf{r}_2} d\ell \, \hat{\boldsymbol{\tau}} \cdot \nabla_S f$$

for any two points $\mathbf{r}_2$ and $\mathbf{r}_0$ on S, and any curve C lying on S and starting at $\mathbf{r}_1$ and ending at $\mathbf{r}_2$. Hence

$$\int_{\mathbf{r}_1}^{\mathbf{r}_2} d\ell \, \hat{\boldsymbol{\tau}} \cdot (\mathbf{v}_S - \nabla_S f) = 0$$

for any curve C on S, so by the remark at the end of subsection 7.4.2, $\mathbf{v}_S - \nabla_S f = \mathbf{0}$. The proof that f is continuously differentiable is omitted. (The easiest way to construct such a proof is to take f constant on each straight line passing through S in the normal direction.) To prove that f is unique up to a constant, note that if $\nabla_S f_1 = \nabla_S f_2$, then $\nabla_S(f_1 - f_2) = 0$, so $f_1 - f_2 = \text{constant}$ by the comment following (7.4.4). $\triangleleft$

7.4.4 Surface Curl

The previous subsection concerned the surface gradient. As might be anticipated, there is also an analog to the curl operator for surfaces. However, the new operator has a number of unexpected and perhaps unintuitive aspects: for example, the ordinary curl has as its domain vector fields, while the new operator works on scalar fields. We remark in passing that the surface gradient and curl represent generalizations of the operators on the sphere ∇_1 and Λ already encountered in subsection 7.2.12; notice, however, that there is a difference of normalization between the two that can be confusing: to obtain those spherical operators from the present operators (when they act on a spherical surface), one must multiply the more general operators by radius of the sphere r.

Let S be a smooth oriented surface with positive unit normal $\hat{n}$. Then the *surface curl* Λ_S is defined by

$$\Lambda_S = \hat{n} \times \nabla_S. \tag{7.4.5}$$

Note that $\Lambda_S f$ depends only on the values of f on S. If we reverse the orientation of S (i.e., change the sign of $\hat{n}$), $\nabla_S = \nabla - \hat{n}(\hat{n} \cdot \nabla)$ so ∇_S does not change sign. Hence Λ_S does change sign. Obviously the surface gradient and the surface curl are both tangent vector fields on S.

Also note that

$$\Lambda_S f = \hat{n} \times \nabla f \tag{7.4.6}$$

if f is defined in an open set containing S.

7.4.5 Applying the FODOs ∇_S and Λ_S to Vector Fields on S

A vector field $\mathbf{v}$ defined on S is called continuously differentiable if its Cartesian components v_i relative to some orthonormal basis are continuously differentiable. As with any vector linear operators on S, we define

$$\nabla_S \cdot \mathbf{v} = (\nabla_S)_i v_i = \hat{x}_i \cdot (\nabla_S v_i)$$
$$\Lambda_S \cdot \mathbf{v} = (\Lambda_S)_i v_i = \hat{x}_i \cdot (\Lambda_S v_i) \tag{7.4.7}$$

where $\hat{x}_1, \hat{x}_2, \hat{x}_3$ is any orthonormal basis for R^3, and $v_i = \hat{x}_i \cdot \mathbf{v}$. Equations (7.4.7) show that $\nabla_S \cdot \mathbf{v}$ and $\Lambda_S \cdot \mathbf{v}$ depend only on the values of $\mathbf{v}$ on S. If $\mathbf{v}$ is defined not only on S but in some open set containing S, we have

$$\Lambda_S \cdot \mathbf{v} = \hat{x}_i \cdot (\Lambda_S v_i) = \hat{x}_i \cdot (\hat{n} \times \nabla v_i) = (\hat{n} \times \nabla)_i v_i$$
$$= (\hat{n} \times \nabla) \cdot \mathbf{v} = \hat{n} \cdot (\nabla \times \mathbf{v}),$$

so

$$\Lambda_S \cdot \mathbf{v} = \hat{n} \cdot (\nabla \times \mathbf{v}). \tag{7.4.8}$$

This equation is the reason for calling Λ_S the surface curl. And this equation also shows that if $\mathbf{v}$ is defined in an open set containing S, then $\hat{\mathbf{n}} \cdot \boldsymbol{\nabla} \times \mathbf{v}$ depends only on the values of $\mathbf{v}$ on S, not the $\mathbf{v}$ values elsewhere.

From this last comment, we can draw two interesting conclusions:

$$\boldsymbol{\Lambda}_S \cdot \hat{\mathbf{n}} = 0 \qquad (7.4.9)$$

$$\boldsymbol{\Lambda}_S \cdot \boldsymbol{\nabla}_S = 0. \qquad (7.4.10)$$

It should be noted that $\boldsymbol{\nabla}_S \cdot \hat{\mathbf{n}}$ does not usually vanish. It is a geometrical property of the oriented surface S, usually called the *total curvature* of S.

Proofs: To prove (7.4.9) and (7.4.10), let f be continuously differentiable in an open set containing S, and set $\mathbf{v} = \boldsymbol{\nabla} f$ in (7.4.8). Then $\boldsymbol{\Lambda}_S \cdot \boldsymbol{\nabla} f = \hat{\mathbf{n}} \cdot (\boldsymbol{\nabla} \times \boldsymbol{\nabla} f) = 0$. But we also have

$$\boldsymbol{\Lambda}_S \cdot \boldsymbol{\nabla} f = \boldsymbol{\Lambda}_S \cdot (\hat{\mathbf{n}} \partial_n f + \boldsymbol{\nabla}_S f)$$

$$= (\boldsymbol{\Lambda}_S \cdot \hat{\mathbf{n}}) \partial_n f + (\hat{\mathbf{n}} \cdot \boldsymbol{\Lambda}_S) \partial_n f + \boldsymbol{\Lambda}_S \cdot \boldsymbol{\nabla}_S f.$$

Since $\hat{\mathbf{n}} \cdot \boldsymbol{\Lambda}_S = 0$, we conclude that

$$(\boldsymbol{\Lambda}_S \cdot \hat{\mathbf{n}}) \partial_n f + (\boldsymbol{\Lambda}_S \cdot \boldsymbol{\nabla}_S) f = 0. \qquad (7.4.11)$$

Now choose f so that $f = 0$ and $\partial_n f \neq 0$ everywhere on S. Since $\boldsymbol{\Lambda}_S$ and $\boldsymbol{\nabla}_S$ see only values on S, $\boldsymbol{\Lambda}_S \cdot \boldsymbol{\nabla}_S f = 0$. Therefore, applying (7.4.11) to such an f yields the conclusion $\boldsymbol{\Lambda}_S \cdot \hat{\mathbf{n}} = 0$. But then (7.4.11) implies that for any f, $\boldsymbol{\Lambda}_S \cdot \boldsymbol{\nabla}_S f = 0$. This is (7.4.10). ◁

The following identities simplify many calculations:

$$\boldsymbol{\nabla}_S \cdot \mathbf{v} = (\boldsymbol{\nabla}_S \cdot \hat{\mathbf{n}}) v_n + \boldsymbol{\nabla}_S \cdot \mathbf{v}_S \qquad (7.4.12)$$

$$\boldsymbol{\Lambda}_S \cdot \mathbf{v} = \boldsymbol{\Lambda}_S \cdot \mathbf{v}_S \qquad (7.4.13)$$

$$\boldsymbol{\Lambda}_S \cdot (\hat{\mathbf{n}} \times \mathbf{v}) = \boldsymbol{\nabla}_S \cdot \mathbf{v}_S \qquad (7.4.14)$$

$$\boldsymbol{\nabla}_S \cdot (\hat{\mathbf{n}} \times \mathbf{v}) = -\boldsymbol{\Lambda}_S \cdot \mathbf{v}_S. \qquad (7.4.15)$$

Proofs:
(7.4.12): Expand $\mathbf{v}$ in normal and tangential parts:

$$\boldsymbol{\nabla}_S \cdot \mathbf{v} = \boldsymbol{\nabla}_S \cdot [\hat{\mathbf{n}} v_n + \mathbf{v}_S]$$

$$= (\boldsymbol{\nabla}_S \cdot \hat{\mathbf{n}}) v_n + (\hat{\mathbf{n}} \cdot \boldsymbol{\nabla}_S) v_n + \boldsymbol{\nabla}_S \cdot \mathbf{v}_S.$$

But $\hat{n} \cdot \boldsymbol{\nabla}_S = 0$, whence (7.4.12).

(7.4.13): Proceed as in the previous demonstration:

$$\boldsymbol{\Lambda}_S \cdot \mathbf{v} = \boldsymbol{\Lambda}_S \cdot [\hat{n} v_n + \mathbf{v}_S]$$
$$= (\boldsymbol{\Lambda}_S \cdot \hat{n}) v_n + (\hat{n} \cdot \boldsymbol{\Lambda}_S) v_n + \boldsymbol{\Lambda}_S \cdot \mathbf{v}_S.$$

But $\boldsymbol{\Lambda}_S \cdot \hat{n} = 0$ by (7.4.9), and $\hat{n} \cdot \boldsymbol{\Lambda}_S = 0$.

(7.4.14): This one requires a lot more work. Since $\hat{n} \times \mathbf{v} = \hat{n} \times \mathbf{v}_S$

$$\boldsymbol{\Lambda}_S \cdot (\hat{n} \times \mathbf{v}) = (\hat{n} \times \boldsymbol{\nabla}_S) \cdot (\hat{n} \times \mathbf{v}_S) = \hat{n} \cdot [\boldsymbol{\nabla}_S \times (\hat{n} \times \mathbf{v}_S)]$$

using (7.2.59) with $\mathbf{Q} = \hat{n} \times \mathbf{v}_S$. By (7.2.68),

$$\boldsymbol{\nabla}_S \times (\hat{n} \times \mathbf{v}_S)$$
$$= (\boldsymbol{\nabla}_S \cdot \mathbf{v}_S + \mathbf{v}_S \cdot \boldsymbol{\nabla}_S)\hat{n} - (\boldsymbol{\nabla}_S \cdot \hat{n} + \hat{n} \cdot \boldsymbol{\nabla}_S)\mathbf{v}_S$$
$$= \hat{n}(\boldsymbol{\nabla}_S \cdot \mathbf{v}_S) + (\mathbf{v}_S \cdot \boldsymbol{\nabla}_S)\hat{n} - \mathbf{v}_S(\boldsymbol{\nabla}_S \cdot \hat{n})$$

since $\mathbf{n} \cdot \boldsymbol{\nabla}_S = 0$. Thus

$$\boldsymbol{\Lambda}_S \cdot (\mathbf{n} \times \mathbf{v}) = \boldsymbol{\nabla}_S \cdot \mathbf{v}_S + \hat{n} \cdot [(\mathbf{v}_S \cdot \boldsymbol{\nabla}_S)\hat{n}] - (\hat{n} \cdot \mathbf{v}_S)(\boldsymbol{\nabla}_S \cdot \hat{n}).$$

But $\hat{n} \cdot \mathbf{v}_S = 0$, and

$$\hat{n} \cdot [(\mathbf{v}_S \cdot \boldsymbol{\nabla}_S)\hat{n}] = n_i(\mathbf{v}_S \cdot \boldsymbol{\nabla}_S)n_i$$

$$= (\mathbf{v}_S \cdot \boldsymbol{\nabla}_S)(\tfrac{1}{2} n_i n_i) = (\mathbf{v}_S \cdot \boldsymbol{\nabla}_S)(\tfrac{1}{2}) = 0.$$

This gives (7.4.14).

(7.4.15): Replace $\mathbf{v}$ by $\hat{n} \times \mathbf{v}$ in (7.4.14) and use $(\hat{n} \times \mathbf{v})_S = \hat{n} \times \mathbf{v}$, $\hat{n} \times (\hat{n} \times \mathbf{v}) = -\mathbf{v}_S$. The identities have now been established. ◁

If we set $\mathbf{v} = \boldsymbol{\nabla}_S f$ in (7.4.15), we obtain

$$\boldsymbol{\nabla}_S \cdot \boldsymbol{\Lambda}_S f = -\boldsymbol{\Lambda}_S \cdot \boldsymbol{\nabla}_S f,$$

so

$$\boldsymbol{\nabla}_S \cdot \boldsymbol{\Lambda}_S = -\boldsymbol{\Lambda}_S \cdot \boldsymbol{\nabla}_S.$$

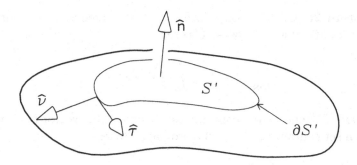

Figure 7.4.3: The setting for the surface form of Gauss' and Stokes' Theorems.

This, together with (7.4.10), gives

$$\nabla_S \cdot \Lambda_S = \Lambda_S \cdot \nabla_S = 0. \qquad (7.4.16)$$

If we set $\mathbf{v} = \nabla_S f$ in (7.4.14) we obtain $\Lambda_S \cdot \Lambda_S f = \nabla_S \cdot \nabla_S f$, so

$$\Lambda_S^2 = \nabla_S^2. \qquad (7.4.17)$$

The second-order differential operator ∇_S^2 or Λ_S^2 is called the *Beltrami operator* on S. We may think of it as the analog to the Laplacian for surfaces.

7.4.6 Surface Forms of the Theorems of Gauss and Stokes

Having followed the development of vector differential calculus for surfaces in the previous subsections, the reader no doubt expects to see a similar edifice for integral calculus; we will not disappoint him or her. These theorems have a number of important and interesting consequences, some of which we set out in this subsection.

Theorem 1: (Stokes, surface form) Let S' be a piece of an oriented smooth surface S, and let $\partial S'$ be the boundary curve of S'. Let $\hat{\tau}$ be the unit vector tangent to $\partial S'$, directed so that the unit vector $\hat{\upsilon} = \hat{\tau} \times \hat{n}$ points out of S'. Let $\mathbf{v}$ be a continuously differentiable vector field on S. Then

$$\oint_{\partial S'} d\ell \hat{\tau} \cdot \mathbf{v} = \int_{S'} d^2 \mathbf{r} \, \Lambda_S \cdot \mathbf{v}_S. \qquad (7.4.18)$$

Theorem 2: (Gauss, surface form) With the same meanings for S', $\partial S'$, $\hat{\tau}$, $\hat{v}$, and $\mathbf{v}$ as in Stokes' Theorem

$$\oint_{\partial S'} d\ell\hat{v} \cdot \mathbf{v} = \int_{S'} d^2\mathbf{r}\, \boldsymbol{\nabla}_S \cdot \mathbf{v}_S. \tag{7.4.19}$$

Proofs: Extend $\mathbf{v}$ to a continuously differentiable vector field in an open set containing S. By the ordinary Stokes' Theorem,

$$\oint_{\partial S'} d\ell\hat{\tau} \cdot \mathbf{v} = \int_{S'} d^2\mathbf{r}\, \hat{\mathbf{n}} \cdot (\boldsymbol{\nabla} \times \mathbf{v}). \tag{7.4.20}$$

But $\hat{\mathbf{n}} \cdot \boldsymbol{\nabla} \times \mathbf{v} = \boldsymbol{\Lambda}_S \cdot \mathbf{v}_S$ by (7.4.8) and (7.4.13). Hence we have (7.4.18). If, in (7.4.18), we replace $\mathbf{v}$ by $\hat{\mathbf{n}} \times \mathbf{v}$, and use (7.4.14), we obtain (7.4.19).   ◁

Theorem 3: If S is a smooth *closed* oriented surface (the surface of a sphere, ellipsoid, torus, or n-holed doughnut) and $\mathbf{v}_S$ is a continuously differentiable tangent vector field on S,

$$\int_S d^2\mathbf{r}\, \boldsymbol{\nabla}_S \cdot \mathbf{v}_S = \int_S d^2\mathbf{r}\, \boldsymbol{\Lambda}_S \cdot \mathbf{v}_S = 0. \tag{7.4.21}$$

Proof: Apply (7.4.18) and (7.4.19) to $S' = S$. By hypothesis, S has no boundary curve ∂S.   ◁

Corollary 3.1: If S is as in Theorem 3 and f and g are scalar fields on S,

$$\int_S d^2\mathbf{r}\, f(\nabla_S^2 g) = \int_S d^2\mathbf{r}\, (\nabla_S^2 f)g. \tag{7.4.22}$$

In other words the Beltrami operator on a closed surface is self-adjoint.

Proof: $f(\nabla_S^2 g) - g(\nabla_S^2 f) = \boldsymbol{\nabla}_S \cdot (f\boldsymbol{\nabla}_S g - g\boldsymbol{\nabla}_S f)$, and by Theorem 3 this integrates to 0 over S.   ◁

Corollary 3.2: If S is as in Theorem 3 and $\nabla_S^2 f = 0$ on S, then $f =$ constant.

Proof: From (7.2.65) we have the following identity

$$\boldsymbol{\nabla}_S \cdot (f\mathbf{u}) = \boldsymbol{\nabla}_S f \cdot \mathbf{u} + f\boldsymbol{\nabla}_S \cdot \mathbf{u}.$$

Then setting $\mathbf{u} = \boldsymbol{\nabla}_S f$ and rearranging we find

$$\boldsymbol{\nabla}_S f \cdot \boldsymbol{\nabla}_S f = \boldsymbol{\nabla}_S \cdot (f \boldsymbol{\nabla}_S f) - f \nabla_S^2 f = \boldsymbol{\nabla}_S \cdot (f \boldsymbol{\nabla}_S f).$$

Thus by Theorem 3, $\int_S d^2\mathbf{r} \, |\boldsymbol{\nabla}_S f|^2 = 0$. Hence $\boldsymbol{\nabla}_S f = 0$. Hence $f = \text{constant}$. ◁

Corollary 3.3: If S is as in Theorem 3 and g is a scalar field on S, then the equation $\nabla_S^2 f = g$ has a solution f if and only if

$$\int_S d^2\mathbf{r} \, g = 0. \tag{7.4.23}$$

If (7.4.23) does hold, then f is uniquely determined by requiring

$$\int_S d^2\mathbf{r} \, f = 0 \tag{7.4.24}$$

as well as

$$\nabla_S^2 f = g. \tag{7.4.25}$$

The unique solution f of (7.4.24) and (7.4.25) will be written $f = \nabla_S^{-2} g$.

Proof: If (7.4.25) does have a solution f, then Theorem 3 with $\mathbf{v}_S = \boldsymbol{\nabla}_S f$ gives (7.4.23). Furthermore, if f and f' are two solutions of (7.4.25), then $\nabla_S^2(f - f') = 0$, so $f = f' + C$, C a constant. Then

$$\int_S d^2\mathbf{r} \, f = \int_S d^2\mathbf{r} \, f' + C|S|$$

where $|S|$ is the area of S. Given a solution f' of (7.4.25), there is only one way to choose C so that $f = f' + C$ satisfies (7.4.24).

It remains to prove that when g satisfies (7.4.23), there is a solution f of (7.4.25). In Chapter 5 we give a complete proof when S is a sphere. Here we give only a *sketch* of a general proof. We want to construct an f on S that satisfies (7.4.25). One way is to choose f as the function that minimizes the quantity

$$L = \int_S d^2\mathbf{r} \, \left[|\boldsymbol{\nabla}_S f|^2 + 2fg \right]. \tag{7.4.26}$$

If $\int_S d^2\mathbf{r}\, g \neq 0$, $-L$ can be made arbitrarily large, by choosing f to be a suitable constant. If $\int_S d^2\mathbf{r}\, g = 0$, there is in fact an f that minimizes L. (This is the nontrivial part of the proof; see, for example, Courant and Hilbert, 1953, vol. II, chap. IV.) If f is this minimizing function, then changing f to $f + \delta f$ produces a first-order change δL that vanishes for all δf. But to first-order in δf,

$$\delta \ell = 2 \int_S d^2\mathbf{r}\, (\boldsymbol{\nabla}_S f \cdot \boldsymbol{\nabla}_S \delta f + g\delta f)$$

$$= 2 \int_S d^2\mathbf{r}\, [\boldsymbol{\nabla}_S \cdot [(\boldsymbol{\nabla}_S f)\delta f] - \delta f \nabla_S^2 f + (\delta f)g]$$

$$= 2 \int_S d^2\mathbf{r}\, \delta f(g - \nabla_S^2 f).$$

Since $\delta \ell = 0$ for all δf, $\nabla_S^2 f - g = 0$.   ◁

Theorem 4: Suppose S is a smooth *simply connected* oriented surface, and $\mathbf{v}_S$ is a continuously differentiable tangent vector field on S. Then $\boldsymbol{\Lambda}_S \cdot \mathbf{v}_S = 0$ if and only if there is a scalar field f such that $\mathbf{v}_S = \boldsymbol{\nabla}_S f$; and f is unique to within an additive constant. Then f is called the *potential* for $\mathbf{v}_S$.
Similarly, $\boldsymbol{\nabla}_S \cdot \mathbf{v}_S = 0$ if and only if there is a scalar field g such that $\mathbf{v}_S = \boldsymbol{\Lambda}_S g$; and g is unique to within an additive constant. Then g is called the *stream function* for $\mathbf{v}_S$.
Proof: By (7.4.16), $\mathbf{v}_S = \boldsymbol{\nabla}_S f$ implies $\boldsymbol{\Lambda}_S \cdot \mathbf{v}_S = 0$, and $\mathbf{v}_S = \boldsymbol{\Lambda}_S g$ implies $\boldsymbol{\nabla}_S \cdot \mathbf{v}_S = 0$.
Conversely, suppose $\boldsymbol{\Lambda}_S \cdot \mathbf{v}_S = 0$. Since S is simply connected, every closed curve C on S consists of the boundary curves $\partial S'$ of one or more patches S' of surface on S. Therefore (7.4.18) implies $\oint_C d\ell \hat{\boldsymbol{\tau}} \cdot \mathbf{v}_S = 0$ for every closed curve C on S. Then the remark at the end of subsection 7.4.2 gives an f such that $\mathbf{v}_S = \boldsymbol{\nabla}_S f$, and shows that f is unique to within an additive constant.
Finally, suppose $\boldsymbol{\nabla}_S \cdot \mathbf{v}_S = 0$. Then, by (7.4.14), $\boldsymbol{\Lambda}_S \cdot (\hat{\mathbf{n}} \times \mathbf{v}_S) = 0$, so there is a scalar field g, unique up to a constant, such that $-\hat{\mathbf{n}} \times \mathbf{v}_S = \boldsymbol{\nabla}_S g$. This equation is equivalent to $-\hat{\mathbf{n}} \times (\hat{\mathbf{n}} \times \mathbf{v}_S) = \hat{\mathbf{n}} \times \boldsymbol{\nabla}_S g$, or $\mathbf{v}_S = \boldsymbol{\Lambda}_S f$.   ◁

7.4.7 Representation of Tangent Vector Fields

Every vector field $\mathbf{v}$ in R^3 can be written in the form $\mathbf{v} = -\boldsymbol{\nabla}\phi + \boldsymbol{\nabla} \times$

$\mathbf{A}$, with scalar potential ϕ and vector potential $\mathbf{A}$; this is Helmholtz's Theorem (see McQuistan, 1965). The corresponding theorem for surfaces is

Theorem: (Helmholtz, surface form) If S is a complete spheroidal surface (continuously deformable into a sphere; a torus is not such a surface, for example) and $\mathbf{v}_S$ is a continuously differentiable tangent vector field on S, there are unique continuously differentiable scalar fields f and g on S such that

$$\mathbf{v}_S = \boldsymbol{\nabla}_S f + \boldsymbol{\Lambda}_S g \qquad (7.4.27)$$

and

$$\int_S d^2\mathbf{r}\ f = \int_S d^2\mathbf{r}\ g = 0. \qquad (7.4.28)$$

Proof: By Theorem 3 of the previous subsection and Corollary 3.3 we can find a scalar f such that $\nabla_S^2 f = \boldsymbol{\nabla}_S \cdot \mathbf{v}_S$. Then $\boldsymbol{\nabla}_S \cdot (\boldsymbol{\nabla}_S f - \mathbf{v}_S) = 0$, so by Theorem 3 there is a scalar g such that $\mathbf{v}_S - \boldsymbol{\nabla}_S f = \boldsymbol{\Lambda}_S g$. This proves that an f and g exist satisfying (7.4.27). We can add constants to them, if necessary, so as to have them also satisfy (7.4.28). To see that f and g are uniquely determined, note that (7.4.27) implies

$$\nabla_S^2 f = \boldsymbol{\nabla}_S \cdot \mathbf{v}_S \qquad (7.4.29)$$

$$\Lambda_S^2 g = \boldsymbol{\Lambda}_S \cdot \mathbf{v}_S. \qquad (7.4.30)$$

These equations and (7.4.28) do determine f and g, according to Corollary 3.3. $\quad\triangleleft$

The surface Helmholtz Theorem can fail if S is not simply connected. For example, suppose S is the torus of revolution obtained by revolving about the z axis a circle in the $x - z$ plane that does not intersect the z axis. Let r, θ, λ be the spherical polar coordinates determined by $\hat{\mathbf{x}}, \hat{\mathbf{y}}, \hat{\mathbf{z}}$. Let $\mathbf{v} = \hat{\boldsymbol{\lambda}}/r \sin\theta$ in R^3. Then $\mathbf{v}|S$ is tangent to S, so $\mathbf{v}_S = \mathbf{v}$ on S. Moreover, $\mathbf{v} = \boldsymbol{\nabla}\lambda$, so $\boldsymbol{\nabla} \times \mathbf{v} = 0$, so $\hat{\mathbf{n}} \cdot \boldsymbol{\nabla} \times \mathbf{v} = 0$ on S. Thus $\boldsymbol{\Lambda}_S \cdot \mathbf{v}_S = 0$. Finally,

$$\boldsymbol{\nabla} \cdot \mathbf{v} = \nabla^2 \lambda = r^{-2}(\sin\theta)^{-2}\partial_\lambda^2 \lambda = 0.$$

But $\boldsymbol{\nabla} = \hat{\mathbf{n}}\partial_n + \boldsymbol{\nabla}_S$, so $\hat{\mathbf{n}} \cdot \partial_n \mathbf{v} + \boldsymbol{\nabla}_S \cdot \mathbf{v} = 0$ on S. Since $v_n = 0$ on S,

$$\boldsymbol{\nabla}_S \cdot \mathbf{v}_S = \boldsymbol{\nabla}_S \cdot \mathbf{v} = -\hat{\mathbf{n}} \cdot \partial_n \mathbf{v}.$$

Since $\hat{\mathbf{n}} \cdot \hat{\boldsymbol{\lambda}} = 0$ on S, and $\partial_n \hat{\boldsymbol{\lambda}} = 0$ on S, $\partial_n \mathbf{v}$ is always in the direction $\hat{\boldsymbol{\lambda}}$ on S, so $\hat{\mathbf{n}} \cdot \partial_n \mathbf{v} = 0$. Thus $\boldsymbol{\nabla}_S \cdot \mathbf{v}_S = 0$ on S. Now we have a tangent vector field $\mathbf{v}_S$ on S with the properties $\boldsymbol{\Lambda}_S \cdot \mathbf{v}_S = \boldsymbol{\nabla}_S \cdot \mathbf{v}_S = 0$. If there were f and g satisfying (7.4.27), then from (7.4.29), (7.4.30), and (7.4.17), $\nabla_S^2 f = 0$ and $\nabla_S^2 g = 0$. From corollary 3.2, $f = $ constant and $g = $ constant. Then from (7.4.27), $\mathbf{v}_S = \mathbf{0}$.

On S, in the preceding example, we have $\mathbf{v}_S = \boldsymbol{\nabla}_S \lambda$. Why is this not already a representation of $\mathbf{v}_S$ in the form (7.4.27) with $f = \lambda$, $g = 0$? The reason is that λ is a multivalued function on S. If we try to make it single-valued, we have to insert a break somewhere with a jump discontinuity of 2π. If we do this, λ is not continuously differentiable, or even continuous, on S. When S is not simply connected, (7.4.27) typically breaks down in this way; it works if one permits either f or g, or both, to be multivalued. We will not pursue this question. For further reading on the subject of tangent vector fields obeying $\boldsymbol{\nabla}_S \cdot \mathbf{v} = \boldsymbol{\Lambda}_S \cdot \mathbf{v} = 0$ on multiply connected surfaces, see Hodge (1959).

Exercises

1. Here is one definition of a vector space.
   A vector space consists of two sets, F and V, and two functions, s and m, with the following properties:

   i. V contains at least one member. (The members of V are called *vectors*.)

   ii. F is a number field (i.e., F has all arithmetic operations). For us F is either the set R of real numbers or the set C of complex numbers. The members of F are called *scalars*.

   iii. If $a \in F$ and $\mathbf{u}$ and $\mathbf{v} \in V$ then s assigns to the ordered pair $(\mathbf{u}, \mathbf{v})$ a value $s(\mathbf{u}, \mathbf{v}) \in V$, and m assigns to the ordered pair $(a, \mathbf{v})$ a value $m(a, \mathbf{v}) \in V$. The values of s and m are usually written thus:

   $$s(\mathbf{u}, \mathbf{v}) = \mathbf{u} + \mathbf{v}, \qquad m(a, \mathbf{v}) = a\mathbf{v} = \mathbf{v}a.$$

   The symbol $\mathbf{u} + \mathbf{v}$ means nothing more nor less than $s(\mathbf{u}, \mathbf{v})$, the value that s assigns to the pair $(\mathbf{u}, \mathbf{v})$; and $a\mathbf{v}$ means simply $m(a, \mathbf{v})$, as does $\mathbf{v}a$. For all scalars a, b

and all vectors **u**, **v**, **w**, the functions s and m obey the
following rules

iv. $\mathbf{u} + \mathbf{v} = \mathbf{v} + \mathbf{u}$ (commutative law for s)

v. $\mathbf{u} + (\mathbf{v} + \mathbf{w}) = (\mathbf{v} + \mathbf{u}) + \mathbf{w}$ (associative law for s)

vi. $a(\mathbf{u} + \mathbf{v}) = a\mathbf{v} + a\mathbf{u}$ (right distributive law)

vii. $(a + b)\mathbf{u} = a\mathbf{u} + b\mathbf{u}$ (left distributive law)

viii. $a(b\mathbf{u}) = ab(\mathbf{u})$ (associative law for m)

ix. $1\mathbf{u} = \mathbf{u}$

x. $0\mathbf{u} = 0\mathbf{v}$

The vector $0\mathbf{u}$ is called the zero vector, and is written **0**. The vector
$(-1)\mathbf{u}$ is written $-\mathbf{u}$, and $\mathbf{u} + (-\mathbf{v})$ is written $\mathbf{u} - \mathbf{v}$.

(a) Show that for any scalar a, $a\mathbf{0} = \mathbf{0}$.

(b) Show that if $a\mathbf{u} = \mathbf{0}$ then either $a = 0$ or $\mathbf{u} = \mathbf{0}$.

(c) Show that for every vector **u**, $\mathbf{u} + \mathbf{0} = \mathbf{u}$.

(d) Show that $\mathbf{u} - \mathbf{u} = \mathbf{0}$.

(e) Show that if $\mathbf{u} + \mathbf{x} = \mathbf{0}$ then $\mathbf{x} = -\mathbf{u}$.

2. Let $F = R$. For each of the following choices of V and the functions s
and m, decide whether F, V, s, m constitute a vector space. Prove
your answer, and if the answer is yes, find the zero vector in the vector
space and for any **u** exhibit $-\mathbf{u}$.

(a) V is the set of all ordered triples (v_1, v_2, v_3) of real numbers. If
$\mathbf{u} = (u_1, u_2, u_3)$, $\mathbf{v} = (v_1, v_2, v_3)$, and $a \in F$ then

$$a\mathbf{u} = (au_1, au_2, au_3)$$

$$\mathbf{u} + \mathbf{v} = (u_1 + v_1, u_2 + v_2, u_3 + v_3).$$

(b) V is the set of all ordered triples of positive real numbers. The
functions m and s are as in (a).

(c) V is as in 3(b). The functions s and m are as follows. If $\mathbf{u} = (u_1, u_2, u_3)$, $\mathbf{v} = (v_1, v_2, v_3)$, and $a \in F$ then

$$a\mathbf{u} = (u_1^a, u_2^a, u_3^a)$$

$$\mathbf{u} + \mathbf{v} = (u_1 v_1, u_2 v_2, u_3 v_3).$$

3. If **P** and **Q** are vector operators such that **P** commutes with **Q**, show
that $\mathbf{P} \times \mathbf{Q} = -\mathbf{Q} \times \mathbf{P}$ and $\mathbf{P} \cdot \mathbf{Q} = \mathbf{Q} \cdot \mathbf{P}$.

4. Suppose that as t increases from 0 to 1, the point $r(t)$ traces out a continuously differentiable curve on $S(b)$, the surface of the sphere of radius b centered at $\mathbf{0}$. Suppose that f is a continuously differentiable scalar field on $S(b)$. Show that at each point on the curve,

$$\frac{d}{dt}g[\mathbf{r}(t)] = \frac{d\mathbf{r}}{dt} \cdot \boldsymbol{\nabla}_s g[\mathbf{r}(t)].$$

REFERENCES

Abramowitz, M., and I. A. Stegun, *Handbook of Mathematical Functions*, National Bureau of Standards, Dover Publications, New York, 1965.

Backus, G. E., The axisymmetric self-excited fluid dynamo, *Astrophysical Journal* 125, 500–524, 1957.

Backus, G. E., A class of self-sustaining dissipative spherical dynamos, *Annals of Physics* 4, 372–447, 1958.

Backus, G. E., Kinematics of geomagnetic secular variation in a perfectly conducting core, *Philosophical Transactions of the Royal Society (London)* A263, 239–266, 1968.

Backus, G. E., Inference from inadequate and inaccurate data, I, *Proceedings of the National Academy of Sciences* 65, 1–7, 1970a.

Backus, G. E., Inference from inadequate and inaccurate data, II, *Proceedings of the National Academy of Sciences* 65, 281–287, 1970b.

Backus, G. E., Inference from inadequate and inaccurate data, III, *Proceedings of the National Academy of Sciences* 67, 282–289, 1970c.

Backus, G. E., Non-uniqueness of the external geomagnetic field determined by surface intensity measurements, *Journal of Geophysical Research, Space Physics* 75, 6339–6341, 1970.

Backus, G. E., The electric field produced in the mantle by the dynamo in the core, *Physics of the Earth and Planetary Interiors* 28, 191–214, 1982.

Backus, G. E., Application of mantle filter theory to the magnetic jerk of 1969, *Geophysical Journal of the Royal Astronomical Society* 74, 713–746, 1983.

Backus, G. E., Poloidal and toroidal fields in geomagnetic field modeling, *Reviews of Geophysics* 24, 75–109, 1986.

Backus, G. E., and J. F. Gilbert, The resolving power of gross Earth data, *Geophysical Journal of the Royal Astronomical Society* 16, 169–205, 1968.

Batchelor, G. K., On the spontaneous magnetic field in a conducting liquid in turbulent motion, *Proceedings of the Royal Society (London)* A213, 349–366, 1950.

Bender, C. M., and S. A. Orszag, *Advanced Mathematical Methods for Scientists and Engineers*, McGraw-Hill, New York, 1978.

Birkhoff, G., and G.-C. Rota, *Ordinary Differential Equations*, Wiley, New York, 1989.

Blackett, P.M.S., The magnetic field of massive rotating bodies, *Nature* 159, 658–666.

Blakely, R. J., *Potential Theory in Gravity and Magnetic Applications*, Cambridge University Press, New York, 1995.

Bloxham, J., D. Gubbins, and A. Jackson, Geomagnetic secular variation, *Philosophical Transactions of the Royal Society (London)* A329, 415–502, 1989.

Bloxham, J., and A. Jackson, Fluid flow near the surface of the earth's outer core, *Reviews of Geophysics* 29, 97–120, 1991.

Booker, J. R., *Geomagnetic Secular Variation and the Kinematics of the Earth's Core*, Thesis (PhD), University of California at San Diego, 1968.

Booker, J. R., Geomagnetic data and core motions, *Proceedings of the Royal Society (London)* A309, 27–40, 1969.

Braginsky, S. I., Structure of the F layer and reasons for convection in the earth's core, *Soviet Physics, Doklady* 149, 8–10, 1963.

Braginsky, S. I., Self-excitation of a magnetic field during the motion of a highly conducting fluid, *Soviet Physics, JETP* 20, 726–735, 1964 (1965 transl.).

Bullard, E. C., The magnetic field within the earth, *Proceedings of the Royal Society of London* A197, 438–453, 1949.

Bullard, E. C., The stability of a homopolar dynamo, *Proceedings of the Cambridge Philosophical Society* 51, 744–760, 1955.

Bullard, E. C., and H. Gellman, Homogeneous dynamos and terrestrial magnetism, *Philosophical Transactions of the Royal Society (London)* A247, 213–278, 1954.

Bullard, E. C., and D. Gubbins, Generation of magnetic fields by fluid motions of global scale, *Geophysical and Astrophysical Fluid Dynamics* 8, 43–56, 1977.

Busse, F. H., Recent developments in the dynamo theory of planetary magnetism, *Annual Reviews of Earth and Planetary Sciences* 11, 241–268, 1983.

Cain, J. C., Z. Wang, C. Kluth, and D. R. Schmitz, Derivation of a geomagnetic model to n=63, *Geophysical Journal of the Royal Astronomical Society* 97, 431–442, 1989a.

Cain, J. C., Z. Wang, D. R. Schmitz, and J. Meyer, The geomagnetic spectrum for 1980 and core-crustal separation, *Geophysical Journal of the Royal Astronomical Society* 97, 433–447, 1989b.

Chapman, S., and J. Bartels, *Geomagnetism* Vols. I & II, Oxford Univerity Press, Oxford, 1962.

Clark, S. P., *Handbook of Physical Constants*, Geological Society of America, New York, 1966.

Clement, B. M., Geographical distribution of transitional VGPs: evidence for non-zonal equatorial symmetry during the Matuyama-Brunhes geomagnetic reversal, *Earth and Planetary Science Letters* 104, 48–58, 1991.

Coddington, E. A., and N. Levinson, *Theory of Ordinary Differential Equations*, McGraw-Hill, New York, 1955.

Courant, R., and D. Hilbert, *Methods of Mathematical Physics* Vols. I & II, Interscience, New York, 1953.

Cowling, T. G., The magnetic field of sunspots, *Monthly Notices of the Royal Astronomical Society* 94, 39–48, 1934.

Cox, A. V., Doell, R. R., and G. B. Dalrymple, Reversals of the earth's magnetic field, *Science* 144, 1537–1543, 1964.

Debnath, L., and P. Mikusiński, *Introduction to Hilbert Spaces with Applications*, Academic Press, New York, 1990.

Elsasser, W. M., Inductions effects in terrestrial magnetism, I. Theory, *Physical Reviews* 69, 106–116, 1946.

Erdelyi, A., W. Magnus, F. Oberhettinger, and F. G. Tricomi, *Higher Transcendental Functions* Vol. 2, McGraw-Hill, New York, 1953.

Gibson, R. D., and P. H. Roberts, The Bullard-Gellman dynamo, in *The Application of Modern Physics to the Earth and Planetary Interiors* (ed. S. K. Runcorn), Interscience, 577–601, 1969.

Gilbert, W., *De Magnete*, translated by P. F. Mottelay, Dover Publications, New York, 1958.

Glen, W., *The Road to Jaramillo: Critical Years of the Revolution in Earth Science*, Stanford University Press, Stanford, CA, 1982.

Gubbins, D., Numerical solution of the kinematic dynamo problem, *Philosophical Transactions of the Royal Society (London)* A274, 493–521, 1973.

Gubbins, D., Can the earth's magnetic field be sustained by core oscillations? *Geophysical Research Letters* 2, 409–412, 1975.

Gubbins, D., and P. H. Roberts, Magnetohydrodynamics of the earth's core, 1-177, in Jacobs, J. A., *Geomagnetism* Vol. II, Academic Press, HBJ, New York, 1987.

Halmos, P. A., *Finite Dimensional Vector Spaces*, van Nostrand, Princeton, NJ, 1958.

Harland, W. B., R. L. Armstrong, A. V. Cox, L. E. Craig, A. G. Smith, and D. G. Smith, *A Geologic Time Scale*, Cambridge University Press, Cambridge, 1989.

Herzenberg, A., Geomagnetic dynamos, *Philosophical Transactions of the Royal Society (London)* A250, 543–585, 1958.

Herzenberg, A., and F. J. Lowes, Electromagnetic induction in rotating conductors, *Philosophical Transactions of the Royal Society (London)* A249, 507–584, 1957.

Hide, R., How to locate the electrically conducting fluid core of a planet from external magnetic observations, *Nature* 271, 640–641, 1978.

Hide, R., and S. R. C. Malin, On the determination of the size of the Earth's core from observations of the geomagnetic secular variation, *Proceedings of the Royal Society (London)* A374, 15–33, 1981.

Hodge, W. V., *The Theory and Applications of Harmonic Integrals*, Cambridge University Press, Cambridge, 1959.

Hoffman, K. A., Dipolar reversal states of the geomagnetic field and core-mantle dynamics, *Nature* 359, 789–794, 1992.

Jackson, A., J. Bloxham, and D. Gubbins, Time-dependent flow at the core surface and conservation of angular momentum in the coupled core-mantle system, in *Dynamics of the Earth's Deep Interior and Earth Rotation* (ed. J.-L. Le Mouël, D. E. Smylies, and T. Herring), American Geophysical Union Monograph, 72, IUGG Vol. 12, 97–107, Washington, DC, 1992.

Jacobs, J. A., *Geomagnetism* Vol. I, Academic Press, HBJ, New York, 1987.

Jault, D., C. Gire, and J.-L. Le Mouël, Westward drift, core motions and exchanges of angular momentum between core and mantle, *Nature* 333, 353–356, 1988.

Johnson, C. L., and C. G. Constable, The time-averaged geomagnetic field as recorded by lava flows over the past 5Ma, *Geophysical Journal International*, 122, 489–519, 1995.

Kahle, A. B., R. H. Ball, and E. H. Vestine, Comparison of estimates of surface fluid motions of the Earth's core for various epochs, *Journal of Geophysical Research* 72, 4917–4925, 1967.

Kellogg, O. D., *Foundations of Potential Theory*, Dover Publications, New York, 1953.

Korevaar, J., *Mathematical Methods*, Academic Press, New York, 1968.

Körner, T. W., *Fourier Analysis*, Cambridge University Press, Cambridge, 1988.

Laj, C., A. Mazaud, R. Weeks, M. Fuller, and E. Herrero-Bervera, Geomagnetic reversal paths, *Nature* 351, 447, 1991.

Laj, C., A. Mazaud, R. Weeks, M. Fuller, and E. Herrero-Bervera, Statistical assessment of the preferred longitudinal bands for recent geomagnetic reversal records, *Geophysical Research Letters* 19, 2003–2006, 1992.

Langel, R. A., The main field, 249–492, in Jacobs, J. A., *Geomagnetism* Vol. I, Academic Press, HBJ, New York, 1987.

Langel, R. A., and R. H. Estes, A geomagnetic field spectrum, *Geophysical Research Letters* 9, 250–253, 1982.

Langel, R. A., R. H. Estes, and G. D. Mead, Some new methods in geomagnetic field modeling applied to the 1960–1980 epoch, *Journal of Geomagnetism and Geoelectricity* 34, 327–349, 1982.

Langel, R. A., R. H. Estes, G. D. Mead, E. B. Fabiano, and E. R. Lancaster, Initial geomagnetic field model from Magsat data, *Geophysical Research Letters* 7, 793–796, 1980.

Lanzerotti, L. J., L. V. Medford, C. G. MacLennan, D. J. Thomson, A. Meloni, and G. P. Gregori, Measurements of the large scale direct current earth potential and possible implications for the geomagnetic dynamo, *Science* 229, 47–49, 1985.

Larmor, J., How could a rotating body like the Sun become a magnet? *Reports of the British Association for the Advancement of Science* 159–160, 1919.

Lorenz, E. N., Deterministic nonperiodic flow, *Journal of the Atmospheric Sciences* 20, 130–141, 1963.

Lortz, D., Exact solutions of the hydromagnetic dynamo problem, *Plasma Physics* 10, 967–972, 1968.

Lowes, F. J., Spatial power spectrum of the main geomagnetic field, and extrapolation to the core, *Geophysical Journal of the Royal Astronomical Society* 36, 717–730, 1974.

Luecke, O., Über Mittelwerte von Energiedichten der Kraftfelder, *Wissenschaftliche Zeitschrift der Pädagogischen Hochschule Potsdam*, Mathematisch-Naturwissenschaftliche Reihe 3, 39–46, 1957.

Madden, T., and J.-L. Le Mouël, The recent secular variation and the motions at the core's surface, *Philosophical Transactions of the Royal Society (London)* A306, 271–280, 1982.

Malin, S. R. C., and E. C. Bullard, The direction of the Earth's magnetic field at London, 1570–1975, *Philosophical Transactions of the Royal Society (London)* A299, 357–423, 1981.

Malkus, W., Reversing Bullard's dynamo, *EOS, Transactions of the American Geophysical Union* 53, 617, 1972.

Malvern, L. E., *Introduction to the Mechanics of a Continuous Medium*, Prentice-Hall, Englewood Cliffs, 1969.

Matsushita, S., and W. H. Campbell (editors), *Physics of Geomagnetic Phenomena* Vols. I & II, Academic Press, New York, 1967.

Mauersberger, P., Das Mittel der Energiedichte des geomagnetischen Hauptfeldes an der Erdoberfläche und seine säkulare Änderung, *Gerlands Beiträge zur Geophysik* 65, 207–215, 1956.

McFadden, P. L., C. E. Barton, and R. T. Merrill, Do virtual geomagnetic poles follow preferred paths during geomagnetic reversals? *Nature* 361, 342–344, 1993.

McFadden, P. L., and R. T. Merrill, Lower mantle convection and geomagnetism, *Journal of Geophysical Research* 98, 6189–6199, 1984.

McQuistan, R. B., *Scalar and Vector Fields: a Physical Interpretation*, Wiley, New York, 1965.

Merrill, R. T., and M. W. McElhinny, Anomalies in the time-averaged paleomagnetic field and their implications for the lower mantle, *Reviews of Physics and Space Physics* 15, 309–323, 1977.

Merrill, R. T., and M. W. McElhinny, *The Earth's Magnetic Field: Its History, Origin and Planetary Perspective*, (2nd printing with corrections) Academic Press, Orlando, FL, 1983.

Merrill, R. T., and P. L. McFadden, Geomagnetic field stability: reversal events and excursions, *Earth and Planetary Science Letters* 121, 57–69, 1994.

Meyer, J., J.-H. Hufen, M. Siebert, and A. Hahn, Investigations of the internal geomagnetic field by means of a global model of the Earth's crust, *J. Geophysics* 52, 71–84, 1983.

Moffatt, K., *Field Generation in Electrically Conducting Fluids*, Cambridge University Press, Cambridge, 1978.

Morrison, L. V., Re-determination of the decade fluctuations in the rotation of the Earth in the period 1861–1978, *Geophysical Journal of the Royal Astronomical Society* 58, 349–360, 1979.

O'Reilly, W., *Rock and mineral magnetism*, Chapman and Hall, New York, 1984.

Panofsky, W. K. H., and M. Phillips, *Classical Electricity and Magnetism*, 2nd ed. Reading, MA, Addison-Wesley Pub. Co., 1962.

Parker, E. N., Hydromagnetic dynamo models, *Astrophysical Journal* 122, 293–314, 1955.

Parker, E. N., *Cosmical Magnetic Fields: Their Origin and Their Activity*, Clarendon Press, Oxford, 1979.

Parker, R. L., The inverse problem of electrical conductivity in the mantle, *Geophysical Journal of the Royal Astronomical Society* 22, 121–138, 1970.

Parker, R. L., The inverse problem of electromagnetic induction: existence and construction of solutions based on incomplete data, *Journal of Geophysical Research* 85, 4421–4428, 1980.

Parkinson, W. D., *Introduction to Geomagnetism*, Academic Press, Edinburgh, 1983.

Phillips, J. D., Time variation and asymmetry in the statistics of geomagnetic reversal sequences, *Journal of Geophysical Research* 82, 835–843, 1977.

Priestley, M. B., *Spectral Analysis and Time Series*, Academic Press, New York, 1981.

Regan, R. D., and B. D. Marsh, The Bangui magnetic anomaly: its geological origin, *Journal of Geophysical Research* 87, 1107–1120, 1982.

Riesz, F., and B. Sz.-Nagy, *Functional Analysis*, Unger, New York, 1955.

Robbins, K. A., A moment equation description of magnetic reversals in the Earth, *Proceedings of the National Academy of Sciences USA* 73, 4297–4301, 1976.

Roberts, G. O., Spatially periodic dynamos, *Philosophical Transactions of the Royal Society (London)* A266, 535–558, 1970.

Roberts, P. H., Dynamo theory, in *Lectures in Applied Mathematics* (ed. W. H. Reid), Vol. 14, 129–206, American Mathematical Society, Providence, RI, 1971.

Roberts, P. H., Kinematic dynamo models, *Philosophical Transactions of the Royal Society (London)* A272, 663–698, 1972.

Roberts, P. H., Origin of the main field: dynamics, 251–303, in Jacobs, J. A., *Geomagnetism* Vol. II, Academic Press, HBJ, New York, 1987.

Roberts, P. H., and D. Gubbins, Origin of the main field: kinematics, 185–246, in Jacobs, J. A., *Geomagnetism* Vol. II, Academic Press, HBJ, New York, 1987.

Roberts, P. H., and S. Scott, On the analysis of the secular variation I. A hydromagnetic constraint: theory, *Journal of Geomagnetism and Geoelectricity* 17, 137–151, 1965.

Roberts, P. H., and M. Stix, The turbulent dynamo: a translation of a series of papers by F. Krause, K.-H. Rädler, and M. Steenbeck. *Tech. Note* 60, NCAR, Boulder, CO, 1971.

Rosenfeld, L., *Theory of electrons*, Series title: Selected topics in modern physics 1, Amsterdam, North-Holland Pub. Co., 1951.

Runcorn, S. K., An ancient lunar magnetic dipole field, *Nature* 253, 701–703, 1975.

Schneider, D. A., and D. V. Kent, The time-averaged paleomagnetic field, *Reviews of Geophysics* 28, 71–96, 1990.

Shure, L., Parker, R. L., and R. A. Langel, A preliminary harmonic spline model from Magsat data, *Journal of Geophysical Research* 90, 11,505–11,512, 1985.

Stacey, F. D., and S. K. Banerjee, *The physical principles of rock magnetism*, Elsevier, New York, 1974.

Steenbeck, M., and F. Krause, On the dynamo theory of stellar and planetary magnetic fields I, *Astr. Nachrichten* 291, 49–84, 1969a. [English translation in Roberts and Stix, 29–47, 1971.]

Steenbeck, M., and F. Krause, On the dynamo theory of stellar and planetary magnetic fields II, *Astr. Nachrichten* 291, 271–286, 1969b. [English translation in Roberts and Stix, 147–220, 1971.]

Steenbeck, M., F. Krause, and K.-H. Rädler, A calculation of the mean electromotive force in an electrically conducting fluid in turbulent motion, under the influence of Coriolis forces. *Zeitschrift für Naturforschung 21a* 369–376, 1967. [English translation in Roberts and Stix, 271–286, 1971.]

Strang, G., *Linear Algebra and Its Applications*, Harcourt, Brace & Jovanovich, San Diego, 1988.

Tanaka, H., M. Kono, and H. Uchimura, Some global features of palaeointensity in geological time, *Geophysical Journal International* 120, 97–102, 1995.

Taylor, J. B., The magneto-hydrodynamics of a rotating fluid and the Earth's dynamo problem, *Proceedings of the Royal Society (London)* A274, 274–283, 1963.

Telford, W. M., L. P. Geldart, R. E. Sheriff, and D. A. Keys, *Applied Geophysics*, Cambridge University Press, Cambridge, 1976.

Thomas, G. B., and R. L. Finney, *Calculus and Analytic Geometry*, Addison-Wesley, Reading, MA, 1984.

Valet, J. P., P. Tucholka, V. Courtillot, and L. Meynadier, Palaeomagnetic constraints on the geometry of the geomagnetic field during reversals, *Nature* 356, 400–407, 1992.

Vine, F. J., and D. H. Matthews, Magnetic anomalies over oceanic ridges, *Nature* 199, 947–949, 1963.

Voorhies, C. V., and E. R. Benton, Pole-strength of the Earth from Magsat and magnetic determination of the core radius, *Geophysical Research Letters* 9, 258–261, 1982.

Weidelt, P., The inverse problem of geomagnetic induction, *Zeitschrift für Geophysik* 38, 257–289, 1972.

Weiss, N. O., The dynamo problem, *Quarterly Journal of the Royal Astronomical Society* 12, 432–446, 1971.

Whaler, K. A., Does the whole of the Earth's core convect? *Nature* 287, 528–530, 1980.

Whittaker, E. T., and G. N. Watson, *A Course of Modern Analysis*, Fourth Ed., Cambridge University Press, Cambridge, 1927.

Wilson, R. L., Permanent aspects of the Earth's non-dipole magnetic field over upper tertiary times, *Geophysical Journal of the Royal Astronomical Society* 19, 417–437, 1970.

INDEX

Taylor, G. I., 278.
Taylor, J. B., 285.
Taylor Theorem, 285.
Telford, W. M., 151.
tesla unit, 3.
Thomas, G. B., 91, 107, 329.
toroidal field, 175.
  in the mantle, 196.
  summary, 181.
toroidal vector field,
  definition of, 177.
toroidal velocity on core, 259, 260.
torque due to magnetic field, 22.
total curvature, 340.
turbulent flow, 277, 278.
two-scale approximation, 278.

unit vectors, 296.

Valet, J. P., 288.
van Allen radiation belts, 5.
vector analysis,
  integral calculus, 328.
vector field, 298.
  continuously differentiable, 299.
  sources, 332.
  confined to a surface, 334.
  on an oriented surface, 336.
  solenoidal, 173.
vector linear operators, 303.

vector linear operators.
  range of, 303.
  dot and cross products of, 308.
  examples of, 304.
  linear combinations of, 306.
  products of, 306.
  commuting, 349.
vector space, definition of, 348.
vector-valued linear operator, 303.
vertical motion below core–mantle
  boundary, 259.
Vestine, E. H., 252, 253, 254.
Vine, F. J., 4.
viscosity in core, 218, 249.
viscous stress tensor, 218.
Voorhies, C. V., 257.
vorticity, 249.

Watson, G. N., 104, 116.
Weidelt, P., 199.
Weierstrass Approximation
  Theorem, 42, 43, 48, 53.
Weiss, N. O., 285.
westward drift, 3, 11.
  of null-flux curves, 255.
Whaler, K. A., 257, 258, 259, 260.
whistlers, 9.
Whittaker, E. T., 104, 116.
Wilson, R. L., 287.
WKBJ approximation, 95, 99.